Limestone in the Built Environment: Present-Day Challenges for the Preservation of the Past

The Geological Society of London
Books Editorial Committee

Chief Editor

BOB PANKHURST (UK)

Society Books Editors

JOHN GREGORY (UK)
JIM GRIFFITHS (UK)
JOHN HOWE (UK)
RICK LAW (USA)
PHIL LEAT (UK)
NICK ROBINS (UK)
RANDELL STEPHENSON (UK)
JONATHAN TURNER (UK)

Society Books Advisors

MIKE BROWN (USA)
ERIC BUFFETAUT (FRANCE)
JONATHAN CRAIG (ITALY)
RETO GIERÉ (GERMANY)
TOM MCCANN (GERMANY)
DOUG STEAD (CANADA)
MAARTEN DE WIT (SOUTH AFRICA)

Geological Society books refereeing procedures

The Society makes every effort to ensure that the scientific and production quality of its books matches that of its journals. Since 1997, all book proposals have been refereed by specialist reviewers as well as by the Society's Books Editorial Committee. If the referees identify weaknesses in the proposal, these must be addressed before the proposal is accepted.

Once the book is accepted, the Society Book Editors ensure that the volume editors follow strict guidelines on refereeing and quality control. We insist that individual papers can only be accepted after satisfactory review by two independent referees. The questions on the review forms are similar to those for *Journal of the Geological Society*. The referees' forms and comments must be available to the Society's Book Editors on request.

Although many of the books result from meetings, the editors are expected to commission papers that were not presented at the meeting to ensure that the book provides a balanced coverage of the subject. Being accepted for presentation at the meeting does not guarantee inclusion in the book.

More information about submitting a proposal and producing a book for the Society can be found on its web site: www.geolsoc.org.uk.

It is recommended that reference to all or part of this book should be made in one of the following ways:

SMITH, B. J., GOMEZ-HERAS, M., VILES, H. A. & CASSAR, J. (eds) 2010. *Limestone in the Built Environment: Present-Day Challenges for the Preservation of the Past*. Geological Society, London, Special Publications, **331**.

RESCIC, S., FRATINI, F. & TIANO, P. 2010. On-site evaluation of the 'mechanical' properties of Maastricht limestone and their relationship with the physical characteristics. *In*: SMITH, B. J., GOMEZ-HERAS, M., VILES, H. A. & CASSAR, J. (eds) *Limestone in the Built Environment: Present-Day Challenges for the Preservation of the Past*. Geological Society, London, Special Publications, **331**, 203–208.

GEOLOGICAL SOCIETY SPECIAL PUBLICATION NO. 331

Limestone in the Built Environment: Present-Day Challenges for the Preservation of the Past

EDITED BY

B. J. SMITH
Queen's University, Belfast, UK

M. GOMEZ-HERAS
Queen's University, Belfast, UK
Universidad Complutense de Madrid; Instituto de Geología Económica (CSIC-UCM), Spain

H. A. VILES
Oxford University Centre for the Environment, UK

and

J. CASSAR
University of Malta, Malta

2010
Published by
The Geological Society
London

THE GEOLOGICAL SOCIETY

The Geological Society of London (GSL) was founded in 1807. It is the oldest national geological society in the world and the largest in Europe. It was incorporated under Royal Charter in 1825 and is Registered Charity 210161.

The Society is the UK national learned and professional society for geology with a worldwide Fellowship (FGS) of over 9000. The Society has the power to confer Chartered status on suitably qualified Fellows, and about 2000 of the Fellowship carry the title (CGeol). Chartered Geologists may also obtain the equivalent European title, European Geologist (EurGeol). One fifth of the Society's fellowship resides outside the UK. To find out more about the Society, log on to www.geolsoc.org.uk.

The Geological Society Publishing House (Bath, UK) produces the Society's international journals and books, and acts as European distributor for selected publications of the American Association of Petroleum Geologists (AAPG), the Indonesian Petroleum Association (IPA), the Geological Society of America (GSA), the Society for Sedimentary Geology (SEPM) and the Geologists' Association (GA). Joint marketing agreements ensure that GSL Fellows may purchase these societies' publications at a discount. The Society's online bookshop (accessible from www.geolsoc.org.uk) offers secure book purchasing with your credit or debit card.

To find out about joining the Society and benefiting from substantial discounts on publications of GSL and other societies worldwide, consult www.geolsoc.org.uk, or contact the Fellowship Department at: The Geological Society, Burlington House, Piccadilly, London W1J 0BG: Tel. +44 (0)20 7434 9944; Fax +44 (0)20 7439 8975; E-mail: enquiries@geolsoc.org.uk.

For information about the Society's meetings, consult *Events* on www.geolsoc.org.uk. To find out more about the Society's Corporate Affiliates Scheme, write to enquiries@geolsoc.org.uk.

Published by The Geological Society from:
The Geological Society Publishing House, Unit 7, Brassmill Enterprise Centre, Brassmill Lane, Bath BA1 3JN, UK

(*Orders*: Tel. +44 (0)1225 445046, Fax +44 (0)1225 442836)
Online bookshop: www.geolsoc.org.uk/bookshop

The publishers make no representation, express or implied, with regard to the accuracy of the information contained in this book and cannot accept any legal responsibility for any errors or omissions that may be made.

© The Geological Society of London 2010. All rights reserved. No reproduction, copy or transmission of this publication may be made without written permission. No paragraph of this publication may be reproduced, copied or transmitted save with the provisions of the Copyright Licensing Agency Ltd, Saffron House, 6–10 Kirby Street, London EC1N 8TS UK. Users registered with the Copyright Clearance Center, 222 Rosewood Drive, Danvers, MA 01923, USA: the item-fee code for this publication is 0305-8719/10/$15.00.

British Library Cataloguing in Publication Data

A catalogue record for this book is available from the British Library.
ISBN 978-1-86239-294-6

Typeset by Techset Composition Ltd, Salisbury, UK
Printed by Antony Rowe, Chippenham, UK

Distributors

North America
For trade and institutional orders:
The Geological Society, c/o AIDC, 82 Winter Sport Lane, Williston, VT 05495, USA
Orders: Tel. +1 800-972-9892
 Fax +1 802-864-7626
 E-mail: gsl.orders@aidcvt.com

For individual and corporate orders:
AAPG Bookstore, PO Box 979, Tulsa, OK 74101-0979, USA
Orders: Tel. +1 918-584-2555
 Fax +1 918-560-2652
 E-mail: bookstore@aapg.org
 Website: http://bookstore.aapg.org

India
Affiliated East-West Press Private Ltd, Marketing Division, G-1/16 Ansari Road, Darya Ganj, New Delhi 110 002, India
Orders: Tel. +91 11 2327-9113/2326-4180
 Fax +91 11 2326-0538
 E-mail: affiliat@vsnl.com

Contents

Preface	vii
SMITH, B. J., GOMEZ-HERAS, M. & VILES, H. A. Underlying issues on the selection, use and conservation of building limestone	1
CASSAR, J. The use of limestone in a historic context – the experience of Malta	13
CALVO, J. P. & REGUEIRO, M. Carbonate rocks in the Mediterranean region – from classical to innovative uses of building stone	27
SIEGESMUND, S., GRIMM, W.-D., DÜRRAST, H. & RUEDRICH, J. Limestones in Germany used as building stones: an overview	37
ESPINOSA-MARZAL, R. M. & SCHERER, G. W. Mechanisms of damage by salt	61
MILLER, A. Z., LEAL, N., LAIZ, L., ROGERIO-CANDELERA, M. A., SILVA, R. J. C., DIONÍSIO, A., MACEDO, M. F. & SAIZ-JIMENEZ, C. Primary bioreceptivity of limestones used in southern European monuments	79
RUIZ-AGUDO, E. & RODRIGUEZ-NAVARRO, C. Suppression of salt weathering of porous limestone by borax-induced promotion of sodium and magnesium sulphate crystallization	93
BECK, K. & AL-MUKHTAR, M. Weathering effects in an urban environment: a case study of tuffeau, a French porous limestone	103
STEFANIDOU, M. A. Approaches to the problem of limestone replacement in Greece	113
DOTTER, K. R. Historic lime mortars: potential effects of local climate on the evolution of binder morphology and composition	119
IOANNOU, I., PETROU, M. F., FOURNARI, R., ANDREOU, A., HADJIGEORGIOU, C., TSIKOURAS, B. & HATZIPANAGIOTOU, K. Crushed limestone as an aggregate in concrete production: the Cyprus case	127
BECK, K., BRUNETAUD, X., MERTZ, J.-D. & AL-MUKHTAR, M. On the use of eggshell lime and tuffeau powder to formulate an appropriate mortar for restoration purposes	137
PÁPAY, Z. & TÖRÖK, Á. Physical changes of porous Hungarian limestones related to silicic acid ester consolidant treatments	147
FIGUEIREDO, C., FOLHA, R., MAURÍCIO, A., ALVES, C. & AIRES-BARROS, L. Pore structure and durability of Portuguese limestones: a case study	157
VAZQUEZ-CALVO, C., VARAS, M. J., ALVAREZ DE BUERGO, M. & FORT, R. Limestone on the 'Don Pedro I' facade in the Real Alcázar compound, Seville, Spain	171
FIGUEIREDO, C., AIRES-BARROS, L. & NETO, M. J. The church of Santa Engrácia (the National Pantheon, Lisbon, Portugal): building campaigns, conservation works, stones and pathologies	183
BUJ, O., GISBERT, J., FRANCO, B., MATEOS, N. & BAULUZ, B. Decay of the Campanile limestone used as building material in Tudela Cathedral (Navarra, Spain)	195

RESCIC, S., FRATINI, F. & TIANO, P. On-site evaluation of the 'mechanical' properties of Maastricht limestone and their relationship with the physical characteristics   203

SEARLE, D. E. & MITCHELL, D. J. The effect of combustion-derived particulates on the short-term modification of temperature and moisture loss from Portland Limestone   209

MOTTERSHEAD, D., FARRES, P. & PEARSON, A. The changing Maltese soil environment: evidence from the ancient cart tracks at San Pawl Tat-Tarġa, Naxxar   219

THORNBUSH, M. J. Measurements of soiling and colour change using outdoor rephotography and image processing in Adobe Photoshop along the southern façade of the Ashmolean Museum, Oxford   231

SASS, O. & VILES, H. A. Two-dimensional resistivity surveys of the moisture contents of historic limestone walls in Oxford, UK: implications for understanding catastrophic stone deterioration   237

Index   251

Preface

There is a general preconception, at least amongst lay observers, that limestones typically decay in a slow, uniform and largely predictable fashion primarily in response to surface dissolution. It is accepted that the rate of dissolution can be accelerated by, for example, an increase in rainfall acidity associated with atmospheric pollution, but the underlying assumption of uniformitarian change still persists. It can be readily demonstrated, however, that the decay of many limestones used in construction is far from uniform and predictable. Instead, in many cases decay is characterized by marked temporal and spatial variability both within individual blocks and often across complete facades. This particularly applies to granular limestones (e.g. bioclastic and oolitic limestones) that are frequently characterized by effective physical breakdown similar to that normally associated with, for example, quartz sandstones. This is commonly accomplished by mechanisms such as granular disintegration, contour scaling and multiple flaking, but in polluted environments may also be linked to the rapid formation of surface crusts composed primarily of gypsum that emphasize the central role of salts as the drivers of decay.

Spatial differentiation of decay at the level of individual blocks may initially be controlled by small-scale variability in key rock properties, most notably the degree of induration, texture and associated porosity/permeability. At the facade scale, however, localized environmental controls on processes such as surface soiling and crust formation may assume a greater significance. In temporal terms, initial gypsum crust formation can stabilize a surface, but over their lifetime, such crusts can be associated with subsurface weakening and act as a source of salts that can eventually penetrate deeply into the underlying stone. Such that, if the crust is breached in any way it is often followed by rapid erosion as pre-weathered subsurface material is lost and 'deep salts' exploited.

It is in the face of such complexity that the idea for the original workshop on granular limestones, on which this volume is based, was conceived under the combined auspices of the Stone Weathering and Atmospheric Pollution NETwork (SWAPNET), the UK Engineering and Physical Sciences Research Council (grant EP/D008603/1) and Heritage Malta. This was on the basis that not only do we need to understand the nature and causes of their decay more fully, but also that we must begin to explore how this improved understanding can be used to inform future conservation strategies. Nowhere is this understanding more urgent than on the limestone island of Malta that provided the ideal venue for the meeting.

This volume would have not been possible without help from numerous colleagues who carried out reviews. Their thorough effort highly improved the level of the papers:

F. J. Alonso	A. McMillan
J. Alvarez	C. McNamara
G. Cultrone	W. Michali
J. Delgado Rodrigues	A. Moropoulou
A. El-Turki	D. Mottershead
T. Gonçalves	S. Nannukutan
A. Gorbushina	F. Puertas
F. Gale	A. Ruffell
R. Grima	C. Sabbioni
C. M. Grossi	G. W. Scherer
C. Hall	S. Siegesmund
C. Hunt	R. Snethlage
J. Hughes	M. Steiger
E. Hyslop	T. Vazquez
G. Leucci	P. Warke
K. Malaga	N. Weiss
N. Masini	G. Wheeler
S. McCabe	W. Wylczynska
J. McKinley	M. Young

Finally we would like to acknowledge the help from the Geological Society staff during the whole process of production of this volume.

B. J. SMITH
M. GOMEZ-HERAS
H. A. VILES
J. CASSAR

Underlying issues on the selection, use and conservation of building limestone

BERNARD J. SMITH[1]*, MIGUEL GOMEZ-HERAS[1,2] & HEATHER A. VILES[3]

[1]*School of Geography, Archaeology and Palaeoecology, Queen's University Belfast, Belfast, UK*

[2]*Departamento de Petrología y Geoquímica, Universidad Complutense de Madrid; Instituto de Geología Económica (CSIC-UCM), Madrid, Spain*

[3]*School of Geography and the Environment, University of Oxford, Oxford, UK*

Corresponding author (e-mail: b.smith@qub.ac.uk)

Abstract: An argument is presented that, despite popular assumptions, many limestones, especially the wide range of clastic and, in general, granular limestones, do not decay in a steady and predictable pattern in response to slow dissolution. Instead these stones, especially when used in construction in polluted environments, invariably decay episodically through physical breakdown. Most commonly this is accomplished through a variety of salt weathering mechanisms that, if unconstrained, can lead to the rapid, catastrophic decay of building blocks and their complete loss – a process that has driven the extensive programmes of stone replacement that are typical of buildings constructed of these stones. In polluted environments, especially those rich in sulphur and particulates, the most common constraint on accelerated decay has been the rapid development of gypsum crusts that, for example, could rapidly 'heal' the scars left by contour scaling. It is ironic, therefore, that any reduction in pollution could conceivably lead to increased erosion by retarding this healing process. Because of this temporal variability of decay and its translation into spatial complexity, it is important that further research is undertaken to understand controls on the decay of these important building stones so that future conservation strategies can be appropriately informed.

On the predictability of decay and the role of solution

Limestones, especially across Europe, have arguably been the building stone of choice for many centuries, if not millennia, and continue to be used extensively in a wide range of structures in a load-bearing capacity and increasingly as cladding (Calvo & Regueiro 2010; Cassar 2010; Siegesmund *et al.* 2010). In addition, crushed stone and processed limestone forms the basis of many mortars and aggregates (Beck *et al.* 2010; Dotter 2010; Ioannou *et al.* 2010). This widespread use is a reflection of their occurrence worldwide within the geological column, the relative ease with which many can be cut and quarried (especially those that harden only on exposure), the readiness with which many can be shaped to produce ashlar blocks and their perceived durability. This perception is, in the UK context, based largely on the survival of many old limestone buildings in cities such as Oxford, Bath and London, and that are assumed to have resisted previously high levels of atmospheric pollution. It should be noted, however, that such interpretations rarely take into account the many buildings that did not survive nor the possible extent of earlier stone replacement (Smith *et al.* 2008*a*).

Limestones are generally homogeneous in their chemical characteristics, being dominated by $CaCO_3$, but can be highly variable in terms of physical characteristics such as hardness, fossil content and porosity (Smith & Viles 2006). It is these physical characteristics, especially porosity and other physical properties (e.g. Figueiredo *et al.* 2010; Rescic *et al.* 2010), that are in turn the major determinants of, for example, patterns of moisture movement (Sass & Viles 2010) and, ultimately, the durability of limestones when they are placed in buildings and exposed to often aggressive environmental conditions (Beck & Al-Mukhtar 2010). So that, for example, dense, impermeable Istrian Limestone has traditionally been used not only as a foundation stone within cities such as Venice but also as an effective barrier to rising groundwater (Simunic Bursic *et al.* 2007). Similarly, a significant number of coarse shell fragments tends to increase durability, whereas the presence of numerous very fine pores can result in a stone that is less durable than those with larger pores (Leary 1983).

Despite the structural differences between limestones, their chemical similarity has lead to the belief, at least amongst lay observers, that their decay in buildings is predominantly through gradual dissolution similar to the so-called karstic erosion of natural limestone outcrops. Karst processes are, as identified by Livingston (1992), those where rainfall acidified with CO_2 is the major agent of deterioration producing dissolution. In the case of natural exposures there is also the general assumption that the rock weathers at a near-constant rate to produce overall surface lowering and distinctive solutional relief forms (such as rillenkarren). Livingston acknowledges, however, that in polluted environments the effects of 'natural' dissolution are enhanced where atmospheric moisture acidified through the presence of sulphur and nitrogen oxides. These react with calcium carbonate to produce, for example, the more soluble calcium sulphate, which can then be removed in solution. In addition, dry deposition can also occur where sulphur and nitrogen oxides in the atmosphere react, under moist conditions, with calcium carbonate to form calcium sulphate – which can accumulate as a gypsum crust. This last process is often associated with the surface deposition of complex mixtures of atmospheric particulates derived from the combustion of fossil fuels and a range of environmental dusts, including marine aerosols, that are themselves rich in carbonates, and a variety of salts. These particulates can themselves be modified by surface reactions with and between acidified atmospheres and the underlying stone to produce secondary salts or simply be included within gypsum crusts as they form to produce their characteristic grey–black tone (McAlister et al. 2006, 2008).

As well as the physico-chemical processes described above, limestones are, like any exposed stone, subject to a wide range of additional weathering regimes. This includes those associated with deteriorative biofilms found under many environmental conditions (Miller et al. 2010), freeze–thaw processes that are important in, for example, many central European locations, and salt weathering (Espinosa-Marzal & Scherer 2010) that is common not just in urban areas, but also in maritime and arid environments characterized by alternate wetting and drying cycles and an adequate supply of salts (Smith & Viles 2006). From this Smith & Viles (2006, p. 192) concluded that:

> limestone deterioration can be more accurately conceived as the result of often complex interactions between many different process regimes. These regimes may be spatially patchy – for example, acid rain processes may dominate one part of a building, whilst other areas may be more sheltered and prone to dry depositional effects. Another complexity is introduced by the fact that, as pollution and climatic conditions have changed, many building stones have been exposed to changing process regimes over their lifetime.

Alternative patterns of decay

The idea that environmental conditions, pollution regimes, and, hence, patterns and rates of weathering and erosion may change over the lifetime of a structure is fundamental to understanding present and future rates of decay. Not just because new processes may come to bear, but because stone carries with it a stress history or 'memory' of past conditions that can make it more or less susceptible to new processes and process combinations. This means that, in general, complex histories are likely to elicit complex future responses, each of them unique not just to a building, but to the micro-environments found on it and even down to the way that previous conditions have been factored through the specific properties of individual stone blocks (McCabe et al. 2007). In the case of limestones, however, it would appear that in many instances there is a belief, translated largely from studies of natural limestone outcrops, that the rate of surface erosion of limestones is gradual and uniform – both spatially and temporally. This has had a major influence on perceptions regarding the nature of the threat that decay poses to buildings, and strongly influenced assumptions underpinning many early scientific investigations into the nature of limestone decay and calculations of erosion rates. Because of this 'uniformitarian' approach it was, for example, assumed that long-term erosion could be readily extrapolated either from the results of downscaled laboratory simulations or from short-term measurements of erosion loss from a limited number of key sites on a building.

Subsequent understanding of the temporal and spatial variability of decay systems has called into question these assumptions, such that Trudgill & Viles (1998) asserted that whilst chemical weathering relevant laboratory rates may be used to predict the order of magnitude of field rates, they cannot be used to ascertain the precise rate. They were even more concerned for physical and biological weathering, and were of the view that 'we are very far from having representative, replicable rate measurements from either field or laboratory studies' (p. 339). Central to their arguments were a series of field measurements and trials, especially repeat micro-erosion measurements of the Portland Limestone of St Paul's Cathedral in London that indicated considerable temporal and spatial variability (Trudgill et al. 1989). These showed that erosion rates can differ significantly from those predicted from a theoretical analysis of potential solution related to rainfall patterns and chemistry. The key to this variability is, they consider, the particular importance of the residence time of water on, and within, the pores of stone in controlling solution rates that is, in turn, controlled by a range of micro-environmental conditions and localized material properties.

This latter point is extremely important, in that when seeking to transfer theories and observations on karstic weathering processes and characteristics to limestone buildings it must be remembered that karstic limestones are always dense, hard and, apart from a well-developed macro-joint system, effectively impermeable. Apart from exceptions such as the aforementioned Istrian Limestone, the majority of building limestones do not fall within this category. Instead, the most widely used limestones are likely to be those that are more easily worked and have not experienced intense deep diagenesis processes. As Leary (1983) observed, these are also likely to be less crystalline, less dense and, therefore, less resilient to many decay processes than many geologically older limestones. Numbered amongst these commonly used stones are a wide range of largely granular limestones that include numerous subvarieties of bioclastic limestones and, in particular, oolitic limestones. Of all the limestones commonly used in construction within the British Isles (and widely across Europe, e.g. Török 2002) it is oolitic limestones that exhibit the greatest variability. By definition they consist primarily of small rounded grains coated with $CaCO_3$ and embedded in calcitic cement, but their durability varies hugely in response to differences in characteristics such as porosity and bioclasts content (Smith & Viles 2006). These characteristics determine a range of secondary properties that include permeability, saturation coefficient and water absorption, that in turn influence the operation of a range of stone decay mechanisms – especially salt weathering (Goudie & Viles 1997).

It is the general susceptibility of oolitic and other granular limestones to physical weathering, especially salt weathering, that sets them apart from dense limestones and from the general preconceptions regarding the pre-eminence of solution loss. Because of this it is more common to see these limestones affected by patterns of decay that one normally associates with sandstones rather than with the surface dissolution of limestone. Although such decay may begin with limited surface loss through granular disaggregation, decay eventually begins to concentrate spatially leading to localized accelerated retreat forming characteristic alveolar weathering or 'honeycombing' of the stone (Fig. 1). Once retreat has been initiated small hollows can amalgamate to form larger cavernous hollows or Tafoni (Fig. 2). Once

Fig. 1. Honeycomb weathering of Globigerina limestone, Malta, controlled by patterns of fossil bioturbation (approximate image width 70 cm).

Fig. 2. Cavernous hollows developing on a Quaternary calcarenite, Mallorca (approximate image width 40 cm).

Fig. 3. Catastrophic decay of an individual block of Globigerina limestone, Malta (approximate image width 2 m).

Fig. 4. Catastrophic decay of a section of wall constructed of Jurassic oolitic limestone, Oxford (approximate image width 4 m).

small-scale cavernous hollows are initiated in oolitic limestones, like those used in cities such as Oxford, there is ample visual evidence that individual blocks (Fig. 3) and, ultimately, whole sections of wall (Fig. 4) can become prone to rapid and, eventually, catastrophic decay. Invariably this is associated with the near-surface presence of salts (e.g. Espinoza-Marzal & Scherer 2010), such as gypsum within the cavernous hollows, and is achieved by a combination of multiple flaking and, especially, granular disaggregation.

Understanding catastrophic decay

Understanding the causes of catastrophic decay is clearly central to understanding the long-term performance of many granular limestones. Likewise, it is apparent that catastrophic decay can only be controlled if the reasons why rapid retreat is triggered in the first place and its continuation is facilitated are understood, and the causes avoided or removed. This is particularly true where inappropriate conservation can accelerate decay, and where choices have to be made regarding stone selection for new build and stone replacement on older structures. To achieve this understanding four questions need to be asked.

- What processes or combination of mechanisms are responsible for rapid retreat?
- What physical, chemical and mineralogical characteristics determine stone susceptibility to rapid retreat, and how do these properties change during decay?
- How do microclimatic conditions at and beneath the stone surface change as stone retreats, and how do these influence decay mechanisms?
- What permits continued weathering despite rapid loss of weathered material in which, for example, salts are concentrated?

Typically, limestones prone to catastrophic decay are clastic or granular and often oolitic in character, but differ from other clastic stone types, such as sandstones, in that they are initially prone to both chemical attack and salt-induced decay as well as severe and/or prolonged freezing. The balance of these processes inevitably changes as decay continues and the retreat of individual limestone blocks progressively shelters them from direct rainfall, but leaves them subject to wetting from dew and frost. Such limestones have been widely used as a building material in England, for example, in many of the large cathedrals (e.g. Wells and Lincoln), historic buildings in Bath and Oxford, and are found in many ecclesiastical ruins. They are diverse in nature, exhibiting a range of petrological fabrics, geochemical characteristics and durabilities (Leary 1983), and often decay through interlinked chemical, physical and biological processes often exacerbated by the presence of atmospheric pollutants.

The first step towards understanding long-term performance is knowledge of what controls the initiation of surface irregularities within which, for example, salts can accumulate. These may originate through slight variations in surface topography produced during the initial working of the stone, but a more general source of spatially concentrated decay may lie in structural and textural variations. In laboratory simulations of salt weathering carried out on oolitic limestones Smith *et al.* (2008b) identified meso-scale structural diversity, such as that associated with distinctive bedding (Fig. 5), fossil bioturbation (Fig. 1) and texturally different inclusions such as clay lenses, as potential focal points for localized decay. In highly polluted environments, however, sulphation (the chemical reaction of $CaCO_3$ with water acidified by SO_2) can be so rapid that it can quickly mask any surface heterogeneity through the development of a surface gypsum crust, especially in areas sheltered from large amounts of surface runoff. In these areas gypsum crusts can act as a hard, relatively impermeable surface, but once breached they may fail

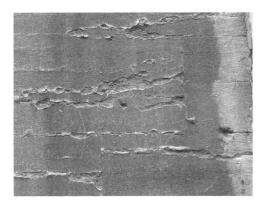

Fig. 5. Decay of a Jurassic oolitic limestone initiated along clay partings within the stone, Oxford (approximate image width 1 m).

catastrophically leading to the production of blisters and scales.

A large body of work has focused on identifying the nature and origin of blackened gypsum crusts on limestone and other calcareous building materials (Sabbioni 1995; Maravelaki-Kalaitzaki & Biscontin 1999; Jimenez de Haro et al. 2004; Sabbioni et al. 2004). Work by Rodriguez-Navarro & Sebastian (1996), Cultrone et al. (2004) and Gomez-Heras et al. (2008) has illustrated experimentally the role of particulate matter in facilitating gypsum formation on limestone. Biofilms and other rock-dwelling organisms can also encourage decay (Schiavon 2002), although some communities may have a more bioprotective role (Carter & Viles 2004). Once initially protective crusts and/or biofilms are breached, salts arising from air pollution, groundwater contamination, road de-icing activities and other sources contribute significantly to limestone decay, producing flaking, disintegration and alveolar forms (Goudie & Viles 1997; Fitzner et al. 2002; Török 2004).

A possible explanation for rapid decay is that once weathering initiates a hollow on the stone surface, accumulated salts are less likely to be washed away by rainfall and reduced moisture availability might enhance salt concentration near the stone surface. However, this does not explain why, with reduced wash-in and near-surface concentration, any salts present are not rapidly lost together with the debris. Neither does such a micro-environmental model explain why only certain stones on an otherwise uniform façade experience rapid, catastrophic retreat. To explain these requires not just an examination of microclimatic conditions, but also subtle variations in the physical characteristics of the stone and the operation of specific decay mechanisms controlled by, for example, available salt types and the balance between salt input, output and storage. One possibility identified by Smith et al. (2002) working on the rapid decay of quartz sandstones in a polluted maritime environment is that the presence of a store of 'deep salts' within stone blocks accumulated over a long period could maintain decay as blocks retreated and tapped into it. Initially this was aided by the presence of the core softening of the stone behind a surface accumulation of gypsum. This phenomenon has been widely observed on oolitic limestones in cities such as Oxford, where rapid retreat has also been associated with areas where dry deposition or mixed process regimes dominate, and where salt deposition and surface formation can continue even in cavernous hollows sheltered from wet deposition of salts and pollutants (Smith & Viles 2006).

Similar studies to those in Oxford have been carried out by Török et al. (2007) in Budapest. Here, they studied ashlar blocks composed of three porous Miocene limestones (a fine-grained limestone, a medium-grained oolitic limestone and a coarse-grained biogranular limestone) from buildings in the city and compared them with quarry blocks of the same lithologies. On the urban stones the most common weathering forms are white and black crusts. Apart from their impact on the visual and thermal properties of stone (Searle & Mitchell 2010; Thornbush 2010), the presence of the surface crusts, together with calcite recrystallization, is also marked by pore occlusion and reduction in microporosity, which was documented using a combination of microdrilling resistance and ultrasonic pulse velocities. These tests also showed degradation of the underlying fine- and medium-grained limestones, with drilling resistances less than those of equivalent quarry stones. Where crusts were detached from this underlying weakened zone the detachment was initiated by the opening up of microfissures that develop below the cemented crust zones. The fine-grained limestone appeared to be less durable than the coarse-grained variety, and more prone to rapid crust formation and detachment. In this particular case, microbiological activity did not appear to play a significant role in crust formation and removal, whereas the latter is probably influenced by freeze–thaw weathering and strongly controlled by the texture and porosity of the limestone substrate.

The 'switching off' of rapid decay

Studies in the city of Oxford have confirmed the episodic nature of blistering on historic limestone walls, its patchiness over wall surfaces, and its complex relation to salt, climatic conditions and pollutant content (Viles 1993; Antill & Viles 1999). Archival investigations of blister development, climate and air pollution over the period 1864–1937 in central Oxford have also indicated particular 'spasms' of blistering (which cannot easily be related to any specific environmental variable) followed by relative stability. Such episodic and complex production of catastrophic decay features poses a severe problem for the conservation and management of historic buildings and monuments, as well as presenting difficult challenges when choosing potential stabilization procedures or replacement stone types. Studies of the rapid decay of non-calcareous stones may again provide some insight into the nature of this decay and what controls its operation. This is because, in the absence of slow and gradual decay in response to surface dissolution, a more appropriate paradigm for the decay of stones such as quartz sandstones is one based on periods of relative surface stability followed by short-lived episodes of rapid surface loss (Smith et al. 1994, 2003). The first episode of rapid loss may very well be associated with the sudden loss of a surface gypsum crust and the rapid erosion of a weakened subsurface layer. After this is removed, the stone appears to be faced by one of two choices, either negative feedbacks come into play or, as described above, the initiation of a localized hollow can trigger a number of positive feedbacks that accelerate decay leading to catastrophic loss (Fig. 6).

The processes by which a scaled surface could be stabilized are numerous. In the case of iron-rich sandstone, for example, surface induration could result from the outward migration of iron to form a surface crust, whereas in a limestone it could take the form of a case-hardened layer formed by the precipitation of calcium carbonate which either migrated outwards or flowed in solution over the surface. The most obvious source, however, of surface stabilization is, as indicated above, the growth of a new gypsum crust. In Oxford, for example, there is widespread observational evidence of the multiple 'healing over' of oolitic limestones following contour scaling through the apparently rapid regrowth of black gypsum crusts (Fig. 7). Similar patterns of multiple black crusts have also been observed on the oolitic limestone of St Matthias Church, which sits above the still polluted city of Budapest (Smith et al. 2003). Regrowth of gypsum crusts on the church is associated with the rapid deposition of gypsum, dust and surface biological colonization by fungal hyphae that help to trap and bind the dust to the surface. As indicated earlier, the presence of the dust appears to catalyse the further precipitation of gypsum in the form of a new, stabilizing surface crust. The importance of dust deposition in a dry environment was highlighted by a comparative analysis of the Hungarian Parliament building that sits below the church on the banks of the Danube. Here, the same stone is exposed to a considerably moister atmosphere and contour scaling appears to be followed more often than not by rapid retreat rather than recrusting.

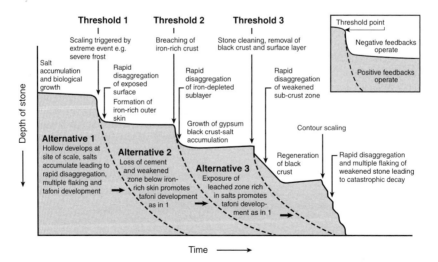

Fig. 6. Hypothetical decay pathways for quartz sandstone used as a building material in a polluted environment. The inset identifies feedback options at critical thresholds of decay associated with rapid surface loss of material. Adapted from Smith et al. (1994) and Smith (2003).

Fig. 7. Multiple gypsum crusts on Jurassic oolitic limestone, Oxford (approximate image width 50 cm).

Summary and conclusions

There is a general preconception, at least amongst lay observers, that limestones used in construction typically decay in a slow, uniform and largely predictable fashion primarily in response to surface dissolution. It is accepted that the rate of dissolution can be accelerated by, for example, an increase in rainfall acidity associated with atmospheric pollution, but the underlying assumption of uniformitarian change still persists. This assumption draws further support from the behaviour in natural outcrops of dense hard limestones that exhibit a distinctive karstic behaviour associated with the formation of persistent surface solutional forms. It can be readily demonstrated, however, that the decay of many limestones used in construction is far from being uniform and predictable. Instead, it has been demonstrated that in many cases decay is characterized by marked temporal and spatial variability, both within individual blocks and often across complete facades. This particularly applies to granular limestones (e.g. bioclastic and oolitic limestones) that are frequently characterized by effective physical breakdown similar to that normally associated with, for example, sandstones. This includes the presence of phenomena such as initial surface roughening, leading to pitting and honeycombing, leading to cavernous weathering, leading in some cases to the rapid, catastrophic decay and complete loss of individual blocks. This is commonly accomplished by mechanisms such as granular disintegration, contour scaling and multiple flaking, but in polluted environments may also be linked to the rapid formation of surface crusts composed primarily of gypsum. The formation of such crusts indicates the significance of dry deposition in the weathering of urban limestones, including surface sulphation and particulate deposition, but also emphasizes the role of salts such as gypsum in driving the various decay mechanisms.

Spatial differentiation of decay at the level of individual blocks may initially be controlled by small-scale variability in key rock properties, most notably the degree of induration, texture and associated porosity/permeability. At the facade scale, however, localized environmental controls on processes such as surface soiling and crust formation may assume a greater significance (Turkington & Smith 2004). In temporal terms, initial gypsum crust formation can stabilize a surface, but over their lifetime such crusts can be associated with subsurface weakening and act as a sources of salts that can eventually penetrate deeply into the underlying stone. As a consequence, if the crust is breached in any way this is often followed by rapid erosion as pre-weathered subsurface material is lost and 'deep salts' exploited. That all such stones (or, indeed, buildings) do not quickly disappear is testimony in many cases to previous and ongoing campaigns of stone replacement by those with a duty of care (Fig. 8), but also to the fact that it is possible for stones to stabilize through the 'switching off' of the positive feedbacks that accelerate.

In the case of cities like Oxford, which were once heavily polluted and in the still polluted city of

Fig. 8. Replacement of a section of wall in Oxford previously destroyed by catastrophic decay of Jurassic oolitic limestone (approximate image width 6 m).

Budapest, it has been shown that stabilization was and is commonly associated with rapid surface sulphation and growth of a new gypsum crust that effectively 'heals' the erosion scars left by surface delamination. During the lifetime of a block this process might be repeated several times, and across a facade one might expect to see a patchwork of stones at different stages of the cycle. This includes the final stages of decay whereby, for whatever combination of circumstances, the rate of regrowth cannot keep pace with the accelerating rate of surface loss and the block (or by this time possibly a collection of adjacent blocks) rapidly disappears. This could be part of a natural progression, in that repeated cycles of scaling and crust development may eventually create a hollow that is deep enough to create a micro-environment that is particularly conducive to, for example, salt weathering. Conversely, there is scope for the potentially ironic situation in which high levels of atmospheric pollution, especially dust in the atmosphere that can catalyse crust formation, may have contributed significantly to the long-term stability of stones that are intrinsically susceptible to processes such as salt weathering, and which can perform relatively badly in durability assessments, such as the standard sodium sulphate crystallization test, used to specify stone for construction.

It is doubly ironic, therefore, that strategies to reduce atmospheric pollution, especially the reduction of sulphur and particulates, could in future curtail the 'healing' process, and that historic stone – often pre-weathered and salt infused over many centuries – might disappear at an accelerated rate following contour scaling. Furthermore, it is possible to envisage that the delicate balance that prevents the triggering of rapid decay could be upset by a number of other interventions and changes. Aggressive removal of black crusts, especially in environments that no longer facilitate their rapid regrowth, could, for example, trigger rapid surface loss. Alternatively, any change in environmental conditions that affected patterns of heating–cooling and wetting–drying, perhaps linked to wider climatic change, might be expected to influence crust stability and/or growth as well as the operation of existing decay processes such as salt weathering, and may even promote new ones.

It is in the face of such complexity and uncertainty, and the demonstrable need for conservation (e.g. Buj et al. 2010; Figueiredo et al. 2010; Pápay & Török 2010; Ruiz-Agudo & Rodriguez-Navarro 2010; Stefanidou 2010; Vazquez-Calvo et al. 2010), that the idea for the original workshop on granular limestones, on which this volume is based, was conceived. This was on the basis that not only do we need to understand the nature and causes of their decay more fully, but also that we must begin to explore how this improved understanding can be used to inform future conservation strategies.

Support for this paper was provided by EPSRC research grant EP/D008603/1.

References

ANTILL, S. A. & VILES, H. A. 1999. Deciphering the impacts of traffic on stone decay in Oxford: Some preliminary observations from old limestone walls. *In*: JONES, M. S. & WAKEFIELD, R. D. (eds) *Aspects of Stone Weathering, Decay and Conservation*. Imperial College Press, London, 28–42.

BECK, K. & AL-MUKHTAR, M. 2010. Weathering effects in an urban environment: a case study of tuffeau. *In*: SMITH, B. J., GOMEZ-HERAS, M., VILES, H. A. & CASSAR, J. (eds) *Limestone in the Built Environment: Present-Day Challenges for the Preservation of the Past*. Geological Society, London, Special Publications, **331**, 103–111.

BECK, K., BRUNETAUD, X., MERTZ, J.-D., BIGAS, J.-P. & AL-MUKHTAR, M. 2010. On the use of eggshell lime and tuffeau powder to formulate an appropriate mortar for restoration purposes. *In*: SMITH, B. J., GOMEZ-HERAS, M., VILES, H. A. & CASSAR, J. (eds) *Limestone in the Built Environment: Present-Day Challenges for the Preservation of the Past*. Geological Society, London, Special Publications, **331**, 137–145.

BUJ, O., GISBERT, J., FRANCO, B., MATEOS, N. & BAULUZ, B. 2010. Decay of the Campanile limestone used as building material in Tudela Cathedral (Navarra, Spain). *In*: SMITH, B. J., GOMEZ-HERAS, M., VILES, H. A. & CASSAR, J. (eds) *Limestone in the Built Environment: Present-Day Challenges for the Preservation of the Past*. Geological Society, London, Special Publications, **331**, 195–202.

CALVO, J. P. & REGUEIRO, M. 2010. Carbonate rocks in the Mediterranean region – from classical to innovative uses of building stone. *In*: SMITH, B. J., GOMEZ-HERAS, M., VILES, H. A. & CASSAR, J. (eds) *Limestone in the Built Environment: Present-Day Challenges for the Preservation of the Past*. Geological Society, London, Special Publications, **331**, 27–35.

CARTER, N. E. A. & VILES, H. A. 2004. Lichen hotspots: raised rock temperatures beneath Verrucaria nigrescens on limestone. *Geomorphology*, **62**, 1–16.

CASSAR, J. 2010. The use of limestone in a historic context – the experience of Malta. *In*: SMITH, B. J., GOMEZ-HERAS, M., VILES, H. A. & CASSAR, J. (eds) *Limestone in the Built Environment: Present-Day Challenges for the Preservation of the Past*. Geological Society, London, Special Publications, **331**, 13–25.

CULTRONE, G., RODRIGUEZ-NAVARRO, C. & SEBASTIAN, E. 2004. Limestone and brick decay in simulated polluted atmosphere: the role of particulate matter. *In*: SAIZ-JIMENEZ, C. (ed.) *Air Pollution and Cultural Heritage*. Balkema, Leiden, 141–145.

DOTTER, K. R. 2010. Historic lime mortars: potential effects of local climate on the evolution of binder morphology and composition. *In*: SMITH, B. J., GOMEZ-HERAS, M., VILES, H. A. & CASSAR, J.

(eds) *Limestone in the Built Environment: Present-Day Challenges for the Preservation of the Past.* Geological Society, London, Special Publications, **331**, 119–126.

ESPINOSA-MARZAL, R. M. & SCHERER, G. W. 2010. Mechanisms of damage by salt. *In*: SMITH, B. J., GOMEZ-HERAS, M., VILES, H. A. & CASSAR, J. (eds) *Limestone in the Built Environment: Present-Day Challenges for the Preservation of the Past.* Geological Society, London, Special Publications, **331**, 61–77.

FIGUEIREDO, C., AIRES-BARROS, L. & NETO, M. J. 2010. The church of Santa Engrácia (the National Pantheon, Lisbon, Portugal): building campaigns, conservation works, stones and pathologies. *In*: SMITH, B. J., GOMEZ-HERAS, M., VILES, H. A. & CASSAR, J. (eds) *Limestone in the Built Environment: Present-Day Challenges for the Preservation of the Past.* Geological Society, London, Special Publications, **331**, 183–193.

FIGUEIREDO, C., FOLHA, R., MAURÍCIO, A., ALVES, C. & AIRES-BARROS, L. 2010. Pore structure and durability of Portuguese limestones: a case study. *In*: SMITH, B. J., GOMEZ-HERAS, M., VILES, H. A. & CASSAR, J. (eds) *Limestone in the Built Environment: Present-Day Challenges for the Preservation of the Past.* Geological Society, London, Special Publications, **331**, 157–170.

FITZNER, B., HEINRICHS, K. & BOUCHARDIERE, D. 2002. Limestone weathering of historical monuments in Cairo, Egypt. *In*: SIEGESMUND, S., WEISS, T. & VOLLBRECHT, A. (eds) *Natural Stone, Weathering Phenomena, Conservation Strategies and Case Studies.* Geological Society, London, Special Publications, **205**, 217–239.

GOMEZ-HERAS, M., SMITH, B. J. & VILES, H. A. 2008. Laboratory modelling of gypsum crust growth on limestone related to soot pollution and gaseous sulphur: implications of 'cleaner' environments for stone decay. *In*: LUKASZEWICZ, J. W. & NIEMCEWICZ, P. (eds) *Proceedings of the 11th International Congress on Deterioration and Conservation of Stone, Torun, September 2008.* Nicolaus Copernicus University Press, Torun, 105–112.

GOUDIE, A. S. & VILES, H. A. 1997. *Salt Weathering Hazards.* Wiley, Chichester.

IOANNOU, I., PETROU, M. F., FOURNARI, R., ANDREOU, A., HADJIGEORGIOU, C., TSIKOURAS, B. & HATZIPANAGIOTOU, K. 2010. Crushed limestone as an aggregate in concrete production: the Cyprus case. *In*: SMITH, B. J., GOMEZ-HERAS, M., VILES, H. A. & CASSAR, J. (eds) *Limestone in the Built Environment: Present-Day Challenges for the Preservation of the Past.* Geological Society, London, Special Publications, **331**, 127–135.

JIMENEZ DE HARO, M. C., JUSTO, A., DURÁN, A., SIGÜENZA, M. B., PÉREZ-RODRIGUEZ, J. L. & BUENO, J. 2004. Crusts formed on different building materials from the cathedral of Seville, Spain. *In*: SAIZ-JIMENEZ, C. (ed.) *Air Pollution and Cultural Heritage.* Balkema, Leiden, 79–84.

LEARY, E. 1983. *The Building Limestones of the British Isles.* BRE Report. HMSO, London.

LIVINGSTON, R. A. 1992. Graphical methods for examining the effects of acid rain and sulphur dioxide on carbonate stones. *In*: DELGADO RODRIGUES, J., HENRIQUES, F. & TELMO JEREMIAS, F (eds) *Proceedings of the 7th International Congress on Deterioration and Conservation of Stone, Lisbon.* Laboratorio Nacional de Engenharia Civil, Lisbon, 375–386.

MARAVELAKI-KALAITZAKI, P. & BISCONTIN, G. 1999. Origin, characteristics and morphology of weathering crusts on Istrian stone in Venice. *Atmospheric Environment*, **33**, 1699–1709.

MCALISTER, J. J., SMITH, B. J. & TÖRÖK, A. 2006. Element partitioning and potential mobility within surface dusts on buildings in a polluted urban environment, Budapest. *Atmospheric Environment*, **40**, 6780–8790.

MCALISTER, J. J., SMITH, B. J. & TÖRÖK, A. 2008. Transition metals and water-soluble ions in deposits on a building and potential catalysis of stone decay. *Atmospheric Environment*, **42**, 7657–7668.

MCCABE, S., SMITH, B. J. & WARKE, P. A. 2007. Preliminary observations on the impact of complex stress histories on sandstone response to salt weathering: laboratory simulations of process combinations. *Environmental Geology*, **52**, 251–258.

MILLER, A. Z., LEAL, N. *ET AL.* 2010. Primary bioreceptivity of limestones used in southern European monuments. *In*: SMITH, B. J., GOMEZ-HERAS, M., VILES, H. A. & CASSAR, J. (eds) *Limestone in the Built Environment: Present-Day Challenges for the Preservation of the Past.* Geological Society, London, Special Publications, **331**, 79–92.

PÁPAY, Z. & TÖRÖK, A. 2010. Physical changes of porous Hungarian limestones related to silicic acid ester consolidant treatments. *In*: SMITH, B. J., GOMEZ-HERAS, M., VILES, H. A. & CASSAR, J. (eds) *Limestone in the Built Environment: Present-Day Challenges for the Preservation of the Past.* Geological Society, London, Special Publications, **331**, 147–155.

RESCIC, S., FRATINI, F. & TIANO, P. 2010. On-site evaluation of the 'mechanical' properties of Maastricht limestone and their relationship with the physical characteristics. *In*: SMITH, B. J., GOMEZ-HERAS, M., VILES, H. A. & CASSAR, J. (eds) *Limestone in the Built Environment: Present-Day Challenges for the Preservation of the Past.* Geological Society, London, Special Publications, **331**, 203–208.

RODRIGUEZ-NAVARRO, C. & SEBASTIAN, E. 1996. Role of particulate matter from vehicle exhaust on porous building stones (limestone) sulphation. *Science of the Total Environment*, **187**, 79–91.

RUIZ-AGUDO, E. & RODRIGUEZ-NAVARRO, C. 2010. Suppression of salt weathering of porous limestone by borax-induced promotion of sodium and magnesium sulphate crystallization. *In*: SMITH, B. J., GOMEZ-HERAS, M., VILES, H. A. & CASSAR, J. (eds) *Limestone in the Built Environment: Present-Day Challenges for the Preservation of the Past.* Geological Society, London, Special Publications, **331**, 93–102.

SABBIONI, C. 1995. Contribution of atmospheric deposition to the formation of damage layers. *Water, Air and Soil Pollution*, **63**, 305–316.

SABBIONI, C., BONAZZA, A., ZAMAGNI, J., GEDNI, N., GROSSI, C. M. & BRIMBLECOMBE, P. 2004. The Tower of London: a case study on stone damage in an urban area. *In*: SAIZ-JIMENEZ, C. (ed.) *Air*

Pollution and Cultural Heritage. Balkema, Leiden, 57–62.

SASS, O. & VILES, H. A. 2010. Two-dimensional resistivity surveys of the moisture content of historic limestone walls in Oxford, UK: implications for understanding catastrophic stone deterioration. *In*: SMITH, B. J., GOMEZ-HERAS, M., VILES, H. A. & CASSAR, J. (eds) *Limestone in the Built Environment: Present-Day Challenges for the Preservation of the Past*. Geological Society, London, Special Publications, **331**, 237–249.

SCHIAVON, N. 2002. Biodeterioration of calcareous and granitic building stones in urban environments. *In*: SIEGESMUND, S., WEISS, T. & VOLLBRECHT, A. (eds) *Natural Stone, Weathering Phenomena, Conservation Strategies and Case Studies*. Geological Society, London, Special Publications, **205**, 195–205.

SEARLE, D. E. & MITCHELL, D. J. 2010. The effect of combustion-derived particulates on the short-term modification of temperature and moisture loss from Portland Limestone. *In*: SMITH, B. J., GOMEZ-HERAS, M., VILES, H. A. & CASSAR, J. (eds) *Limestone in the Built Environment: Present-Day Challenges for the Preservation of the Past*. Geological Society, London, Special Publications, **331**, 209–218.

SIEGESMUND, S., GRIMM, W.-D., DÜRRAST, H. & RUEDRICH, J. 2010. Limestones in Germany used as building stones: an overview. *In*: SMITH, B. J., GOMEZ-HERAS, M., VILES, H. A. & CASSAR, J. (eds) *Limestone in the Built Environment: Present-Day Challenges for the Preservation of the Past*. Geological Society, London, Special Publications, **331**, 37–59.

SIMUNIC BURSIC, M., ALJINOVIC, D. & CANCELLIERE, S. 2007. Kirmenjak Pietra d'Istria: a preliminary investigation of its use in Venetian architectural heritage. *In*: PRIKRYL, R. & SMITH, B. J. (eds) *Building Stone Decay: From Diagnosis to Conservation*. Geological Society, London, Special Publications, **271**, 63–68.

SMITH, B. J. 2003. Background controls on urban stone decay: lessons from natural rock weathering. *In*: BRIMBLECOMBE, P. (ed.) *The Effects of Air Pollution on the Built Environment*. Air Pollution Reviews, **2**. Imperial College Press, London, 31–61.

SMITH, B. J. & VILES, H. A. 2006. Rapid, catastrophic decay of building limestones: thoughts on causes, effects and consequences. *In*: FORT, R., ALVAREZ DE BUERGO, M., GOMEZ-HERAS, M. & VÁZQUEZ-CALVO, C. (eds) *Heritage Weathering and Conservation*. Taylor & Francis, London, 191–197.

SMITH, B. J., GOMEZ-HERAS, M. & MCCABE, S. 2008*a*. Understanding the decay of stone-built cultural heritage. *Progress in Physical Geography*, **32**, 439–461.

SMITH, B. J., GOMEZ-HERAS, M., MENEELY, J., MCCABE, S. & VILES, H. A. 2008*b*. High resolution monitoring of surface morphological change of building limestones in response to simulated salt weathering. *In*: LUKASZEWICZ, J. W. & NIEMCEWICZ, P. (eds) *Proceedings of the 11th International Congress on Deterioration and Conservation of Stone, Torun, September 2008*. Nicolaus Copernicus University Press, Torun, 275–282.

SMITH, B. J., MAGEE, R. W. & WHALLEY, W. B. 1994. Breakdown patterns of quartz sandstone in a polluted urban environment: Belfast, N. Ireland. *In*: ROBINSON, D. A. & WILLIAMS, R. B. G. (eds) *Rock Weathering and Landform Evolution*. Wiley, Chichester, 131–150.

SMITH, B. J., TÖRÖK, A., MCALISTER, J. J. & MEGARRY, Y. 2003. Observations on the factors influencing stability of building stones following contour scaling: a case study of oolitic limestones from Budapest, Hungary. *Building and Environment*, **38**, 1173–1183.

SMITH, B. J., TURKINGTON, A. V., WARKE, P. A., BASHEER, P. A. M., MCALISTER, J. J., MENEELY, J. & CURRAN, J. M. 2002. Modelling the rapid retreat of building sandstones. A case study from a polluted maritime environment. *In*: SIEGESMUND, S., WEISS, T. & VOLLBRECHT, A. (eds) *Natural Stone, Weathering Phenomena, Conservation Strategies and Case Studies*. Geological Society, London, Special Publications, **205**, 339–354.

STEFANIDOU, M. A. 2010. Approaches to the problem of limestone replacement in Greece. *In*: SMITH, B. J., GOMEZ-HERAS, M., VILES, H. A. & CASSAR, J. (eds) *Limestone in the Built Environment: Present-Day Challenges for the Preservation of the Past*. Geological Society, London, Special Publications, **331**, 113–117.

THORNBUSH, M. J. 2010. Measurements of soiling and colour change using outdoor rephotography and image processing in adobe photoshop along the southern façade of the Ashmolean Museum, Oxford. *In*: SMITH, B. J., GOMEZ-HERAS, M., VILES, H. A. & CASSAR, J. (eds) *Limestone in the Built Environment: Present-Day Challenges for the Preservation of the Past*. Geological Society, London, Special Publications, **331**, 231–236.

TÖRÖK, A. 2002. Oolitic limestone in a polluted atmospheric environment in Budapest: weathering phenomena and alterations in physical properties. *In*: SIEGESMUND, S., WEISS, T. & VOLLBRECHT, A. (eds) *Natural Stone, Weathering Phenomena, Conservation Strategies and Case Studies*. Geological Society, London, Special Publications, **205**, 363–380.

TÖRÖK, A. 2004. Leithalkalk-type limestones in Hungary: an overview of lithologies and weathering features. *In*: PRIKRYL, R. & SIEGL, P. (eds) *Architectural and Sculptural Stone in Cultural Landscape*. Karolinum Press, Prague, 157–172.

TÖRÖK, A., SIEGESMUND, S., MÜLLER, C., HÜPERS, A., HOPPERT, M. & WEISS, T. 2007. Differences in texture, physical properties and microbiology of weathering crust and host rock: a case study of the porous limestone of Budapest (Hungary). *In*: PRIKRYL, R. & SMITH, B. J. (eds) *Building Stone Decay: From Diagnosis to Conservation*. Geological Society, London, Special Publications, **271**, 261–276.

TRUDGILL, S. T. & VILES, H. A. 1998. Field and laboratory approaches to limestone weathering. *Quarterly Journal of Engineering Geology*, **31**, 333–342.

TRUDGILL, S. T., VILES, H. A., INKPEN, R. & COOKE, R. U. 1989. Remeasurement of weathering rates, St Paul's Cathedral, London. *Earth Surface Processes and Landforms*, **14**, 175–196.

TURKINGTON, A. V. & SMITH, B. J. 2004. Interpreting spatial complexity of decay features on a sandstone wall: St Matthew's Church, Belfast. *In*: SMITH, B. J.

& TURKINGTON, A. V. (eds) *Controls and Causes of Stone Decay*. Donhead Press, Shaftesbury, Dorset, 149–166.

VAZQUEZ-CALVO, C., VARAS, M. J., ALVAREZ DE BUERGO, M. & FORT, R. 2010. Limestone on the 'Don Pedro I' facade in the Real Alcázar compound, Seville, Spain. *In*: SMITH, B. J., GOMEZ-HERAS, M., VILES, H. A. & CASSAR, J. (eds) *Limestone in the Built Environment: Present-Day Challenges for the Preservation of the Past*. Geological Society, London, Special Publications, **331**, 171–182.

VILES, H. A. 1993. The environmental sensitivity of blistering of limestone walls in Oxford, England: A preliminary study. *In*: THOMAS, D. S. G. & ALLISON, R. J. (eds) *Landscape Sensitivity*. Wiley, Chichester, 309–326.

VILES, H. A. 2002. Implications of future climate change for stone deterioration. *In*: SIEGESMUND, S., WEISS, T. & VOLLBRECHT, A. (eds) *Natural Stone, Weathering Phenomena, Conservation Strategies and Case Studies*. Geological Society, London, Special Publications, **205**, 407–418.

The use of limestone in a historic context – the experience of Malta

JOANN CASSAR

Department of the Built Heritage, Faculty for the Built Environment, University of Malta, Msida, MSD 2080, Malta (e-mail: joann.cassar@um.edu.mt)

Abstract: The Maltese Islands measure only 316 km^2, have a population of just over 405 000 and are situated in the central Mediterranean. They are composed of sedimentary rocks, of which the Globigerina and Coralline Limestones have been used as building materials since prehistoric times. This paper gives an overview of the use of these materials, and other imported materials, for building from prehistoric times to the present day, and also looks at the exploitation of the underground environment through the ages.

The history of the Maltese Islands has long been closely associated with their strategic location in the central Mediterranean, whilst the Islands' natural and built environments have largely been shaped by their geology and their surface rock sequence of sedimentary rocks. Narrow dry valleys, karstland, cliffs and a varied coastline typify the Maltese landscape, which also includes innumerable quarries, caves, hypogea and catacombs, megalithic structures, dolmens, wayside chapels and baroque churches, kilometres of fortifications, walled cities, palaces, as well as humble farmhouses and other dwellings all built, or carved, out of the local limestone: Globigerina Limestone, known locally as *franka* (freestone), or softstone, and Coralline Limestone, or hardstone. Globigerina Limestone, an easily cut, carved and shaped stone, is one of the few natural resources of the Islands, and has been the main local building stone used for millennia; Coralline Limestone, on the other hand, has also been widely used but somewhat less exploited. Both stones were also exported in the past: Globigerina Limestone, often referred to simply as 'Malta stone', was sent to areas such as North African countries, Greece and Turkey (Bruno 2004). Thus, it was recorded that 'important foreign buildings constructed with Malta stone are the Royal Palace at Athens, St Louis Cathedral at Carthage, and a well-known Protestant Church in Naples, besides many others in Greece, Egypt and Northern Africa' (Mamo 1936). Often referred to as 'Maltese marble', Coralline Limestone was also shipped abroad, primarily during the British period, when it was documented that exports 'to England and Belgium have exceeded 400 tons in the last 12 months' (Mamo 1936).

In this paper the case will be made that, since prehistoric times, the Maltese builder and sculptor have overwhelmingly used the local limestones as the main building and decorative materials, supplementing these since earliest days with the importation of materials not found locally, for specific uses. When employing local materials, the inhabitants often chose to use Globigerina Limestone and Coralline Limestone differently – using the softer, easily carved Globigerina Limestone as the material of choice for buildings and decoration, whilst using the harder Coralline Limestone for situations where greater durability was required.

It will also be shown how, from the earliest times, even when only stone tools were available, humans also chose to excavate and use underground sites, fully exploiting the properties of the soft Globigerina Limestone (Pace 2000), whilst also utilizing naturally occurring caves that widely occur in this limestone country (Buhagiar 2005). This tradition of living in underground locations was revived during the heavy bombardments of World War II, when numerous shelters and tunnels were dug in the soft Globigerina Limestone, and were used not only to accommodate the local population, but also to house the British Military Operations Rooms in Valletta.

Materials

Globigerina Limestone

Globigerina Limestone forms part of the Oligo-Miocene 'soft limestones' widely found in the Mediterranean Basin, including Turkey, Israel, Tunisia, Spain and Italy. This formation is made up of three members, the Lower, Middle and Upper Globigerina Limestone (Table 1), separated from each other by a band of phosphatic nodules. It is the Lower Globigerina Member that has been used as the main local building material. It can be described as a pure limestone (calcite >92%), containing small amounts of quartz, feldspars, apatite, glauconite and clay minerals; detailed composition and property data have been published in Cassar (2004). The porosity is very high, reaching values of up to 40.7% (Cassar *et al.* 2008). Other test

Table 1. *Lithostratigraphical subdivisions on the Maltese Islands (Oil Exploration Directorate 1993)*

Formation/Member	Thickness (m)	Lithologies
Upper Coralline Limestone Formation		
Ġebel Imbark Member	4–25	Hard pale grey carbonates with sparse faunas. Deposits now restricted to erosional outliers and synclinal cores
Tal Pitkal Member	30–50	Pale grey and brownish-grey coarse-grained wackestones and packstones containing significant coralline algae, mollusc and echinoid bioclasts
Mtarfa Member	12–16	Massive–thickly bedded carbonate mudstones and wackestones, yellow in their lower levels and unconformable upon Greensand in western outcrops. Carbonates become white and chalky in the upper two-thirds of eastern outcrops
Għajn Melel Member	0–13	Massive bedded dark–pale brown foraminiferal packstones containing glauconite occur above a basal Upper Coralline Limestone erosion surface in western Malta
Greensand Formation	0–11	Friable, brown–greenish glauconite-rich sands occur above a marked erosion surface truncating the Blue Clay Formation
Blue Clay Formation	15–75	Medium grey and soft, pelagic marls, typically with well-developed pale bands rich in planktonic foraminifera, but lower clay content
Globigerina Limestone Formation		
Upper Globigerina Limestone Member	8–26	A tripartite, fine-grained planktonic foraminiferal limestone sequence comprised of a lower cream-coloured wackestone, a central pale grey marl and an upper pale cream-coloured wackestone
Middle Globigerina Limestone Member	15–38	A planktonic foraminifera-rich sequence of massive, white, soft carbonate mudstones locally passing into palegrey marly mudstones
Lower Globigerina Limestone Member	0–80	Pale cream–yellow planktonic foraminiferal packstones rapidly becoming wackestones above the base
Lower Coralline Limestone Formation		
Il Mara Member	0–20	Tabular beds of palecream–palegrey carbonate mudstones, wackestones and packstones in 1–2 m-thick units
Xlendi Member	0–22	Planar–cross-stratified, coarse-grained limestones (packstones) with abundant coralline algal fragments
Attard Member	10–15	Grey limestones (wackestones and packstones) are typical throughout Malta
Magħlaq Member	>38	Massive bedded, pale yellowish–grey carbonate mudstones are dominant, benthonic foraminifera alone are frequent

data relative to Lower Globigerina Limestone are given in Table 2.

Freshly quarried Globigerina Limestone is pale yellow in colour, and is fine-grained and generally homogeneous in texture. Sections where bioturbation is concentrated also occur, and in other areas concentrations of yellow or brown stains can be found. Traditionally, however, the unblemished variety has normally been chosen as building material. Quarry workers have long recognized that two facies occur, locally called *franka* and *soll*. *Ġebla tal-franka* (freestone) is the generic name for building stone, which is also generally reputed to weather well, whereas stone obtained from the *soll* facies is alleged to be less durable and will deteriorate in a few years, depending on exposure (Cassar 2002). These two types of Globigerina Limestone differ in geochemical and mineralogical composition, and also in physical properties (Cassar 2002). The *soll* variety has a lower overall porosity and a higher proportion of small pores (Farrugia 1993; Fitzner *et al.* 1996;

Table 2. *Properties of Lower Globigerina, Upper Coralline and Lower Coralline Limestone*

	Specific gravity	Porosity (%)	Saturation coefficient	Compressive strength uniaxial (N mm^{-2})	Flexural strength (N mm^{-2})
Lower Globigerina Limestone	2.35–2.61[1]	23.9–40.7[1-8]	0.66–0.95[1,8]	15.0–32.9[2] (dry) 7.95–16.3[2,3] (sat)	1.1–4.7[1,2] (dry) 1.2–3.7[2] (sat)
Upper Coralline Limestone	2.05–2.52[1]	2.4–32.3[1]	0.52–1.0[1]	8.8–67.2[1] (dry)	NA
Lower Coralline Limestone	2.43–2.65[1]	1.8–28.3[1]	0.61–1.0[1]	6.8–105[1] (dry) 7.7–23.8[9] (sat)	5.90–11.23[9] (dry) 5.96–11.23[9] (sat)

References: [1]Bonello (1988); [2]Cachia (1985); [3]Sammut (1991); [4]Farrugia (1993); [5]Fitzner *et al.* (1996); [6]Rothert *et al.* (2007); [7]Cassar *et al.* (2008); [8]Building Research Station (1963); [9]Tampone *et al.* (1994).
(sat), saturated; NA, not available.

Cassar 2002; Rothert *et al.* 2007; Cassar *et al.* 2008). The highest percentage porosity measured for Globigerina Limesone (40.7%) has been for a *franka* sample (Farrugia 1993), whereas the lowest measured (23.9%) has been for a *soll* sample (Sammut 1991). *Soll* samples also tend to have a higher average dry compressive strength when compared to *franka* samples (Cachia 1985; Sammut 1991; Cassar *et al.* 2008).

In the fresh state, the two types cannot be distinguished visually; however, abandoned quarry faces show the two types as bands of stone differing in weathering intensities. The difference in weathering has also been observed through salt weathering tests (Fitzner *et al.* 1996; Rothert *et al.* 2007; Cassar *et al.* 2008). Frequently, the main deterioration phenomenon is alveolar weathering, often leading to back-weathering (Fig. 1). The primary weathering process responsible for the deterioration of Globigerina Limestone has been recognized as salt crystallization in the pores of this very porous limestone (Cassar 2002). The weathering forms, and intensity, vary because of local differences in salt content and types, which are mostly, but not exclusively, of marine origin (Torfs *et al.* 1996), and also because of the differing quality and weathering resistance of the stone (Cassar 2002).

Coralline Limestone

Both Upper and Lower Coralline Limestone have been used locally since Prehistoric times, albeit to a lesser extent than the Globigerina Limestone. The Coralline Limestone Formations are composed of strata that vary widely in appearance and strength. They include reef limestones, foraminiferal limestones, lime muds and cross-bedded lime sands (Pedley *et al.* 2002). The Ġebel Imbark Member of the Upper Coralline Limestone, and the Xlendi and Attard members of the Lower Coralline Limestone, have been used for construction purposes in Valletta (Pedley & Hughes Clarke 2002), whereas the Għajn Melel Member, also of the Upper Coralline Limestone, was used to build the outside of the Citadel in Gozo (Table 1). Test data pertaining to these two types of Coralline Limestone are given in Table 2. Also interesting to note is that the majority of caves associated with the cave-dwelling phenomenon in Malta from the Late Medieval Period onwards (see later) are located in the Mtarfa Member of the Upper Coralline Limestone Formation (Buhagiar 2007).

Buildings

The Maltese Islands are heavily built-up, with 23.22% of the land being used for buildings and related purposes; this includes built-up land, industrial land, and land used for quarries and for infrastructural purposes (NSO 2007). Existing structures range from complex megalithic prehistoric temples, Medieval wayside chapels, impressive Baroque buildings and fortifications built by the Knights of St John, and vernacular architecture, to

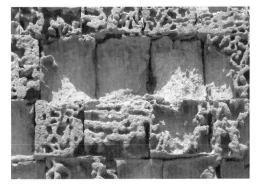

Fig. 1. Deterioration forms on a Globigerina Limestone wall, showing predominantly alveolar weathering, as well as back-weathering (author's photograph).

military constructions built at the time when Malta was a British colony, to modern edifices.

The oldest buildings on the Islands, the prehistoric megalithic temples, were built during the period 4000–2500 BC, also known as the 'Temple Period' (Pace 2000). In the Mediterranean this is the period that falls between the Neolithic and the Bronze Age. The best-preserved temple complexes are those of Ħaġar Qim, Mnajdra, Tarxien and Ġgantija (Fig. 2), which are recognized as UNESCO World Heritage Sites. Here, extensive use is made of both Globigerina and Coralline Limestone (see below).

Numerous towns and villages dot the Islands of Malta and Gozo; traditionally, these were built wherever a good supply of Globigerina Limestone was locally available. This has resulted in a wealth of vernacular architecture still present over most of the Islands. Most towns and villages also boast impressively large village churches, often built from stone excavated from the basement or crypt of the very church. The countryside is also dotted with innumerable wayside churches.

Another striking aspect of the Maltese Islands is an imposing Baroque architecture, a legacy of the occupation by the Order of St John, and a major characteristic of the walled cities of Valletta and Mdina, respectively, the present and past capital cities of the Islands. These conurbations house magnificent palaces and churches, as well as kilometres of fortifications, all built of the local limestone.

The British legacy consists mostly of forts and fortifications, which were built primarily in three areas: on the Great Fault that divides the northern and southern parts of Malta; around the harbour area; and on the SE coast (Spiteri 1991). Also during this period other buildings and monuments were built in Valletta and elsewhere, 'the Doric Revival [...] used by the British to assert their presence in the island and as a symbol of their imperial might and glory' (Mahoney 1996, p. 211). Ample use was made here of the readily available and easily cut and shaped Globigerina Limestone, although this period of great innovation in building materials also introduced into Malta cast iron and concrete. This led the way to the modern era, where, although Globigerina Limestone is still widely employed in the erection of domestic dwellings, reinforced concrete is also one of the main materials used in the construction of modern, especially high-rise, buildings (see later).

But we will start this journey through the millennia with structures that were built 'around 3500 BC, [when] no-one else was raising free-standing buildings in stone anywhere else in the world' (Trump 2002, p. 69).

Prehistoric sites

The megalithic temples

Both Globigerina Limestone and Coralline Limestone were used in the construction of the prehistoric

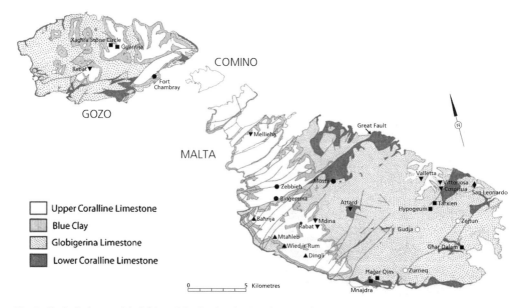

Fig. 2. Geological map of the Maltese Islands, showing locations mentioned in the text. Adapted by J. Bonello from Bowen Jones *et al.* (1961) and Buhagiar (2007). (■, Prehistoric sites; ▲, areas with cave dwellings; ○, towns with Late Medieval churches; ▼, important towns; ●, locations with forts.)

temple complexes. The mode of construction is megalithic, with stones up to 6.40 m long being used to build a series of adjacent curved apses apparently, as can currently be observed, without the use of any bedding mortar. Although built over a period of 1000 years, the temples have a number of consistent characteristics. These include the use of freestanding megaliths that form 'trilithon' portals along the main axis of the temples, whilst other freestanding uprights were used to build semi-circular apses, arranged along the axis, with horizontal megaliths laid in 'courses' above (Torpiano 2004).

Studies carried out by Grima (2004) concluded that the type of underlying rock did not determine the building of a temple structure. However, it is clear that the temples are built mainly of stone available in the immediate vicinity, and only stone for very specific purposes was carried for significant distances (Grima 2004). There is also a strict correlation with the use of a particular material and its location within the structure. In the case of Mnajdra, for example, Lower Coralline Limestone, abundantly present in the immediate vicinity, was used for the construction of the external walls, whereas Globigerina Limestone, brought in from further afield, was used for the internal walls. Other temples, such as that of Ħaġar Qim (Fig. 3), are built entirely of Globigerina Limestone, as no Coralline Limestone is readily available nearby. The Globigerina Limestone was also crushed to make the *torba* (beaten earth) flooring of some of the temples (Evans 1971). *Torba:* 'A plaster-like material much used in the temples flooring. It was produced by crushing Globigerina limestone, then spreading it over a rubble foundation. After repeated wetting and pounding, it set hard and could be polished. The result can easily be mistaken for bedrock, and in excavations it frequently is' (Trump 2002, p. 77). At Ġgantija in Gozo, however, very little use was made of Globigerina Limestone, practically the entire complex being built of Upper Coralline Limestone cropping out nearby (Tampone *et al.* 1987).

Most decorative elements, including altars, animal reliefs and spiral motifs, were carved out of the soft Globigerina Limestone, as were a number of statues and statuettes – representing both seated and standing figures – found in several of the megalithic temples, ranging in size from a few centimetres to a colossal statue from the Tarxien temples, which must have been about 2 m high when intact (Evans 1971). Of great importance are six Globigerina Limestone stick figures, approximately 16 cm high, with human heads on plaque-like bodies, some unfinished, which were found during excavations at the Xagħra (Brochtorff) Stone Circle (see later) in Gozo (Stoddart *et al.* 1993). Functional elements, such as tools, on the other hand, were generally made out of Coralline Limestone (Grima 2004).

Excavations of prehistoric sites have also unearthed numerous artefacts made of materials not naturally available in the Islands. These include obsidian from the neighbouring islands of Pantalleria and Lipari, 'greenstone' from Calabria and other 'exchange products from as far afield as the Alps' (Stoddart *et al.* 1993). Brown *et al.* (1995) identified the lithologies used for 'ground stone artefacts' found in the Xagħra Stone Circle as falling 'into three broad petrological groups', these include: 'greenstones' (a nephrite group, green serpentinite, white tremolite, eclogite, metagabbro); volcanic rocks (basaltic); and fine-grained limestones, limonitic iron-carbonate and sandstone. Possible sources include Calabria, Sicily and Pantelleria, and even northern Italy or the western Alpine zone (Brown *et al.* 1995). Flint, probably from Sicily, as well as local chert are also mentioned. Another material found in Maltese prehistoric sites is pumice from Lipari, mentioned by Trump (2002). Evans (1971) also records artefacts made of 'alabaster' and 'lava'. Decoration and ritual made wide use of red ochre, possibly originating, at least in part, from Sicily (Stoddart *et al.* 1993). Malone (1995) mentioned finding 'lumps of ochre', as well as 'ochre-stained material', in the Brochtorff Circle excavations, in addition to a 'bowl full of ochre' (Stoddart *et al.* 1993).

Underground locations

The earliest known underground location occupied by the Neolithic people who first arrived in Malta is Għar Dalam cave (Fig. 2), a natural cave located in the Lower Coralline Limestone (Zammit Maempel 1989*a*). This 'natural water-worn cave on the south eastern part of Malta [...] has a length of about 145 m' (Zammit Maempel 1989*a*, p. 13). This site is important for the numerous

Fig. 3. Façade of Ħaġar Qim megalithic temples (author's photograph).

remains of dwarf animals (hippopotami, deer and, to a lesser extent, elephants) dating back to the Pleistocene (Zammit Maempel 1989a). This cave was inhabited around 5000 BC, and gives its name to the first phase of Maltese prehistory, the Għar Dalam Phase. Here, human occupation is evidenced by the discovery of the earliest pottery in the Maltese Islands, as well as flakes of obsidian and worked flint and chert (Evans 1971).

The Ħal Saflieni Hypogeum

During the temple building period in the Maltese Islands, not only were extraordinary and complex megalithic temples constructed, but underground sanctuaries also developed. These include the Ħal Saflieni Hypogeum in Malta and the Xagħra Stone Circle in Gozo.

The Ħal Saflieni Hypogeum (Fig. 2), a unique underground site carved out of the Globigerina Limestone Formation, dates to the period 4000–2500 BC (Pace 2000). It is thought to have been used for religious purposes whilst also having a funerary function (Trump 2002).

This site consists of a series of irregular intercommunicating halls, chambers, passages, recesses and niches spreading out over three levels, covering an area of approximately 500 m^2 and extending to a depth of 10.6 m. The uppermost level, at ground level, is in part built and in part excavated; this may have consisted of natural caves, altered at some point by the site's users (Pace 2002). The other two levels are located entirely underground. Of the three levels, the middle one is the most ornate, with elaborately carved rooms that are embellished with rock-cut replicas of architectural features to be found in the local contemporaneous megalithic temples (Fig. 4). Many areas are also decorated with red ochre designs that range from a simple wash over the walls to elaborate spirals and hexagons, which also extend over the carved ceiling (Cassar 1996).

Fig. 4. Chambers, recesses and niches carved out of the middle level of the Ħal Saflieni Hypogeum (photograph by D. Cilia).

Xagħra Stone Circle

An equally intriguing prehistoric underground site is that of the Xagħra Stone Circle (Fig. 2) in Gozo. Excavated from 1987 to 1994, it was also primarily used for burial purposes (Trump 2002). This site was not dug by humans, but consists of a series of natural caves adapted for burial. Situated in the Upper Coralline Limestone, there is a lack of elaborate carved decorations found in the Ħal Saflieni Hypogeum, this material being much harder and difficult to carve and shape. Site furniture, including altars and trilithon doorways, were made from blocks of Globigerina Limestone transported from about 1 km away (Trump 2002), and also included 'curved screens of pitted and painted globigerina architecture' (Stoddart et al. 1993, p. 10).

Phoenician and Roman times

After the magnificent temple building period, a succession of cultures followed in the Maltese Islands, none of which, however, left anything comparable as regards construction. The Bronze Age saw the appearance for the first time on the Islands of copper and bronze, appearing as axes and daggers in local sites (Trump 2002). Extant structures dating to this period are primarily dolmens dotting the islands of Malta and Gozo. The stones used here are usually completely unworked, and 'were probably slabs which had weathered out naturally on the surface' made of Coralline Limestone or Globigerina Limestone (Evans 1971, p. 193), depending on the materials present in the neighbouring area.

The succeeding colonizers were the Phoenicians, a consistent group of whom had established themselves on the Islands at 700 BC at the latest (Bonanno 2005). No significant structural remains survive from this period, archaeological evidence consisting mainly of rock-cut tombs (Bonanno 2005). From the Punic period, which dates from 500 BC, to the arrival of the Romans, we still find an 'outstanding surviving structure [...] [a] curious square building [...] [which] consists of a single room and stands about five metres high preserving a crowning cornice of Egyptian inspiration [...]' (Bonanno 2005, p. 91). This is built of Globigerina Limestone (N. Vella 2008 pers. comm.).

Malta became part of the Roman Empire in 218 BC. Some of the earliest writings of the period, which mention the islands of 'Melite' (Malta) and 'Gaulos' (Gozo), also mention local buildings. One such writer, Diodorius Siculus, writing between 60 and 30 BC, writes about a city and harbours, and also says 'The dwellings on the island are worthy of note, being ambitiously constructed with cornices and finished in stucco with unusual workmanship' (Bonanno 2005, pp. 182–184).

Surviving remains of buildings dating to the Roman period range from town houses, to country villas, to baths. Rock-cut tombs and necropoleis also survive from this period. Some of the finest remains are those of the Roman *domus* at Rabat, Malta (Fig. 2), dated stylistically to around 80 BC. Here, columns and entablature were made of the local (Globigerina) limestone (Gouder 1983; Bonanno 2005). This stone was also used for statuary. Conversely, the Romans used Coralline Limestone, once again, for utilitarian objects, such as olive-pippers. There is also much evidence for the importation and use of non-local materials. This is in keeping with the great fascination the Romans had for exploiting marble, including coloured marbles, from all over the Mediterranean and transporting them to colonies not only in the Mediterranean, but also as far away as Britain (Lazzarini 2007). In Malta, we find primarily the use of white marble, used for statuary and inscriptions; coloured marbles, and other stones, were imported and used to fashion impressive mosaics, found in several of the Roman remains in Malta and Gozo. Those of the Roman *domus* in Rabat are dated to 'a time span of 70 years from *c*. 125 to *c*. 50 BC' (Bonanno 2005, p. 164). Here we find used 'black and white marble and a soft green stone', as well as red, yellow and green tesserae (Gouder 1983, p. 4).

The Arab period

The chief remaining tangible legacy of the Arab period in Malta is the building system, which seems to have been wholly imported to the Maltese Islands (Mallia 2005). There is no written evidence for when the system appeared here; however, the use of Arab technical terms, including several archaic ones, indicates that this could have occurred during the Arab period of 870–1091 AD (Mallia 2005). Also possibly dating to this period are irrigation systems used locally before the late nineteenth century, when water management policies changed, which could have been part of 'a new horticultural and technological package introduced during the Muslim and post-Muslim period between the 11th and 13th centuries AD' (Buhagiar 2007, p. 103). These include 'narrow rock-cut tunnels tapping the perched aquifer' (Buhagiar 2007, p. 103).

Traditionally, buildings in Malta make ample use of Globigerina Limestone. These consist of thick walls and flat, or in the case of churches, shallow pitched roofs, all in stone (Hughes 1993). Transverse stone arches support large, flat stone slabs [*xriek*: 'a thin roofing slab' (Mahoney 1996)] to form roofs and floors; a thin roofing Globigerina Limestone slab can, at most, span a width of 2 m without cracking. Greater widths (up to about 2.75 m at ground level) were achieved by sloping the walls slightly inwards and adding corbels just below the roofing slabs (Mahoney 1996). The roofs were traditionally covered in the local lime-based mortar enriched with crushed pottery [*deffun*: '*Deffun*' mortars are part of the Maltese building industry history. [...] The "Deffun" is based on a mix of lime, crushed pottery and water, which is applied in a specific way leading to a high performance hydraulic mortar used for the protection of roofs against the action of rainwater' (Chetcuti 2003)] or pozzolana to make it waterproof (Chetcuti 2005). This technique was still utilized locally until the introduction in the twentieth century of modern waterproofing materials.

The Late Medieval Period

The earliest post-Muslim remaining structures are churches that dot the Maltese countryside. Of these, the better preserved ones still surviving today are the Church of the Annunciation of the Virgin at Ħal Millieri, limits of Zurrieq, Sta Marija ta' Bir Miftuħ, limits of Gudja, and the church of St Catherine of Alexandria, in Zejtun (Fig. 2). The earliest written references to the Church of the Annunciation date to 1495 (Luttrell 1976); Buhagiar (2005) reported that the more important countryside churches were enlarged and generally structurally modified in the fifteenth and early sixteenth centuries, resulting in the development of 'an essentially Maltese style of church' based on a technique that made ample use of stone, and very little timber.

In all three churches, extensive use is made of Globigerina Limestone, with transverse arches, and limestone roofing slabs resting on them. Mahoney (1996) stated that this roofing system was first observed in the fourth–seventh century Christian churches of the Hauran district in Syria, where, like Malta, there was a scarcity of timber and a plentiful supply of good building stone. This is reiterated by Buhagiar (2005) who also explained that stone, the exclusive building material, determined both the physical appearance of the building and its limitations.

Noteworthy in the churches at Ħal Millieri and Bir Miftuħ are wall paintings. Buhagiar (2005, p. 99) stated that the Ħal Millieri frescoes 'are the best-preserved and most complete cycle of church paintings to survive from the Late Middle Ages'. He also informs us that 'cheap and simple colours' were used in the execution of these murals. These were burnt and raw sienna, azurite, indigo, lime white, yellow, red and green earths, verdigris and charcoal (?) black (Zanolini 1976, translated by R. De Angelis 2008). The technique was in use in Italy from approximately 1250 to 1500, from where it, and possibly the non-local materials, were probably imported to Malta. The wall paintings of Bir Miftuħ, on the other hand, are post-medieval (Buhagiar 2005).

Underground locations

It is very common in the Mediterranean to use natural caves as places of habitation, as well as to excavate, or artificially enlarge, underground locations and use them as dwellings (Fiorini 1993). Other uses include burial places and cultic shrines (Buhagiar 2007). Cave dwellings were diffused in the Late Medieval Period, when troglodytic living 'became a principal mode of habitat and enjoyed a permanency that lasted well into the modern period' (Buhagiar 2005, p. 40). Jean Quintin d'Autin's *The Earliest Description of Malta* (1536) says 'There are many troglodytes in Malta; they dig caves, and these are their houses' (Vella 1980, p. 39). A *Descrittione di Malta*, dated 1716, also mentions cave-dwelling: 'La gente di campagna abitava anticamente nelle grotte, e ce ne vedono ancora' (translation: 'country folk in the past lived in caves, and this can still be seen') (Mallia-Milanes 1988, p. 57). Underground locations were also often used as places of worship, with troglodytic churches abounding, especially in the Mellieħa and Rabat areas of Malta (Mahoney 1996).

Cave dwellings

Troglodytic dwellings in Malta were of two distinct types in medieval Malta, those within natural karst depressions and those within cliff faces, which occur primarily in the Upper Coralline Limestone formation (Fig. 2), with cliff-face settlements generally found in the Mtarfa Member (Buhagiar 2007). The caves were often enlarged by their occupiers. Dry-stone walling was often used to partition interiors into separate living and storage areas. The entrance was commonly screened by dry- and wet-stone walls; these, and parts of the caves' interior, were sometimes plastered and whitewashed (Buhagiar 2007). Shafts leading to the exterior let in light and air, but were devised to exclude rain and wind.

Two types of roofing were in use: dry-stone screening walls were bridged and sealed with rough, thin slabs of stone (*xriek*) and a hydraulic mortar mix (*deffun*). Buhagiar (2007, p. 114) states that 'the materials utilized and the construction methods employed are probably similar to the late-medieval roofing methods'. Larger gaps were covered over with lighter materials, including dead vegetal material such as twigs, branches, bamboo reeds and hay (Buhagiar 2007).

Rock-cut cemeteries and churches

In his first chapter in his opus *The Late Medieval Art and Architecture of the Maltese Islands*, Buhagiar (2005, p. 2) stated 'The story of art in Malta [...] starts in the rock-cut cemeteries that honeycombed the area of the city of Melite, and several outlying country districts. These display a variety of tomb architecture and a richness of detail that make them unique in the Early Christian world'. Carved out of the soft Globigerina Limestone, these catacombs possess features such as tables, couches, niches, screens, recesses, pilasters and columns (Buhagiar 2005). Besides these architectural features, few of these catacombs are richly carved: 'the few wall paintings and wall-carvings that survive are [...] humble and primitive' (Buhagiar 2005, p. 3).

Several rock-cut churches occur near the old city of Mdina (Fig. 2), and scattered in the countryside, where these were 'an essential aspect of the troglodytic phenomenon' (Buhagiar 2005, p. 68); there is a major concentration in the north of Malta. Some were adapted or re-cut from Early Christian and Byzantine burial places; others made use of natural caves or grottoes. Many of these troglodytic churches were still in use in 1575, mentioned in the Apostolic Visitation Report of Mgr. Pietro Dusina (Buhagiar 2005). Some are adorned with wall paintings.

The Knights of St John

With the arrival of the Knights of St John from Rhodes in 1530, a great building phase, unprecedented in local history, started. This included the building of the fortified city of Valletta (Fig. 5), and the remodelling and enlarging of the walled city of Mdina. The Knights amply used locally available materials for buildings, but also imported and used great amounts of foreign material for

Fig. 5. Aerial view of Valletta, the capital city of Malta built by the Knights of St John (photograph by D. Cilia).

decorative purposes, as seen in the magnificent interior of St John's Co-Cathedral in Valletta. This included 'coloured ornamental sediments, serpentine breccias, and brown marble claddings to column bases, from Italy, Spain and Greece' (Pedley & Hughes Clarke 2002, p. 16). Lazzarini (2007) mentions the use of the *breccia di Aleppo* in the tomb of Grand Master Perellos in the Co-Cathedral. Black basalt from Sicily (Mount Etna) was used for street paving (Pedley & Hughes Clarke 2002).

In *The Earliest Description of Malta* published in Lyons in 1536 by Jean Quintin d'Autin (Vella 1980, pp. 37–39) we are told that in this period 'Masons [...] make good use of the island's stone for building purposes, or turn it into lime. The Maltese stone is white and remarkable for its softness; it is sawed more easily than wood. Often huge blocks of stones are prised loose by wedges from the solid rock. Hence it is worked easily, but it is not strong enough against moistures and sea breezes; and because of its dryness it is not strong enough for mortar and cements'. Another *Descrittione di Malta*, this time dating to 1716, mentions the local stone: 'ora tutti hanno case di pietra bianca, che è dolcissima al taglio' (translation: 'now all have houses made of white stone, which is very easy to cut') (Mallia-Milanes 1988, p. 57).

Most of the stone used by the Knights for building the fortifications was excavated on site, came largely from the ditches of the respective fortifications and was mainly (with a few exceptions) Lower Globigerina Limestone (Spiteri 2008). However, the Knights also made ample use of other materials obtained from other locations on the Islands – Upper Coralline Limestone was obtained from San Leonardo, and Lower Coralline Limestone from Cospicua, Attard and Mosta (Pedley & Hughes Clarke 2002). The Upper and Lower Coralline Limestones were more expensive, as they were more difficult to cut (Spiteri 2008). The Lower Coralline Limestone, known to be resistant to sea-spray, began to be employed, from the eighteenth century onwards, in coastal works of fortification (Spiteri 2008). The *tal-franka* (Globigerina) stone, of which seemingly unlimited supplies were available, was considered 'ideal for building, white in colour, easy to cut and specially suitable for use in the erection of fortress walls as it is not easily crushed by artillery', although, conversely, it was reputed not to stand up very well to humidity, and was rather soft (Spiteri 2008, p. 398; de Giorgio 1985).

By the late seventeenth century it had become established practice to construct, as much as possible, those sections of the fortifications exposed to 'harmful hot winds' out of the best quality stone. It was also the custom to build the lower courses of both buildings and ramparts out of hardstone (Coralline Limestone) (Spiteri 2008).

Fig. 6. Main gate of Mdina, the old capital city of Malta (author's photograph).

For the building of the main gateway at Mdina (Fig. 6), strict instructions were detailed regarding the materials to be used. A 1724 document specifies that 'all the stones must be quarried only from the superficial, as opposed to the lower profiles of the quarry' and 'if, for one reason or another, approval of the quarried stone is not forthcoming, the contractor [...] is obliged to abandon the quarry in favour of a more satisfactory one' (De Lucca undated).

An interesting and even more detailed insight into the materials used in construction comes from the archives relating to the building of Fort Chambray in Gozo in the mid-eighteenth century (Buhagiar & Cassar 2003; Spiteri 2008). The engineer of the Order initially spent 3 weeks on site to ascertain the quality of the ground and its suitability for building, as well as the workability of the stone that could be quarried there. One of the advantages of the site was said to be the rock itself, which was such that it was both easy to quarry and was strong on exposure to the atmosphere (Buhagiar & Cassar 2003). During the building of the fort, individual master masons were brought in occasionally to report on the quality of stone being quarried, amongst other things. Independent experts were brought in to estimate the amounts due to workers, and to certify the number and quality of stones that could be extracted from a particular area (Buhagiar & Cassar 2003). Stone quality was graded according to hardness and was used in different parts of the works according to the local strength requirements, and this also determined the price to be paid. Four main categories were distinguished: strong stone (*pietra forte*); hard stone (*pietra dura*); soft stone (*pietra molla*); and very soft stone (*pietra più molla*) (Buhagiar & Cassar 2003). For exposed areas, the better quality stone was used as much as possible. It was sometimes recommended that the first courses were to be of hard stone in areas said to be exposed to what was referred to as 'a harmful hot wind'. Areas facing west and SW needed to be built of 'seasoned' Globigerina Limestone, using the summer heat to 'dry out' the stone that was very damp when first quarried (Buhagiar & Cassar 2003).

British period

Under British rule, many publications appeared on the geology of the Maltese Islands, and on its primary building stone, Globigerina Limestone. The first systematic description was published in 1843 by Captain T. Spratt, who can in fact be considered as the founder of studies on the geology of the Maltese Islands (Zammit Maempel 1989*b*). Spratt referred to the formation nowadays called 'Globigerina Limestone' as 'freestone'. He stated that 'This is the stone which is commonly used for building in the two islands' and added that it was '[...] used extensively in all the public and private edifices of Malta and Gozo, and is an article of considerable export to all parts of the Mediterranean' (Spratt 1843, p. 227).

Murray (1890) was the first to use the name 'Globigerina Limestone'. Colson, writing in the same publication (Murray 1890, p. 472), said of what is today referred to as the Lower Globigerina Limestone: 'It is from this bed that the greatest quantity of building stone used in the island is obtained. It is a pale yellow limestone [...] is soft and easily worked, but hardens somewhat when exposed to air, weathers very well, and turns after a time to a light reddish brown colour [...]'. Colson also made what is probably the first written reference to the less durable *soll*, calling it '*saul*' (Murray 1890, p. 472).

Cooke was one of the most prolific nineteenth-century writers on the geology of the Maltese Islands. In his 'Observations on the geology of the Maltese Islands' [Cooke (Pt VIII) 1892] he described the Globigerina Limestone as '[...] a fine-grained freestone which, owing to its many excellent qualities, is largely used for building purposes [...]'.

Galea (1915, p. 152) made the first written reference to *franca* (*franka*), and also spoke of *soll*, which was said to be 'much inferior in quality, weather rather badly, and scale off especially when facing the south'.

Hyde (1955, pp. 72–73) referred to building stone that varied in quality. He stated that 'Anyone who strolls through Valletta and Malta generally can soon observe the difference in weathering due to the choice and quality of the rock. Whilst some of the fortifications and monuments are in perfect condition even after 300 years and more, other rock surfaces are corroded and show a "honey-comb" like appearance caused by some portions weathering out more quickly than others'.

The first studies on the durability of the local stone took place during the period 1958–1964, when the Building Research Station in the UK produced three internal notes on the building limestones of the Maltese Islands (Building Research Station 1958, 1963, 1964). In the 1958 report (p. 6), it was stated that 'The survey confirmed the statements recorded in already published work that Globigerina limestone varies considerably in natural durability', and 'there seemed little doubt that differences in quality did exist; indeed it would be unusual if this were not so'. The report went on to say that '[...] poor quality stones may be of the type known as "soll" or "*saul*"'.

Although many buildings built by the British largely utilized the local Globigerina Limestone, and hence their interest in the properties and durability of this material, the British period was also a time when much innovation in building materials was introduced. Structural steel was introduced, with the Naval Bakery at Vittoriosa being one of the first buildings where masonry and steel were combined structurally (Mahoney 1996). Built between 1842 and 1845, the Bakery saw the novel combination of '[...] cast-iron columns in parts of the building as supports for subsequent roofing structures and of rolled steel joists instead of timber beams for roofing various floors' (Magro Conti 2006, p. 5). Soon after this rolled steel joists also started being used in domestic buildings instead of traditional timber beams, used in conjunction with Malta stone until World War II. This system of construction disappeared with the introduction of reinforced concrete (Mahoney 1996).

Cast-iron pillars and girders were also used in the construction of the Valletta Market (commenced 1859) (Mahoney 1996, p. 238). A Maltese innovation here involved the roof cladding, where 'red deal boards [were] covered over with *torba* and *deffun*' providing a cheap and effective way of waterproofing and insulating the roof.

Following research into the production of hydraulic mortars for engineering works in the eighteenth century, Portland cement was first patented by Joseph Aspdin in England in 1824 (Weaver 1993). The earliest British Standard for cement (BS 12) was published in 1904 (Wright & Kendall 2008). This was the time when the Royal

Engineers were building forts and fortifications in Malta, for the defence of the Islands. 1844 saw the building of the first fort (Spiteri 1991); whereas from 1872, and for the next 30 years, fortifications were built primarily in three areas: on the Great Fault, which divides the northern and southern parts of Malta; around the harbour area; and on the SE coast (Spiteri 1991). Although ample use was made of local building materials, especially Globigerina Limestone – but also Coralline Limestone – 'concrete' was also utilized (Spiteri 2008 pers. comm.)

Powter (1978) informed us that whereas in the 1930s British military construction used almost exclusively brick or stone masonry, in 'the 1860's concrete was making significant inroads, and by the 1870's and 1880's it almost monopolized fortress construction' (p. 62). He continues to add that 'the first British fortress to make extensive use of concrete was built at Newhaven in Sussex' (p. 65). In Malta, cement was definitely in use in Malta by the mid-1870s (Spiteri 2008 pers. comm.). Among the first to include this material were the forts at Mosta and Bingemma, and the Dwejra Lines (Spiteri 2008 pers. comm.). Spiteri also observed that the earlier forts had a more fragile concrete than that used after the 1890s. Powter (1978, p. 69) in his paper gives as one of the conditions that can influence the deterioration of concrete fortifications and gun batteries the fact that 'The concrete is possibly not based on Portland cement' – this could have been the case also in the earlier forts in Malta. Spiteri mentions the use of Portland cement in Victorian forts, particularly those built around the turn of the century. Unfortunately, to date, this subject has not been studied locally, and it is not even known exactly when Portland cement was introduced in the Islands (Torpiano 2008, pers. comm.).

Today

Today, building in Malta continues at a fast pace. Production of softstone 'from the mid 1990s was around 0.7 million tonnes [...] per annum' and of hardstone 1.9 million tonnes (Mt); there are at present 22 hardstone quarries and 66 softstone quarries in the Maltese Islands (Cromie *et al.* 2003); the total area of the Islands is only 316 km^2 and the population in 2005 was 405 387 (NSO 2006). Coralline Limestone is used primarily as aggregate in the manufacture of concrete products, building and civil engineering projects, and road building and maintenance (Cromie *et al.* 2003). For these purposes, the importation of cement into the Maltese Islands reached an all-time high of 393 kt in 2006 and 359 kt in 2007 (Cembureau 2007). Globigerina Limestone is still much sought after for the building of housing and for cladding purposes; it tends to be used for buildings that are up to six or eight storeys high. Aesthetic considerations based especially on a higher demand for whiter stone means that there is much wastage (Cromie *et al.* 2003). The quarrying sector produced almost 2.2 Mt of waste in 2004 and almost 1.2 Mt of waste in 2005 (NSO 2006).

During the last 10–15 years, the construction industry has shown a reduced demand for Globigerina Limestone and an increased use of concrete products manufactured from hardstone (Cromie *et al.* 2003). This is resulting in a change in the face of the Islands, especially certain areas of Malta, where the number of concrete buildings greatly outweighs the number of buildings in Globigerina Limestone, which, as this paper has shown, was the building material of choice for millennia, starting from the first structures built in the Maltese Islands by their prehistoric inhabitants.

The author wishes to thank R. M. Cachia and N. Vella for reading and commenting on the original draft of the paper, and R. De Angelis for the translation of Italian terms. J. Bonello is warmly thanked for adapting Figure 2; the valuable help of M. Coleiro is also here acknowledged. D. Cilia was the photographer for Figures 4 and 5.

References

BONANNO, A. 2005. *Malta Phoenician, Punic, and Roman*. Midsea Books, Malta.

BONELLO, S. 1988. *Engineering Properties of Rocks and Soils of the Maltese Islands*. B.E. & A. dissertation, University of Malta.

BOWEN JONES, H., DEWDNEY, J. C. & FISHER, W. B. 1961. *Malta: A Background for Development*. Durham University Press, Durham.

BROWN, C., DIXON, J. E. & LEIGHTON, R. 1995. Stone axes and stone axe pendants. *In*: MALONE, C., STODDART, S., BONANNO, A., GOUDER, T. & TRUMP, D. (eds) *Mortuary Ritual of 4th Millennium BC Malta: The Zebbug Period Chambered Tomb from the Brochtorff Circle at Xaghra (Gozo). Proceedings of the Prehistoric Society*, **61**, 303–345.

BRUNO, B. 2004. *L'Arcipelago Maltese in Eta' Romana e Bizantina*. Edipuglia, Bari.

BUHAGIAR, K. 2007. Water management strategies and the cave-dwelling phenomenon in late-medieval Malta. *Medieval Archaeology*, **51**, 103–131.

BUHAGIAR, K. & CASSAR, J. 2003. Fort Chambray: the genesis and realisation of a project in eighteenth century malta. *Melita Historica*, **13**(4), 347–364.

BUHAGIAR, M. 2005. *The Late Medieval Art and Architecture of the Maltese Islands*. Fondazzjoni Patrimonju Malti, Malta.

BUILDING RESEARCH STATION. 1958. *The Maltese Islands: use of limestone for building*. Unpublished. Department of Scientific and Industrial Research, BRS, Watford.

BUILDING RESEARCH STATION. 1963. *Maltese limestones: relation of durability to laboratory-measured properties and efficacy of silicone treatments*. Note no. C965 (unpublished). Department of Scientific and Industrial Research, BRS, Watford.

BUILDING RESEARCH STATION. 1964. *The Maltese Islands: properties and behaviour of local limestone*. Internal note 6 (unpublished). Department of Scientific and Industrial Research, BRS, Watford.

CACHIA, J. 1985. *The Mechanical and Physical Properties of the Globigerina Limestone as used in Local Masonry Construction*. A. & C.E. dissertation, University of Malta.

CASSAR, J. 1996. Deterioration of a prehistoric hypogeum in Malta. *In: Proceedings of the XIII Congress of the International Union of Prehistoric and Protohistoric Sciences (UISPP)*, Volume **6**. ABACO Edizioni, Forli', Italy, Tome 1, 707–713.

CASSAR, J. 2002. Deterioration of the Globigerina Limestone of the Maltese Islands. *In*: SIEGESMUND, S., WEISS, T. & VOLLBRECHT, A. (eds) *Natural Stone, Weathering Phenomena, Conservation Strategies and Case Studies*. Geological Society, London, Special Publications, **205**, 33–49.

CASSAR, J. 2004. Composition and property data of Malta's building stone for the construction of a database. *In*: PŘIKRYL, R. & SIEGL, P. (eds) *Architectural and Sculptural Stone in Cultural Landscape*. Karolinium Press, Prague, 11–28.

CASSAR, J., MARROCCHI, A., SANTARELLI, M. L. & MUSCAT, M. 2008. Controlling crystallization damage by the use of salt inhibitors on Malta's limestone. *Materiales de Construcción*, **58**(289–290), 281–293.

CEMBUREAU. 2007. *Activity Report 2007*. The European Cement Association, Available at http://www.cembureau.be/Documents/Publications/Activity%20Report%202007.pdf (viewed on 16 August 2008).

CHETCUTI, F. 2003. *Deffun – Analysis of a Local Hydraulic Mortar*. B.Cons. dissertation, University of Malta.

CHETCUTI, F. 2005. *The Conservation of Deffun Roof Mortars*. M.Cons. dissertation, University of Malta.

COOKE, J. H. 1892. Observations on the Geology of the Maltese Islands. [Pt. VIII]. *Mediterranean Naturalist*, **1**(10), 152–154.

CROMIE, I., COLE, M. & MEPA. 2003. *Malta Environment & Planning Authority – Malta. Minerals Subject Plan for the Maltese Islands 2002*. Entec UK Limited. Available at http://www.mepa.org.mt/Planning/factbk/SubStudies/MineralsSS/Minerals_SP.pdf (viewed on 11 August 2008).

DE GIORGIO, R. 1985. *A City by an Order*. Progress Press Co. Ltd., Malta.

DE LUCCA, D. UNDATED. Architectural interventions in Mdina following the earthquake of 1693. *In*: AZZOPARDI, J. (ed.) *Mdina and the Earthquake of 1693*. Heritage Books, Malta, 45–76.

EVANS, J. D. 1971. *The Prehistoric Antiquities of the Maltese Islands*. Athlone Press, London.

FARRUGIA, P. 1993. *Porosity and Related Properties of Local Building Stone*. B.E. & A. dissertation, University of Malta.

FIORINI, S. 1993. Malta in 1530. *In*: MALLIA-MILANES, V. (ed.) *Hospitaller Malta 1530–1798*. Mireva Publications, Malta, 111–198.

FITZNER, B., HEINRICHS, K. & VOLKER, M. 1996. Model for salt weathering at Maltese Globigerina Limestones. *In*: ZEZZA, F. (ed.) *Origin, Mechanisms and Effects of Salt on Degradation of Monuments in Marine and Continental Environments. Proceedings of the European Commission Research Workshop on Protection and Conservation of the European Cultural Heritage, Bari, Italy*. Research Report, **4**, 333–344.

GALEA, R. V. 1915. Geology of the Maltese Archipelago. *In*: MACMILLAN, A. (ed.) *Malta and Gibraltar Illustrated*. Collingridge, London, 173–182.

GOUDER, T. C. 1983. *The Mosaic Pavements in the Museum of Roman Antiquities at Rabat, Malta*. Department of Museums, Malta.

GRIMA, R. 2004. The landscape context of megalithic architecture. *In*: CILIA, D. (ed.) *Malta Before History*. Miranda Publishers, Malta, 327–345.

HUGHES, Q. 1993. The architectural development of Hospitaller Malta. *In*: MALLIA-MILANES, V. (ed.) *Hospitaller Malta 1530–1798*. Mireva Publications, Malta, 483–507.

HYDE, H. P. T. 1955. *Geology of the Maltese Islands*. Lux Press, Valletta, Malta.

LAZZARINI, L. 2007. *Poikiloi Lithoi, Versicvlores Macvlae: I marmi Colorati della Grecia Antica*. Fabrizio Serra Editore, Pisa.

LUTTRELL, A. (ed.) 1976. *Hal Millieri: A Maltese Casale, its Churches and Paintings*. Midsea Books, Malta.

MAGRO CONTI, E. 2006. *The Malta Maritime Museum Vittoriosa*. Insight Heritage Guides Series, **11**. Heritage Books, Malta.

MAHONEY, L. 1996. *5000 Years of Architecture in Malta*. Valletta Publishing, Malta.

MALLIA, D. 2005. *Building Technology Transfer Between Malta and the Middle East: A Two Way Process*. Paper presented at the *International Seminar on the Management of the Shared Mediterranean Heritage 5th Conference on the Modern Heritage*. ISMARMED, Alexandria. Available at: http://www.unesco.org/archi2000/pdfmallia.pdf (viewed on 8 August 2008).

MALLIA-MILANES, V. 1988. *Descrittione di Malta Anno 1716 – A Venetian Account*. Bugelli Publications, Malta.

MALONE, C. 1995. General description. *In*: MALONE, C., STODDART, S., BONANNO, A., GOUDER, T. & TRUMP, D. (eds) *Mortuary Ritual of 4th Millennium BC Malta: The Zebbug Period Chambered Tomb From the Brochtorff Circle at Xaghra (Gozo)*. *Proceedings of the Prehistoric Society*, **61**, 303–345.

MAMO, J. 1936. Marbles and limestones of Malta. *Sands, Clays and Minerals*, **2**(4), 83–88.

MURRAY, J. 1890. The Maltese Islands, with special reference to their geological structure. *Scottish Geographical Magazine*, **6**, 449–488.

NSO. 2006. *Environment Statistics 2006*. National Statistic Office, Government of Malta. Available at: http://www.nso.gov.mt (viewed on 16 August 2008).

NSO. 2007. *Malta in Figures 2007*. National Statistics Office, Government of Malta. Available at: http://www.nso.gov.mt (viewed on 16 August 2008).

OIL EXPLORATION DIRECTORATE. 1993. *Geological Map of the Maltese Islands*. Office of the Prime Minister, Malta.

PACE, A. 2000. The Prehistoric Hypogeum at Hal Saflieni. *In*: PACE, A. (ed.) *The Hal Saflieni Hypogeum 4000 BC–2000 AD*. PEG Ltd, Malta.

PEDLEY, M. & HUGHES CLARKE, M. 2002. *Geological Itineraries in Malta and Gozo*. PEG Ltd, Malta.

PEDLEY, M., HUGHES CLARKE, M. & GALEA, P. 2002. *Limestone Isles in a Crystal Sea*. PEG Ltd, Malta.

POWTER, A. 1978. History, deterioration, and repair of cement and concrete in nineteenth century fortifications constructed by the Royal Engineers. *Bulletin of the Association for Preservation Technology*, **10**(3), 59–77.

ROTHERT, E., EGGERS, T., CASSAR, J., RUEDRICH, J., FITZNER, B. & SIEGESMUND, S. 2007. Stone properties and weathering induced by salt crystallization of Maltese Globigerina Limestone. *In*: PŘIKRYL, R. & SMITH, B. J. (eds) *Building Stone Decay: From Diagnosis to Conservation*. Geological Society, London, Special Publications, **271**, 189–198.

SAMMUT, A. 1991. *An Assessment of Globigerina Limestone Resources*. B.E. & A. dissertation, University of Malta.

SPITERI, S. C. 1991. *The British Fortifications: An Illustrated Guide to the British Fortifications in Malta*. Malta.

SPITERI, S. C. 2008. *The Art of Fortress Building in Hospitaller Malta*. BDL Publishing, Malta.

SPRATT, T. 1843. On the geology of the Maltese Islands. *Proceedings of the Geological Society of London*, **4**, Part II, (97), 225–230.

STODDART, S., BONANNO, A., GOUDER, T., MALONE, C. & TRUMP, D. 1993. Cult in an island society: prehistoric Malta in the Tarxien Period. *Cambridge Archaeological Journal*, **3**(1), 3–19.

TAMPONE, G., VANNUCCI, S. & CASSAR, J. 1987. Nuove Ipotesi sull'Architettura del Tempio Megalitico di Ggantija a Gozo. *Bollettino Ingegnieri*, **3**, 3–21.

TAMPONE, G., VANNUCCI, S. ET AL. 1994. I Templi Megalitici Preistorici delle Isole Maltesi: Determinazione delle Propieta' Meccaniche dei Materiali ed Interpretazione dei Dissesti. [Prehistoric megalithic temples of the Maltese Arcipelago: determination of mechanical properties of materials and interpretation of structural failures.] *In*: FASSINA, V., OTT, H. & ZEZZA, F. (eds) *Conservation of Monuments in the Mediterranean Basin, Proceedings of the 3rd International Symposium, Venice, Italy*. Soprintendenza dei Beni Artistici e Storici di Venezia, Italy, 567–575.

TORFS, K., VAN GRIEKEN, R. & CASSAR, J. 1996. Environmental effects on deterioration of monuments: case study of the Church of Sta. Marija Ta' Cwerra, Malta. *In*: ZEZZA, F. (ed.) *Origin, Mechanisms and Effects of Salt on Degradation of Monuments in Marine and Continental Environments. Proceedings of the European Commission Research Workshop on Protection and Conservation of the European Cultural Heritage, Bari, Italy*. Research Report, **4**, 441–451.

TORPIANO, A. 2004. The construction of the megalithic temples. *In*: CILIA, D. (ed.) *Malta Before History*. Miranda Publishers, Malta, 347–365.

TRUMP, D. H. 2002. *Malta Prehistory and Temples*. Midsea Books, Malta.

VELLA, H. C. R. 1980. *The Earliest Description of Malta (Lyons 1536)*. Translation of Jean Quintin d'Autun *Insulae Melitae Descriptio*. DeBono Enterprises, Malta.

WEAVER, M. E. 1993. *Conserving Buildings. A Guide to Techniques and Materials*. Wiley, New York.

WRIGHT, A. & KENDALL, P. 2008. The listening mirrors. A conservation approach to concrete repair techniques. *Journal of Architectural Conservation*, **14**(1), 33–54.

ZAMMIT MAEMPEL, G. 1989a. *Għar Dalam Cave and Deposits*. PEG, Malta.

ZAMMIT MAEMPEL, G. 1989b. *Pioneers of Maltese Geology*. Mid-Med Bank, Malta.

ZANOLINI, P. 1976. Operazioni di Restauro. *In*: LUTTRELL, A. (ed.) *Hal Millieri: A Maltese Casale, its Churches and Paintings*. Midsea Books, Malta, 106–107.

Carbonate rocks in the Mediterranean region – from classical to innovative uses of building stone

JOSÉ P. CALVO* & MANUEL REGUEIRO

Instituto Geológico y Minero de España, c/Ríos Rosas 23, 28003 Madrid, Spain
*Corresponding author (e-mail: jose.calvo@igme.es)

Abstract: Carbonate rocks are present in many geological formations of the Mediterranean region, thus having favoured their common use as building stone for the many civilizations that inhabited the area throughout history. The wide presence of carbonate rocks has been supplemented by a large variety of rock types that can be found in monumental, funerary and normal constructions. Five main carbonate rock types used as building stone can be differentiated: metamorphic marble; banded fine-grained limestone; shell limestone; travertine; and brecciated–nodular carbonate rocks. In most cases these carbonate rocks have been traditionally used, although new rock types are currently marketed. Nowadays, several Mediterranean countries are major producers of carbonate building stone that is marketed for its ornamental and decorative characteristics. Italy, Spain and Turkey are placed in high rank positions of the global carbonate ornamental stone market, whilst Greece, France, Croatia, Israel, Morocco, Egypt, Algeria and Tunisia produce mainly for internal consumption.

Carbonate rocks have been used as building stone by many Mediterranean civilizations. This is not surprising as calcareous geological formations crop out widely in most of the countries around the Mediterranean Sea, yielding a variety of carbonate rocks with appropriate petrophysical characteristics for building commodities. Exploiting carbonate rocks from neighbouring natural outcrops, successive civilizations from the Ancient times (i.e. Etruscans, Phoenicians, Greeks, Romans, inhabiting the Mediterranean coasts, architects during the Middle Age, the Renaissance and more recent centuries) had no difficulty in finding source rocks for use in funerary works, churches, cathedrals, palaces and other constructions.

Production of carbonate building stone, mainly as ornamental and decorative stone, constitutes an important industry in many Mediterranean countries nowadays (Fig. 1), especially Italy, Spain and Turkey, which are, respectively, the second, third and sixth building stone producers in the world (Table 1). Other countries, such as Greece, France, Croatia, Israel, Egypt, Tunisia, Algeria and Morocco, also produce carbonate ornamental stone but they are less important in terms of production value and/or presence in the global stone market. Nonetheless, all these countries together contribute significantly to the world carbonate building stone production, which accounted for 55 million tonnes (Mt) (54% of the total world stone production) in 2006 (*Roc Máquina* 2006).

From Ancient times, some carbonate rocks from Italy, Spain, Greece and Turkey have been highly prized for decorative and sculptural uses. In particular, metamorphic marble from Carrara (northern Italy), Macael (SE Spain), the Greek hinterland and islands (Thassos, Paros, Skyros, etc.), and the western Turkish coast was widely traded across the Mediterranean. White metamorphic marble from these various provenances was used as raw material for some of the most famous sculptures and monuments in the history of mankind.

In the modern commercial literature the term 'marble' is used as a generic name for all carbonate rocks that are dimensioned, processed and marketed as pieces of varied shape and size for ornamental purposes in buildings. This introduces considerable confusion with the common use in geological terminology where 'marble' is only used for metamorphic carbonate rocks. Moreover, commercial use of the term 'marble' also includes serpentine and other green rocks of non-carbonate composition. No differentiation between limestone and dolostone is recognized in commercial literature dealing with carbonate building stone.

In this paper an overview of the main carbonate rock types used as building stone in the Mediterranean region is presented. Both metamorphic marble and limestone have been considered, although they are clearly differentiated throughout the text. Some of these carbonate rocks are the basis for the traditional architecture of the different Mediterranean countries. Moreover, new rock types and processing methods have enlarged the list of ornamental stone types available worldwide from this relatively small area.

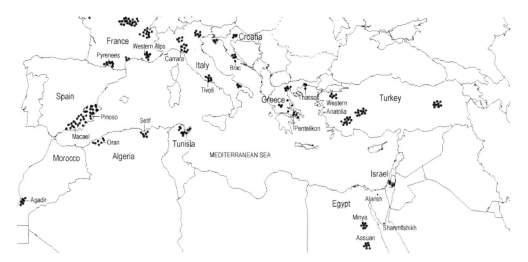

Fig. 1. Schematic map of the Mediterranean showing the most important productive zones (dotted areas) for carbonate building stone.

Table 1. *List of the main building stone producers in 2006, as well as data of imports/exports (adapted from Roc Máquina 2006). Values include marble, granite and slate*

Main producers (Mt)		Main importers (Mt)		Main exporters (Mt)	
China	15.72	China	4.11	China	10.60
India	11.20	USA	3.84	Italy	3.18
Italy	10.40	Italy	2.49	India	3.15
Spain	8.57	South Korea	1.81	Turkey	2.67
Turkey	7.72	Germany	1.77	Spain	2.48
Brazil	6.40	Taiwan	1.62	Brazil	1.81
Portugal	2.83	Japan	1.60	Croatia	1.15
Egypt	2.20	Spain	1.42	Portugal	1.13
Greece	2.10	Holland	1.06	Belgium	0.81
USA	1.30	Belgium	1.02	South Africa	0.68
France	1.20	France	0.96	Iran	0.50
Mexico	1.10	UK	0.89	Germany	0.45
Poland	1.10	Saudi Arabia	0.77	Canada	0.39

Geological features of the Mediterranean region

Southern Europe is flanked by Mesozoic–Cenozoic orogenic systems resulting from the convergence and collision between the European and African–Arabian plates. The Alpine–Mediterranean orogenic belts have resulted from a complex evolutionary history of extension, subsidence, subduction, collision, topographic build-up, extensional collapse and back-arc spreading that started in the early Mesozoic (Giese 2005). During the Mesozoic, widespread carbonate platforms formed in passive continental margins flanking the Tethys Ocean. By the late Cretaceous, the Mediterranean area was dominated by subduction zones that inverted the previous extensional regime.

The main Cenozoic orogenic zones in the Mediterranean are the Alps–Betics, the Apennines–Maghrebides and the Dinarides–Hellenides–Taurides (Carminati & Doglioni 2005). Extensional basins superimposed on these orogenic belts during the Neogene. On the western side of the Mediterranean, the Valencia, Provencal, Alborán, Algerian and Tyrrhenian basins developed, and the Aegean Basin on the eastern side.

Main development of carbonate rocks in the Mediterranean was closely linked to periods of extensional tectonics during the Mesozoic. These rocks underwent more or less metamorphic transformation depending on their location relative to the orogenic belts created by subsequent subduction and collision. This is the case for the Triassic metamorphic marble in the Apennines and SE Spain.

During the Neogene carbonate platforms, including huge volumes of bioclastic sediments, developed that were later uplifted, mainly during the Pliocene.

Throughout the Quaternary formation of carbonate rocks in the Mediterranean region was mostly restricted to small lake and river basins that were closely related to carbonate massifs. In this continental setting, cold tufa and travertine are the most typical carbonate deposits.

Carbonate rock types used as building stone

A review of the carbonate rocks that have been and/or are mostly used as building stone in the Mediterranean region allows five main types to be defined whose petrographical features may be summarized as follows.

Metamorphic marble

Marble is a metamorphic crystalline rock derived from either limestone or dolostone. Most priced metamorphic marble is a uniform, cold white rock, formed of interlocking fine crystals of calcite or dolomite. These features give these rocks maximum density and strength with a minimum pore space. Coarser-grained varieties are less favourable for monumental, memorial or statuary stone (Winkler 1997). The colour of metamorphic marble may vary from white to pink, grey, green and yellow. Impurities due to the presence of clays and other accessory minerals may result in bands and streaks. Tectonic strength causes cataclastic and strongly folded structures that give an attractive appearance to the carbonate rock.

Several marble quarries in the Mediterranean region have been famous since historical times. Carrara, in NW Italy, is the best-known locality for metamorphic marble in the world, having attracted famous sculptors throughout human history. Macael, in SE Spain, has also provided metamorphic marble since Roman times. Pentelikon marble has largely been used in Greece and is still currently extracted near Athens, for example, Dionissos marble.

Banded fine-grained limestone

Several types of cream–greyish limestone showing centimetre-thick banding are included in the catalogues of many building stone producers. The banded structure is the result of changes in rock fabric, from mudstone to wackestone, where peloids and/or small bioclasts are the main constituents. The band limits may be emphasized by stylolites caused by interstatal pressure solution. Local crenulation resulting in crinkly beds is also a relatively common feature in fine-grained limestones.

These carbonate rocks are used as thin wall and floor panels. This limestone facies is typical of inner carbonate platforms, and occurs mainly in Mesozoic formations from Italy, Croatia and Greece.

Shell limestone

The limestone consists of coarse grainstone (Dunham 1962; 'rudstone' in Embry & Klovan's 1971 classification) formed of numerous, tightly packed fossil seashells showing varied sizes and shapes. The origin of these deposits relates to development of shallow, strongly agitated carbonate platforms where molluscs, bryozoans and algae skeletons were deposited with a marked orientation. Most commonly used shell limestone in the Mediterranean occurs in Neogene formations from SE Spain, southern Italy, Turkey, Israel and Egypt.

Shell limestone may be well cemented, although more frequently it shows high pore space, which enables this stone to resist freezing and salt crystallization. This latter characteristic can explain the relatively good performance of shell limestone in historical monuments from the Magna Greece period and cathedrals located in sea-coast cities around the Mediterranean (Fig. 2).

Travertine

The term 'travertine' is somewhat misleading as, in a commercial sense, it is used for continental carbonate rocks of different origin (Pentecost 2005). Thus, cold-water tufa, travertine '*sensu stricto*' and speleothems are included under this definition.

Carbonate tufa comprises large accumulation of cold-water plant remains on which calcium

Fig. 2. Ancient quarries in the Cave di Chiusa area, southern Sicily. Pliocene shell limestone was carved out directly as cylinders, about 1.50 m in diameter, to be used as columns for the Magna Greece temples.

carbonate precipitated after abrupt degassing in streams and/or springs. Tufa deposits can be massive or roughly bedded, and they have largely been used because of their ability to be easily cut and carved. Carbonate tufa is a common building stone for castles, cathedrals and public works in general in several cities of Spain. Travertine 'sensu stricto' is a product of hot natural springs and related ponds. In this setting chemical precipitation of carbonate mimics delicate bacterial textures, giving a very decorative appearance. The evenly cream-coloured travertine of Tivoli is one of the best examples of this rock type and was the favourite building stone in ancient Rome.

Some speleothems filling voids in karstic carbonate systems are also considered to be 'travertine' by commercial building stone producers. Carbonate is banded, and shows a variety of colours that are mostly caused by iron and other impurities. The carbonate bands are usually translucent when thinly cut, which gives an opal-like appearance to the rock.

Brecciated–nodular carbonate

A variety of limestone and dolostone types is included in this group. The common feature of these rocks is their inhomogeneous structure caused by several diagenetic processes. The most common process is brecciation of the primary carbonate rock owing to burial and/or tectonic deformation. This causes the stone to show a combination of 'matrix' – parent material and veins and/or voids that are filled by coarser crystalline carbonate cement. In other cases, the inhomogeneous fabric is the result of differential recrystallization of the carbonate; this process may be related to dolomitization of the primary limestone.

Red and greyish nodular limestone has been used as building stone in many countries of the Mediterranean region. The most typical nodular limestone comes from the so-called 'Ammonitico Rosso' formations that represent condensed limestone facies at several levels of the Tethyan Jurassic.

Miscellanea

The five main types of carbonate building stone described here do not include all the traditionally used and/or presently offered stones on the market. Several types of massive, homogeneous limestone can be found in both the monumental and popular architecture. For instance, oolitic limestone is relatively common in Middle Jurassic formations and its appearance after polishing is much prized. Yellow fine-grained dolomitic marble of Triassic age showing Mn-oxide stringers from SE Spain is marketed in the Far East. These are just two examples of carbonate rocks that are not included in by the five-type classification.

Carbonate building stone by countries

Italy

Italy is the third largest world stone producer (Table 1) and is the foremost producer of carbonate rocks. Metamorphic marble excavated in Carrara, northern Italy (Fig. 3), shows both white and veined varieties, and accounts for two-thirds of the total Italian carbonate building stone production.

The Apuan Alps are the source for more than 40 metamorphic marble varieties, exploited since the first century BC. From a geological point of view, the area belongs to the autochthonous part of the Tuscanian Apennines. The Apuan area is a tectonic window in which the allochthonous Ligurid terranes are not present, so allowing the original basement to crop out. The formation containing the Carrara metamorphic marble is of lowermost Jurassic age (Carmignani et al. 2005). The texture of the marble is typically saccaroidal. Colour is usually

Fig. 3. Exploitation of metamorphic marble of Lower Jurassic age in Carrara, Apuan Alps, northern Italy. The height of the working benches reaches up to 8 m.

white, but the metamorphic marble locally shows grey banding. The purest, fully recrystallized marble is known as 'statutario' marble. Other varieties include 'Marmo ordinario', 'Marmo venato', 'Marmo nuvolato', 'Marmo arabescato' and 'Marmo calacata'. Some of these varieties are not texturally homogeneous metamorphic marble rocks but metabreccia formed of clasts and matrix of contrasted colours.

In addition to the carbonate rocks exploited in the Apuan region, Italy produces several coloured varieties such as the beige-coloured, banded fine-grained limestone ('Serpeggiante', 'Trani') from Puglia, the 'Botticino' from Lombardy or the yellow limestone ('Giallo Siena' and 'Giallo Reale') from central Italy. The Venetto region hosts several red and pink carbonate rocks like 'Rosso Rubino' in Tuscany and 'Portoro' in Liguria. 'Tuscan Arabescatos', 'Orobici' and 'Fior di Pesco' are well-known coloured varieties as well.

Travertine is also a well-known, famous Italian building stone that has been extensively used for ornamental and construction purposes in many cities of the country. Most important travertine deposits are found around Tivoli, Rome, Grosseto and Siena, all in central Italy.

Spain

Spain is today, after Italy, the second largest world producer of carbonate building stone with over 5 million tonnes per year (Mt year^{-1}). Metamorphic marble, interbedded with mica- and calc-schist, crops out in the Betic Range, SE Spain (Lombardero & Regueiro 2002). Metamorphic marble beds are tightly folded and very irregular, with variations in thickness, number of beds, lateral continuity, etc. Most extensive exploitation of metamorphic marble takes place in Sierra de los Filabres, Almería, where the prized Blanco Macael (Fig. 4) and other white, grey, yellow and cream varieties are mined in many large quarries. In central-western Spain (Iberian Massif), metamorphic marbles of Cambrian age are exploited in a few areas (García-Guinea & Martínez-Frías 1992).

Major carbonate building stone potential in Spain is represented by Jurassic limestone and dolostone deposits located in the External Zones of the Betic Range (Lombardero & Regueiro 2002). White-beige oolitic and micrite limestone is mined in Granada, Seville and Córdoba provinces. Nodular red limestone is quarried in Murcia, Alicante, Granada and Málaga. Brecciated dolostone showing varieties such as 'Beige Serpiente' and 'Marrón Imperial' are exploited in Alicante and Murcia.

In Tertiary formations, the variety 'Crema Marfil' is extensively exploited from the Eocene

Fig. 4. Use of metamorphic marble sourced from the Macael quarries (Almería, SE Spain) for cladding the external walls of the mosque in Madrid.

of the Betic Range, SE Spain (Fig. 5). The Iberian Range is the second major geological structure containing calcareous ornamental formations. Quarries in Valencia and Tarragona supply pink, brown and beige Liassic and Upper Cretaceous limestone building stone. The Basque–Cantabrian Basin in northern Spain provides black ('Negro Marquina'), as well as grey, beige and red, limestone that is usually related to Urgonian reefs. Moreover, late Tertiary formations from several continental basins are formed of lake massive and travertine limestone deposits.

In summary, Spain provides a large variety of carbonate rocks that have been and/or are used as building stone. Lombardero & Quereda (1992) have reported more than 60 carbonate rock types quarried in the country. Current commercial catalogues from Spanish companies offer a similar number of 'marble' types from domestic sources.

Fig. 5. Large quarry (the largest marble quarry in Europe) near Pinoso, Alicante province, SE Spain, producing the Eocene limestone 'Crema Marfil'.

Turkey

The potential for carbonate building stone of Turkey is remarkable as the country possesses reserves reaching around 13 800 Mt, which represent 35% of the world total reserves (*Roc Máquina* 2002) and produces more than 100 varieties from around 500 quarries. Total annual production is around 1.62 Mt. Carbonate building stone formed part of Turkey's history as it is present in all constructions of the various civilizations of Anatolia.

Quarries for carbonate building stone are located in almost all regions of Anatolia and Tracie, although 90% of the quarries are located in the western part of the country, particularly in the regions of the Aegean and Marmara seas.

Among the best-known commercial varieties, the following types can be highlighted because of their features and quality: Süpren, which shows an amazing cataclastic texture with red and yellow patches in a white grey mass, Elazıg Cherry, a mixed carbonate–serpentine rock, Black Aksehir, Manyas White, Bilecik, Afyon Tigerskin, with brecciated structure, Denizli Travertine, Aegean Bordeaux, Crimson Milas, Leopard (Salome), showing cataclastic structure, White Mustafa Kemalpasa, a coarse-grained metamorphic marble, Gemlik Diabase, Karacabey Black, Vize Pink, Sazara White, Muğla White, Marmara White, Milas Kavaklıdere and Afyon Sugar.

Greece

Greece is the cradle of metamorphic marble. The word 'marble' derives from the Greek term 'marmaro', which derives from the Marmara Sea. Metamorphic marble and other carbonate rock types have been used in the country for institutional and sacred buildings since earliest times. The famous temple of Zeus in Olympia and the temple of Apollo in Delphi were built using metamorphic marble from Paros. The Parthenon and the Erechtheum in the Acropolis of Athens (Fig. 6) used carbonate rocks from the Pentelikon Mountains around the city. Statues such as Aphrodite of Milos and Hermes of Praxiteles are also examples of the use of metamorphic marble during the golden age of the Greek classical period.

Today, carbonate building stone is exploited in several regions of Greece (Laskaridis 2004). In eastern Macedonia, the Drama–Kavala region presents limestone-bearing beds, whereas in the Thassos region the Cape Vathy is the most important limestone-quarrying area in Greece, producing up to 80% of the total exported carbonate building stone of the country. Well-known white varieties are produced in Thassos (e.g. Limenas Thassos, Prinos, white of Salaria Thassos, and Crystallina

Fig. 6. View of the Parthenon, in the Acropolis of Athens, built using metamorphic marble from Pentelikon, Attica region.

of Thassos). In western Macedonia, major production is confined to the white and coloured limestone of the regions of Kozani and Veria.

In Ipiros, Ioannina region, a beige limestone is exploited, whereas the area of Larissa and Volos in Thessalia offers a varied set of white, whitish, pink and coloured limestone.

The area of Pentelikon, in the region of Attika, is still an important production centre, with renowned rocks such as the Bianco di Penteli or the Marmo Greco Fino. Today, exploitation of carbonate building stone in the Penteli Mountain is restricted for environmental reasons to its northern part in the area of Dionyssos and Agia Marina. Here, more than 40 500 t of metamorphic marble are produced annually from underground operations.

The area of Leadia–Domvrena, in Sterea Hellas, is rich in pink-white, whitish and black limestone. Varieties such as Black of Levadia, Pink of Levadia and Whitish of Helikona are marketed.

Apart from the above-mentioned regions, Greece produces carbonate building stone from several islands in the Aegean Sea such as Naxos, Tinos, Paros, Evia and Crete, as well as in the Argolis region (Peloponnese).

France

Production of carbonate building stone in France reached up 1.17 Mt in 2006 and reserves are abundant. The number of active quarries exceeds 100, being located mainly along the northern Pyrenees, the Rhone Valley and the Montagne Noire region.

Metamorphic marble is scarce in France, except for some Devonian marble deposits in the Pyrenees. Several limestone types displaying pronounced diagenetic features and/or strong brecciation by tectonism are recognized in Palaeozoic formations, especially from Devonian and Mississippian

(Carboniferous) deposits cropping out in the Languedoc and Boulonnais regions (Perrier 1993).

Middle Jurassic platform carbonate deposits provide most of the carbonate building stone in France. Oolitic and reefal limestones are the more common carbonate types used as building stone. Crinoidal and other bioclastic carbonate rocks in Upper Jurassic formations are also worked. Several brecciated carbonate deposits that display features characteristic of karst diagenesis supplement this market. Cretaceous formations, in particular Urgonian reefs and several carbonate breccia beds, also provide a variety of carbonate building stone that have been used in local construction.

Both marine and continental limestone of Tertiary age has been exploited in the Paris Basin and in areas located in SE France. In the Paris Basin, quarries are active in Eocene (mainly Lutetian) and Oligocene formations. As an example, the Chartres Cathedral was built with Oligocene lacustrine limestone from Beauce. Marine bioclastic limestone of Miocene age is used as building stone in the subalpine regions of Savoie, Languedoc and Provence, and it is also extracted in quarries from the Aquitania Basin. Some Quaternary travertine is extracted locally in France.

Croatia

The dimension stone industry in Croatia is small, reaching up to $135\,000\,t\,year^{-1}$, and is almost exclusively carbonate rocks (Crnkovi & Jovicic 1993). Croatia's main calcareous ornamental stone production is located in the regions of Istria and Dalmatia, in the Outer (karstic) Dinarides and in the Island of Brac. Limestone from this island, which is almost completely formed of carbonate rocks of Turonian–Senonian and Eocene age, adorns the Statue of Liberty and the White House in Washington, DC. Market varieties produced in the island are Adria Grigio Macchiato (Adria grey mottled), showing uniform grey colour, and Adria Grigio Venato (Adria grey veined) with darker-grey veins and oval stains.

The region of Istria also contains abundant dimension stone resources from Jurassic, Upper Cretaceous and Eocene limestone formations. The most renowned regional material is called 'Orsera' (also known as 'Kirmenjak'), a hard, dense Upper Jurassic limestone showing characteristic parallel to bedding stylolites and varied tones, from whitish to brown-greyish and green to blue. The Lower Cretaceous limestone variety called Istria Giallo is extracted from both underground and opencast operations.

The variety of Badenian limestone called Lithothamnion (derived from the presence of red algae as a main carbonate component of the rock), cropping out in the hills surrounding Zagreb, was employed in the thirteenth century to build the cathedral of the city. The limestone was later used in the Renaissance and in the Baroque period in the construction of the main city buildings, and is still used today.

In Split, several companies exploit well-known varieties of limestone, such as Zeleni Jadran, Red Alkasin, Light Alkasin, Rozalit, Vrsine, Seget and Plano (Steblez 1998).

Croatia possesses an immense architectural Roman legacy in stone (Daniel 2002). Many cities had their own amphitheatre, but the most impressive is that from Pula, built in the first century using the stone from the village of Vinkuran, the oldest quarry in Croatia. Diocletian's palace complex in Split, built in the fourth century BC, employing the limestone from the island of Brac, has been described as a 'symphony in stone'. The Eufrasian Basilica in Istria, northern Croatia, was built with metamorphic marble and other stones imported from Greece and Bosnia, and remains as standing proof of a flamboyant international stone market in the fifth and sixth centuries BC.

Israel

Some 75% of Israelian outcrops are carbonate rocks that represent the main source of natural stone of the country (Shadmon 2000). Domestic production of stone tiles for cladding reaches up to $600\,000\,m^2$.

The famous limestone 'Jerusalem Stone', a dolomitic limestone quarried around the cities of Jerusalem and Bethlehem as well as in other parts of the country, was excavated by the Israelites as early as 1000–2000 BC and was used to construct religious sites such as Solomon's Temple. Since 1918 the use of 'Jerusalem Stone' has been strictly regulated, as it must be used for the external walls of all new buildings constructed in the city. In this way Jerusalem preserves its glory as the 'City of Gold', produced by the effect that the reflecting sunlight of dusk and dawn has upon the buildings and walls. Today, the stone is produced in areas of the Jordan River and from the southern Port of Eilat in the Red Sea.

Besides Jerusalem Stone, other varieties of carbonate building stone in Israel are the Hebron limestone, exploited in the hills of Judea (Hebron White and Shell), and the Makhtesh Ramon, from quarries in Negev that supply a limited limestone production under the commercial names of Ramon Grey and Ramon Gold (*Roc Máquina* 2000).

Egypt

The millenary Egyptian culture has given testimony to the magnificent constructions using carbonate rocks as raw material. Egypt has abundant carbonate deposits with a great variety of qualities and colours. The main quarries are located in the south of the country and in the Sinai Peninsula (Alarish, Shanmflishikh) (Fig. 1). Egypt produces cream and yellowish limestone (*Roc Máquina* 2007). In 1986 fiscal restrictions imposed on marble imports led to remarkable discoveries in the area of Minya of the Filetto, El Hassana, Galala, Sunny varieties Golden Cream and Sylvia varieties that changed the structure of the Egyptian building stone market.

Algeria

The majority of quarries for carbonate building stone in Algeria are located in the northern part of the country, mainly in the regions of Oran and Setif. Main carbonate deposits in the Oran area are Terga and Kristel. According to the statistics of the Office National de Recherches Géologiques et Minières (ORGM) (Martín de Bernardo & Castelos 2007), natural stone reserves in Algeria are around 64.8 Mt, although resources might reach 151.2 Mt. Algeria has 50% of its potential natural stone resources yet to be exploited.

Tunisia

Tunisia is not recognized as a large carbonate building stone producer, although its ex-works total production is around 4 Mt year^{-1} from 21 quarrying companies (UNIDO Tunisia 2005). Italy is the main client of Tunisia, receiving 56% of its exports (around 50 000 t year^{-1}). Imports of carbonate building stone for internal consumption reaches up 87 000 t year^{-1}.

Carbonate building stone in Tunisia is mainly exploited in a radius of 30 km from the capital of the country, Tunis, and also in El Kef and Kasserine.

The geology of the Tunisian calcareous deposits is varied (Gaied *et al.* 2000). Most important sedimentary formations furnishing carbonate building stone in Tunisia are grey and black Jurassic limestone from the Aziza massif, and yellow–reddish limestone and dolostone from the Jurassic of the Hairesh massif, this having been exploited since Roman times. In western Tunisia, Albian–Aptian black limestone and bioclastic beige Jurassic limestone are exploited in the region of Tataouine. South of Tunisia, beige–greyish homogeneous sublithographic dolomite of Turonian age is quarried under the commercial name of Matmata. Whitish, pink, red and yellowish Cenomanian massive reefal limestone (Keddel type) is intensely exploited in Jebel Keddel, 20 km south of Tunis, where reserves are estimated at 8.1 Mt. In central western Tunisia, the Campanian–Maastrichtian Abiod Formation includes varied carbonate rock types that are commercialized as 'Thala' marble.

Eocene nummulite-rich limestone is common in Tunisia. The limestone shows different colours, from brown to red and beige, known as the commercial type 'Kesra'.

Tunisia is also known for its decorative 'False onyx' exploitations (travertine). The onyx of Sejnane consists of ribbon or streaked bands of calcite and brown iron carbonate. The onyx deposit of Jebel Mzar, located 10 km south of Bir Lahmar, is a travertine of possible hydrothermal origin (Gaied *et al.* 2000).

Morocco

Reserves of natural building stone in Morocco are estimated to reach up 2700 Mt (*Roc Máquina* 2000). Seventy per cent of those reserves are located in the centre of the country and in the coastal plain, particularly in Agadir. The construction of the Hassan II mosque in Casablanca (1986–1993) absorbed the entire carbonate building stone production for several years, especially that coming from the southern part of the country.

Concluding remarks

Carbonate rocks are present in many geological formations of the Mediterranean region. Extensive carbonate platforms formed throughout Jurassic and Cretaceous times constitute a main source for many of the carbonate rock types, both limestone and dolostone, that were used historically as building stone in the area and those that are currently traded in the global market of ornamental stone. Metamorphic marble from several quarried areas in Italy (Carrara), Spain (Macael), Greece (Pentalikon, Paros, etc.) has furnished some of the most famous monuments and sculptures. Nowadays, the Mediterranean countries contribute significantly to the world carbonate building stone production.

Metamorphic marble, banded fine-grained limestone, shell limestone, travertine and brecciate–nodular carbonate rocks can be differentiated as the most common carbonate building and ornamental stone quarried in the Mediterranean region. At present, these carbonate stones are traded under a wide variety of commercial names. Major production corresponds to Italy, Spain and Turkey, where the stone industries are highly developed and a significant part of the natural stone production is exported. Other countries, such as Greece, France, Croatia, Israel, Egypt, Algeria, Tunisia and Morocco, are also important building stone producers but their production is limited and, thus, is mainly traded to the domestic market.

We acknowledge the editors of this volume (Drs B. J. Smith, M. Gómez-Heras, H. A. Viles and J. Cassar) for their trust on the expertise of the authors for providing a well-based overview of carbonate building stone in the Mediterranean. Thanks are given to Dr A. McMillan and Dr S. Siegesmund for their useful comments that have significantly improved the quality of an initial version of the paper.

References

CARMIGNANI, L., MECCHERI, M. & PRIMAVORI, P. 2005. Marbles and other ornamental stones from the Apuane Alps (northern Tuscany, Italy). *Giornale di Geologia Applicata*, **1**, 233–246.

CARMINATI, E. & DOGLIONI, C. 2005. Mediterranean Tectonics. *In*: SELLEY, R. C., COCKS, L. R. M. & PLIMER, I. R. (eds) *Encyclopedia of Geology*, Volume **2**. Elsevier/Academic Press, Amsterdam, 135–146.

CRNKOVI, B. & JOVICIC, D. 1993. Dimension stone deposits in Croatia. *Rudarsko-geolosko-naftni zbornik*, **5**, 139–163.

DANIEL, P. 2002. La industria de la piedra en Croacia, en expansión. *Litos*, **59**, 24–38.

DUNHAM, R. J. 1962. Classification of carbonate rocks according their depositional texture. *In*: HAM, W. E. (ed.) *Classification of Carbonate Rocks*. AAPG Memoirs, **1**, 108–121.

EMBRY, A. F. & KLOVAN, J. E. 1971. A late Devonian reef tract on northeastern Bank Island, North-west Territories. *Canadian Petroleum Geology Bulletin*, **19**, 730–781.

GAIED, M. E., BEN HAJ ALI, M., CHAABANI, F. & TAAMALLAH, H. 2000. Les potentialités en pierres marbrières et ornementales de Tunisie. *In*: *Les Pierres Marbrières de Tunisie*. Mines and Geology Review, **38**. Edition of the Geological Survey of Tunisia, Office of Mines, Tunis.

GARCÍA-GUINEA, J. & MARTÍNEZ-FRÍAS, J. (eds) 1992. *Recursos Minerales de España*. Consejo Superior de Investigaciones Científicas. Colección Textos Universitarios, **15**, Madrid.

GIESE, P. 2005. Moho Discontinuity. *In*: SELLEY, R. C., COCKS, L. R. M. & PLIMER, I. R. (eds) *Encyclopedia of Geology*, Volume **3**. Elsevier/Academic Press, Amsterdam, 645–659.

LASKARIDIS, K. 2004. Greek marble through the ages: an overview of geology and the today stone sector. *In*: PRIKRYL, N. (ed.) *Dimension Stone 2004*. Taylor & Francis, London.

LOMBARDERO, M. & QUEREDA, J. M. 1992. La Piedra Natural para la construcción. *In*: GARCÍA-GUINEA, J. & MARTÍNEZ-FRÍAS, J. (eds) *Recursos Minerales de España*. Consejo Superior de Investigaciones Científicas. Colección textos Universitarios, **15**, Madrid.

LOMBARDERO, M. & REGUEIRO, M. 2002. Industrial Minerals and Rocks. *In*: GIBBONS, W. & MORENO, T. (eds) *The Geology of Spain*. Geological Society, London, 485–494.

MARTÍN DE BERNARDO, J. & CASTELOS, I. 2007. *El mercado del mármol y granito de Argelia*. Notas sectoriales. Oficina Económica y Comercial de la Embajada de España en Argel. ICEX, Madrid.

PENTECOST, A. 2005. *Travertine*. Springer, Heidelberg.

PERRIER, R. 1993. Les roches calcaires de France. La pierre en France. *Mines et Carrières, Les Techniques*, **75**, 54–69.

ROC MÁQUINA. 2000. *Mercados. El sector de la piedra natural en Marruecos*. Reed Business Information, Bilbao, September, 52–61.

ROC MÁQUINA. 2002. *Turquía, cuna de antiguas civilizaciones y del uso del mármol*. Reed Business Information, Bilbao, May, 110–116.

ROC MÁQUINA. 2006. *Natural Stone in the World. 2006 Directory*. Reed Business Information, Bilbao.

ROC MÁQUINA. 2007. *Mercados. Egipto. Uno de los diez principales exportadores de piedra*. Reed Business Information, Bilbao, May–June, 38–41.

SHADMON, A. 2000. Israel building stone resources and production 1950–2000. *GSI Current Research*, **12**, 71–74.

STEBLEZ, W. G. 1998. *The Mineral Industry of Croatia – 1998*. US Geological Survey, Mineral Commodities Summaries.

UNIDO TUNISIA. 2005. *Marble in Tunisia*. United Nations Industrial Development Organization, Vienna.

WINKLER, E. M. 1997. *Stone in Architecture. Properties, Durability*. Springer, Berlin.

Limestones in Germany used as building stones: an overview

SIEGFRIED SIEGESMUND[1]*, WOLF-DIETER GRIMM[2], HELMUT DÜRRAST[3] & JOERG RUEDRICH[1]

[1]*Göttinger Zentrum Geowissenschaften, Universität Göttingen, Goldschmidtstrasse 3, 37077 Göttingen, Germany*

[2]*Zamboninistrasse 25, 80638 München, Germany*

[3]*Geophysics Group, Department of Physics, Faculty of Science, Prince of Songkla University, HatYai 90112, Thailand*

*Corresponding author (e-mail: ssieges@gwdg.de)

Abstract: Germany has an enormous number of different carbonate rock units, which vary widely in their geological age and sedimentary depositional environment. Limestones quarried from these exposures have a wide range of usages and applications, such as dimension and ornamental stones, floor tiles and panelling, and for use as paving stones and massive stones. Since antiquity, limestones were used as building materials in areas where they were naturally available and abundant. Limestones exhibit a relatively good weathering resistance, which is mainly controlled by the mineralogical composition and the rock structure. The susceptibility of limestones to weathering and alteration is only secondarily related to the stone's contact with rainwater and its exposure to frost. In this situation the pore space is the main controlling factor. Industrialization and the subsequent increase in air pollutants, which started at the end of the nineteenth century, led to the formation of dark and unsightly crust deposits on the limestones surfaces. These crusts, being the result of man-made activities, are the main weathering problem for carbonate dimension stones.

Limestone occurrences: a geological and stratigraphical record

In Germany, different types of carbonate rocks ranging in age from Precambrian to Quaternary are exposed at the surface (Fig. 1). The rocks outside the Alps area can be divided into three groups. The first group and oldest formations consist of folded and schistose sedimentary and metamorphic rocks with magmatic rocks intruded at varying extent. These Precambrian–Upper Carboniferous rocks form the basement. The second group lies discordant on the basement and consists mainly of sediments that have experienced only low tectonic stresses. This sediment cover comprises rocks from the Upper Carboniferous to the Tertiary. The third group are young Quaternary deposits, unconsolidated sediments from the last ice age and from the Holocene, which are mainly found in northern Germany and in the foreland of the Alps.

During the Variscan orogeny the basement rocks were metamorphosed and deformed. Carbonate rocks were only sporadically interfolded into these formations. The more authentic carbonate rocks can be found at the northern margin of the Rhenohercynikum orogen, because these rocks, mainly Devonian in age, experienced lesser amounts of deformation. In the Lower and Middle Devonian, Germany was mainly a marginal sea that was progressively filled with large amounts of sediments from the Old Red Continent to the north. As the sedimentation rate decreased from the Middle to the Upper Devonian, barrier-related reef carbonates appeared. These massive limestones (Massenkalk) were formed mainly by the reef-building stromatopora, rugose and tabulate corals.

Clastic sediments and erosional detritus from the Variscan Mountains, named the Rotliegend, characterize the Lower Permian in central Europe. The first marine sediments that stratigraphically overlie the Rotliegend belong to the Zechstein Epoch. They reflect a transgression from the north into the Germanic Basin, through a narrow strait between the Shetland Platform and Scandinavia (Ziegler 1990). Several marine-evaporitic cycles of deposition followed, with limestone and dolostone formations being deposited at the beginning of these cycles. These carbonate rocks are essentially lacking in fossils owing to the hypersaline conditions during the sedimentation.

During the Lower Triassic central Europe experienced a regression combined with a continuous subsidence. This led to massive, partly cyclical, sand-, silt- and mudstone depositions from a meandering fluviatile river system that led into a central playa sea (Paul 1982, 1999; Olsen 1988). Only in regions with a lagoonal sedimentary facies was an oolitic limestone formed, called the Rogenstein,

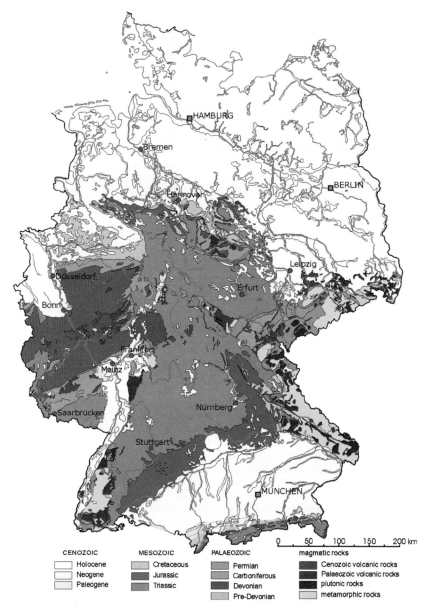

Fig. 1. Geological map of Germany showing the main distribution of limestone occurrences (for details see pp. 37–39) modified from Federal Institute for Geosciences and Natural Resources 1:2000000, support@geoshop-hannover.de, 2008 (modified).

which was widely used as dimension stone in the region of occurrence.

During the Middle Triassic the central European area was characterized by an extensive marine inundation. The first deposits were massive Wellenkalk limestone formations of the Lower Middle Triassic (Muschelkalk). The Wellenkalk fauna in general is small in size and shows a lower species diversity. This is seen as an effect of reduced water circulation with lower oxygen content and a slightly higher water salinity. Graded shell banks might be interpreted as a result of storm events. In contrast, extensive circulation events in-between have resulted in the formation of peloidal, oolitic and bioclastic carbonate rocks, key beds that today represent the main dimension stone horizons. In the Middle Triassic the water circulation was restricted and disconnected from the open marine area. These changes in the environmental conditions resulted in the deposition of marl, gypsum and rock-salt-bearing sedimentary formations. Later, in the Upper Middle Triassic, open marine conditions were realized again, with layers of oolitic shell limestones and tempestite-dominated lime-marl formations.

In the Upper Triassic the sediments again had a non-marine character, as the Germanic Basin represented a restricted lake with large fluctuations in the water level, temporarily leading to complete evaporation. The sediments during this period were mainly mudstones and marlstones, and terrestrial sandstones. Only sporadic, thin limestone and dolostone layers were interbedded, with no significance for use as dimension stones.

The marine environment was realized once more and showed a greater extent in the Jurassic (Lias). The development of an east–west-oriented barrier in the central part of the Germanic Basin led to the division of a Northern and Southern German Basin. A strait temporarily connected both basins. The Lower Jurassic (Lias) is characterized by the deposition of mudstones and marlstone series with greater thicknesses in the northern basin. Reduced circulation in the vertical water column resulted in anoxic conditions and the formation of black shales. In the Middle Jurassic (Dogger) mainly brown Fe-rich sediments were deposited, whereas in the Upper Jurassic (Malm) deposition of dominantly limestone and marlstone sequences occurred. In the Northern German Basin the thickness of these formations can vary greatly owing to uplift and subsidence tectonics and associated halokinetic movements in the subsurface. The sedimentary deposits are mainly bioclastic and oolitic limestones. At that time limestone formations were also deposited in the warm and well-aerated waters of the Southern German Basin, a shallower sea with a large spatial extension at the margin of the former Tethys Ocean. The shallower parts are characterized by reef and reef detritus limestone, whereas in the deeper parts massive and platy limestones were deposited. Today, both areas represent important dimension stone deposits.

Parts of the Northern German Basin further subsided at the beginning of the Cretaceous, and were filled with detritus eroded and transported from the central German geanticline or high (Mitteldeutsche Schwelle). The main deposits were sandstones, and mudstones with coal beds. In southern Germany, Lower Cretaceous sediments are limited to the foothill areas of the Alps. The further uplift of this orogenic zone resulted in flysch deposits in the foothill valleys. Then, in the Upper Cretaceous, an extensive transgression increased the areas of sedimentation, mainly with glauconitic and calcareous sandstones, and limestones and marlstones. Outcrops of these formations can be found in the Cretaceous basin near Münster, in the NW part of the Harz Mountains, and in the area around the city of Dresden in east Germany.

The continental conditions at the end of the Cretaceous period were superseded by a transgression at the beginning of the Tertiary. Later, in the Miocene, a regression followed that left most parts of Germany in the Pliocene under continental conditions once again. In contrast to the marine carbonate formations in the Cretaceous, the marine and continental sediments in the Tertiary period are mainly clastic.

The youngest limestones of Quaternary age are the travertine and calc-sinter (tufa) deposits, which are mainly geographically restricted local deposits with no regional extension. Only the travertine has relevance for the dimension stone industry and is still quarried today. Calc-sinter currently has no significance today, but was used and mined in the past. For example, it was mined near Helmstedt and Heiligenstadt in central northern Germany, and in particular in the pre-alpine moraine regions in Bavaria, southern Germany. The calc-sinter or tufa represents a deposition from oversaturated stream waters near spring areas.

Limestone fabrics

The use of carbonate rocks as dimension stones for buildings and sculptures dates back to early antiquity. The aesthetic value of limestones with their varieties in colours and colour distributions, together with their hardness, as well as softness and their wider availability, led to its widespread use (Fig. 2). However, from a geological standpoint, limestones are very heterogeneous. This mainly depends on the depositional environment in various facies, and on the diagenetic and tectonic processes (cementation, dolomitization, recrystallization, fracturing, etc.), that are associated with limestones or other types of carbonate rocks (Murray & Pray 1965; Wardlaw 1965). These heterogeneities are scale dependent; at a scale of 10 m–10 km they represent major lithostratigraphic boundaries, large faults and extensive fractures (Weber 1986). At a smaller scale (millimetres–metres) the heterogeneities are caused by the rock fabric. In limestones four major rock fabric types can be distinguished (Fig. 3): (a) major constituents consisting of fossils, ooides, peloides and crystals; (b) pore space; (c) fractures; and (d) stylolites. The distribution and orientation of the various microfabrics influence the physical and technical properties as well as the weathering processes. If there is a spatial distribution (e.g. pores) or a preferred

Fig. 2. Examples of some well-known limestones used for monuments and sculptures. (**a**) Lion at the Emperor's Cathedral in Königslutter made from Elmkalkstein. (**b**) Petri Church in Braunschweig built from Rogenstein. (**c**) Historical quarry of the Devonian Lahn marble (Unica, Villmar), which is used as an open-air museum today. This material was also used in the construction of the Empire State Building. (**d**) Western portal of the Halberstädter Cathedral made from Huy limestone (photo from U. Kalisch). (**e**) A figure group at the main entrance of the Naumburg Cathedral constructed from Schaumkalk, a Triassic limestone.

orientation of the fabric elements (e.g. fractures), the properties of a limestone show an anisotropic behaviour (see Schön 1996; Siegesmund 1996). If two or more of these parameters appear together, their effect on the properties can be more complex. The various heterogeneities of carbonate rocks or limestones led to a number of different classification systems, most of which originated from and are used in the petroleum industry. Limestones and dolomites have been classified based on mineralogy, texture, composition and physical parameters, like pore types and porosity (see Roehl & Choquette 1985; Mazullo *et al.* 1992). Füchtbauer (1959) used the chemical and mineralogical composition for a classification. Folk (1959, 1962) and Dunham (1962) presented classifications based on the relative amounts of four textural and diagenetic components in the rocks (particles, lime-mud matrix, cement and

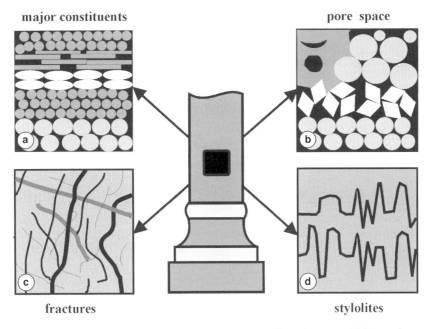

Fig. 3. Schematic diagram of the major fabric types of carbonate rocks as dimension stones. Major constituents: mainly lime mud and various carbonate components (their size, shape and orientation), the spatial distribution of the various components, the preferred orientation of crystals, and the preferred orientation related to the shape. Pore space: different types of pores, their size, the pore radii distribution, the spatial distribution of the different pores in the rock and the interconnectivity of the pores. Fractures: different sizes, from micro to macro cracks, their width and spatial extension in the dimension stone, the state of cracks and fractures (like open, closed and mineralized), the interconnectivity of the cracks and fractures, naturally originated cracks in the rock mass and man-made induced cracks during the processing of the dimension stones. Styloliths: occurrence and spatial distribution of styloliths, their size, thickness, and orientation in the rock mass and their composition.

pores), as well as the characteristics of particles v. matrix. These classifications combine descriptive and genetic aspects (Mazullo *et al.* 1992). Furthermore, classification systems, where pore space is considered with regards to porosity and permeability, were also developed (Archie 1952; Choquette & Pray 1970; Lucia 1983, 1995).

In the petroleum industry the Dunham classification is currently used, which Embry & Klovan (1972) later modified. Dunham's (1962) classification is based on the depositional texture of the carbonate rocks, and uses the grain-/mud-support principle, with a grain-supported and a mud matrix-supported framework/rock as the main types. 'Mud' is defined as components with a diameter of less than 20 μm. The relationship between the components, the matrix and the cement is considered here with a textural point of view. The following five classes were originally defined: mudstone (mud-supported, with less than 10% of components with a diameter of more than 2 mm); wackestone (mud-supported, with more than 10% of components with a diameter of more than 2 mm, components float in the matrix); packstone (grain-supported with mud, components support each other); grainstone (grain-supported without mud, components support each other, with and without cement); and boundstone (original components bound together). Embry & Klovan (1972) later modified this and added more classes. For limestones with 10% of particles larger than 2 mm, they added to wackestone the floatstone (mud-supported, with more than 10% of components having a diameter of greater than 2 mm). To packstone and grainstone, they included the rudstone (grain-supported, with more than 10% of components showing a diameter of greater than 2 mm). Furthermore, they separated the boundstone into three classes: bafflestone (original components bound together by sediment-entrapping organisms, like corals); bindstone (original components bound together by sediment-binding organisms, like blue algae); and framestone (original components bound together by frame-building organisms, like reef corals or sponges). This classification system has been slowly adapted in the dimension stone industry, and will be used throughout this paper. Although the pore space properties are of primary interest for hydrocarbon exploration, they are also important for the weathering behaviour of carbonate rocks. In the following section, the major constituents of limestones will be explained in more detail (see also Fig. 3).

Major constituents

Limestones are mainly composed of fossils, ooides, peloides, pellets, intraclasts, crystals and lime mud. Calcite and dolomite are the main carbonate minerals. Secondary constituents of importance consist of sulphate and sulphide minerals as well as organic material. The arrangement and distribution of the components result in various bedding and layering features, which are basically controlled by the deposition and sedimentation. These features are characterized by their texture and structure. This includes variations in composition, size, shape and orientation of the particles and components, and in their packing (Collinson & Thompson 1989). Furthermore, arrangement and distribution of the components can be modified by secondary processes (e.g. tectonics), resulting in a crystallographic preferred orientation (texture) of the rock-forming minerals (see Schaftenaar & Carlson 1984; O'Brien et al. 1993; Ratschbacher et al. 1994).

Pore space

In carbonate rocks various pore types can be distinguished. All of them are the result of the primary distribution of different components in a matrix influenced by secondary diagenetic processes. During the early water expulsion stage, the mechanical compaction resulting from the overburden pressure is the dominant mechanism for porosity reduction (Schlanger & Douglas 1974; Kim et al. 1985). Archie (1952) was the first one who classified carbonate rocks using their pore space. Choquette & Pray (1970) presented a fabric-selectivity concept. Lucia (1983, 1995, 1999) added a more petrophysical view to the classification of the pore space: the pore size distribution controls the porosity, the permeability and the saturation, and it is related to the rock fabrics. These considerations and investigations led to two major pore space groups: (a) the interparticle pore space; and (b) the vuggy pore space, which is divided into separate-vug pores and touching-vug pores. For the first group, Lucia (1995) characterized three rock fabric/petrophysical classes defined by certain permeability and water-saturation fields.

Fractures

In the pore space classification of Lucia (1995), fractures are one type of touching-vug pores. The fractures in this study are separated from the pore space because they can contribute to anisotropy as well as to porosity and permeability. The fracture intensity in dolomites is higher than in limy dolomites and in limestones (Stearns 1967; Sinclair 1980), and with decreasing grain size the fracture intensity in dolomites increases (Sinclair 1980). For the examination of fractures van Golf-Racht (1996) presented a descriptive classification scheme defining the following categories: (a) open/closed fractures; (b) macro/micro fractures; and (c) natural/induced fractures. Induced fractures can occur at every processing step of dimension stones, from quarrying and transportation to cutting and carving. As the applied processing techniques and technologies have been changed and improved over time, the occurrence and distribution of processing-induced fractures and cracks in the dimension stones has been reduced.

Stylolites

Stylolites are a common diagenetic feature in carbonate rocks, and their origin is independent of the rock facies and the geological age (Nelson 1985; van Golf-Racht 1996). Generally, they are irregular planes of discontinuity with an orientation from horizontal to vertical (Fig. 5b and e). The stylolite irregularities are displayed by the stylas, columns and rows of different width and height. The stylolites themselves are characterized by the concentration of relatively insoluble constituents of the enclosing rock (Park & Schot 1968). The (continuous) presence of relatively insoluble material can act as barriers to the fluid dynamic system of the rock. It is generally accepted that stylolites are the result of a concentration–pressure process or a pressure–dissolution process. Both processes are mainly controlled by the solubility of the components and particles in the rock, and by the regional stress field (Nelson 1985). The solution plane is more or less perpendicular to the maximum stress direction, which is commonly the overburden pressure direction in basin regions with less significant tectonics. A classification of stylolites in carbonate rocks is given by Logan & Semeniuk (1976) with respect to configuration, arrangement, fabric and structure of the pressure-solution phenomena.

Limestone examples from Germany

Mudstones

The mudstones are composed mainly of lime mud with less than 10% grains. All of the five different stones investigated show no visible pore space, although porosity can be measured (see Table 1).

Table 1. Rock physical properties of selected limestones from Germany (data are from Clemens et al. 1990; Grimm 1990; Katzschmann & Lepper 1999; Katzschmann et al. 2006; Wehinger unpublished data and own unpublished data)

	Dimension stone or trade name	Locality	Stratigraphic system	Series	Bulk density (g cm^{-3})	Porosity (%)	Degree of saturation (dimensionless)	Water uptake (atmosphere) (weight%)	Water uptake (vacuum) (weight%)	Compressive strength (MPa)	Tensile strength (MPa)	Flexural strength (MPa)	Young's modulus (MPa)	Abrasion (cm^3/ 50 cm^2)
1	Deutsch-Rot-Kalkstein	Horwagen	Devonian	Late	2.72	0.47	0.76	0.12	0.17					
2	Wallenfelser Kalkstein	Wallenfels	Devonian	Late	2.72	0.13	0.43	0.02	0.05					
3	Weinberg "Marmor"	Kerpen	Devonian	Middle	2.70	0.50	0.70	0.19	0.70	110.6		11.70		19.17
4	Vilmarer Kalkstein (Bongard)	Villmar	Devonian	Middle (Givet)	2.70	0.59	0.71	0.16	0.22	123.3				19.44
5	Vilmarer Kakstein (Unika)	Villmar	Devonian	Middle (Givet)	2.71	0.65	0.77	0.19	0.24	123.3				19.44
6	Aachener Blaustein	Hahn (Aachen)	Devonian	Late (Frasne)	2.70	0.74	0.65	0.11	0.17	80.0		5.00		
7	Saalburger Marmor	Tegau	Devonian	Late	2.67	3.96	0.89	0.21	1.41	127.0		11.40	80.99	22.30
8	Harzer Dolomit (Nüxeier Stein)	Steina (Bad Sachsa)	Permian	Late (Zechstein, (Werra/Stassfurt)	2.74	3.84	0.77	1.07		219.0				19.20
9	Braunschweiger Rogenstein	Nußberg (Braunschweig)	Triassic	Early (Skyth)	2.58	5.26	0.78	1.58	2.05	179.5				16.90
10	Füssener Steinbruchkalk	Alterschrofen	Triassic	Middle (Ladin)	2.74	0.74	0.80	0.21	0.27					
11	Osnabrücker Wellenkalk	Osnabrück	Triassic	Middle (Anis/Ladin)	2.60	4.54	0.79	1.38	1.75					
12	Elmkalkstein	Königslutter	Triassic	Middle (Anis/Ladin)	2.04	24.64	0.38	4.07	12.06	27.5	4.70			18.90
13	Crailsheimer Muschelkalk	Crailsheim-Satteldorf	Triassic	Middle (Anis/Ladin)	2.44	11.04	0.40	1.82	4.65	30.3		6.50		
14	Kirchheimer Muschelkalk (Kernstein)	Kirchheim	Triassic	Middle (Ladin)	2.64	2.93	0.68	0.76	1.11					
15	Kirchheimer Muschelkalk (Blaubank and Goldbank)	Kirchheim	Triassic	Middle (Ladin)	2.67	1.85	0.73	0.51	0.70	80.0		12.00		24.00

(*Continued*)

Table 1. Continued

	Dimension stone or trade name	Locality	Stratigraphic system	Series	Bulk density (g cm^{-3})	Porosity (%)	Degree of saturation (dimensionless)	Water uptake (atmosphere) (weight%)	Water uptake (vacuum) (weight%)	Compressive strength (MPa)	Tensile strength (MPa)	Flexural strength (MPa)	Young's modulus (MPa)	Abrasion (cm^3/ 50 cm^2)
16	Poller Trochitenkalk	Polle	Triassic	Middle (Ladin)	2.70	0.89	0.66	0.21	0.33					
17	Kuacker Muschelkalk	Kuacker (Kirchheim)	Triassic	Middle	2.67	4.64		0.35		82.5	5.5	12.66		23.78
18	Eichestädt	Eichestädt	Triassic	Middle	2.58	3.73	0.88	0.34	0.20	79.2	7.5	12.68		28.54
19	Tegernseer Kalkstein	Enterbach	Jurassic	Late (Malm)	2.70	0.53	0.80	0.17						
20	Solnhofener Kalkstein	Solnhofen	Jurassic	Late (Malm)	2.58	4.77	0.86	1.47	1.85	215.0		15.60		15.00
21	Treuchtlinger Kalkstein	Treuchtlingen	Jurassic	Late (Malm)	2.62	3.38	0.75	1.10	1.29	149.0		14.10		21.60
22	Hoheneggelsener Korallenoolith	Hoheneggelsen	Jurassic	Late (Malm)	2.21	18.30		6.25	8.29					
23	Salzhemmendorfer Korallenoolith	Salzhemmendorf	Jurassic	Late (Malm)	2.46	13.32	0.91	2.54	5.42					
24	Thüster Kalkstein	Thüster Berg bei Salzhemmendorf	Cretaceous	Lower Cretaceous	2.08	23.43	0.57	6.35	11.28	25.0	7.00			55.48
25	Anröchter Grünsandstein	Anröchte	Cretaceous	Late (Turon)	2.45	10.54	0.94	4.03	4.30	85.0				25.80
26	Baumberger Sandstein	Baumberge	Cretaceous	Late (Campan)	2.18	19.08	0.73	6.42	8.76	50.0		11.00		19.08
27	Enzenauer Nummulitenkalk	Unterenzenau	Tertiary	Eocene	2.67	2.09	0.80	0.62	0.78					
28	Rosenheimer Kalkstein	Rohrdorf	Tertiary	Eocene	2.61	3.91	0.88	1.31	1.50					
29	Gönninger Kalktuff	Gönningen	Quarternary	Holocene	1.70	35.12	0.72	14.83	20.67					
30	Travertin	Bad Langensalza	Quarternary	Holocene	2.25	17.58	0.79	3.60		35.0–74.0		8.30– 7.60	44.61– 49.72	19.80

The Upper Devonian Deutsch-Rot-Kalkstein is dense, mainly light red in colour with lump-like structures (Knollenkalkstein) and has a flaser structure. The stylolites are easily weathered after a short exposure time. Although the stone is resistant to weathering, the colours are clearly bleached. The Wallenfelser Kalkstein, also Upper Devonian, is more dark grey–grey-black with a cloudy flaser structure, calcite-filled fissures and some fossil fragments. It has a good–moderate weathering resistance; however, it shows bleaching and increasing surface roughness after a short exposure time. The Middle Triassic Osnabrücker Wellenkalk is a fine-grained and dense yellow-grey limestone, mainly fossil free, seldom with a very small amount of shell fragments. It shows slightly larger pore spaces in parts where larger shell fragments occur. Further characteristics are intensive bioturbation with burrows approximately 2 cm in diameter and numerous stylolites, where the stone can be easily broken apart. In general, this stone shows a good resistance to weathering. The Tegernseeer Kalkstein is an Upper Jurassic dense limestone with no components visible, and has a red-brown–dark-red colour. The stone is intensively crinkled with crease-like structures, as well as white calcite fissures and stylolites. After a certain exposure time the stone shows bleaching and an increased surface roughness, as well as weathering along the stylolites. The Solnhofene Kalkstein is a pale brown, extremely dense, micitric and homogenous limestone. It shows in parts dispersed Fe- and Mn-mineralization, especially along the bedding planes and cracks, but also in the form of dots. This stone is famous for the well-preserved Upper Jurassic (Malm Zeta) fauna.

Wackestones

The key characteristic of a wackestone is that it is still lime-mud-supported but has more than 10% of carbonate grains in total. Therefore, original components are not bound together during the deposition; however, the depositional texture can usually be recognized. All of the six different stones investigated show no visible pore space, although their porosity can also be determined in the laboratory (see Table 1). The Middle Devonian Weinberg 'Marmor' is a dense, grey–pink-grey coloured, fine–medium and coarse detrital reef limestone. The stone shows irregular fissures cemented with calcite, and dark violet stylolites, as well as some fossil fragments. Although the stone shows a good weathering resistance in general, polished surfaces with outside exposure are not durable, sometimes exhibiting holey weathering structures. The Aachener Blaustein has a dark blue-grey–black colour, which gives it the name 'bluestone'. The Upper Devonian dense limestone contains fossils, and shows white, nearly parallel, calcite veins (millimetres in thickness), as well as sparitic-filled voids ranging up to centimetres in size. Stylolites are irregularly distributed but widely found. Their occurrence affects the generally good weathering resistance, with dissolution and break apart along the stylolites. The Lower Triassic Braunschweiger Rogenstein (Figs 4b & 5b) mainly consists of ooides ranging up to 1 cm in diameter, giving it its name 'Roestone', and it consists partly of stromatolithes in layers parallel to the ooide-containing layers. It is a dark-red–grey layered and sandy limestone, and has a good weathering resistance. Single ooides rarely break out. The light- to medium-grey Füssener Steinbruchkalk shows darker coloured bioturbation features, resulting in finger-like structures. These more porous parts have a higher water intake, resulting in more intensive weathering than the parts with no bioturbation. Furthermore, the stone shows irregularly distributed fine calcite fissures and stylolites. In general, this limestone has a good–moderate weathering resistance. The Treuchtlinger Kalkstein (Fig. 4d) is a grey–blue-grey or yellow-grey dense limestone, with poorly sorted fossils in a fine-grained matrix. Fossils consist of algae, Tubiphytes, parts of porifera, belemnites and others. The white foraminifer Tubiphytes give the limestone an irregular appearance. Small holes between the fossil parts are filled with micritic matrix. Several varieties exhibit extensive bioturbation patterns. In general, the stone has a good weathering resistance; however, not necessarily at freezing temperatures. The Anröchter 'Grünsandstein' (or 'green sandstone') is actually a blue-green, dense limestone, with fossil fragments up to 3 cm. The grain size varies from very-fine to coarse grained, and visible layering is the result of the alignment of elongated fossil pieces. The colour is generated by the homogeneous distribution of glauconite, which gives the rock its name. Generally, the weathering resistance is moderate, but shows dissolution, the break out of particles, and the exfoliation of small pieces and, rarely, larger ones. Gypsum efflorescence can also be found but is an infrequent occurrence.

Packstone

The packstone is generally grain-supported by carbonate grains and shows an observable fraction of mud. In some stones investigated the pore spaces are macroscopically visible, but not in all (see Table 1). The Villmarer Kalkstein (Bongard variety) is a dense, poorly sorted limestone, varied in colour, with organogenic structures, and some stylolites, joints and fissures. Some of them are

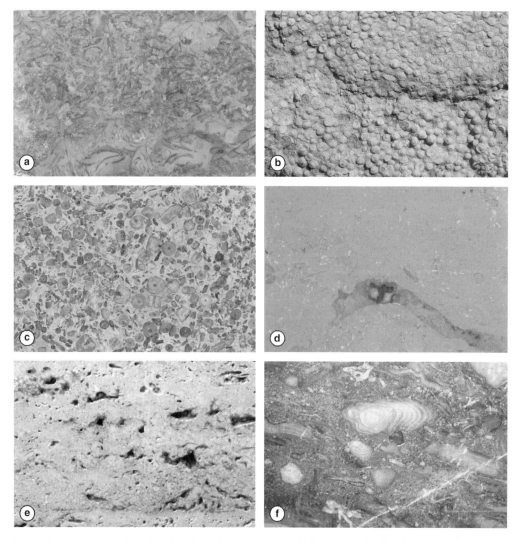

Fig. 4. Macroscopic overview of the limestones from (**a**) Kuacker, (**b**) Rogenstein, (**c**) Crailsheimer Muschelkalk, (**d**) Treuchtlinger, (**e**) Bad Langensalza travertine and (**f**) 'Lahn marble'.

filled with calcite. The stone has a very-good–good weathering resistance; however, in an outdoor environment the polish shows an accelerated deterioration. The break out of parts can be only observed after a long period of exposure. The Crailsheimer Muschelkalk (Figs 4c & 5c) is a mainly light grey and grey limestone with Trochite parts and shell fragments. It also comes as a blue-grey and yellow-grey variety. The limestone is relatively homogenous and has a dense structure. The pore space is indistinctly distributed, mainly interparticle pores, voids and molds, with an average pore size of 0.5 mm (meso pores). In general, the weathering resistance is good. However, surface dissolution can led to a degeneration of the polished surface and microkarst features, causing a loosening of the fossil parts out of the matrix. The Poller Trochitenkalk is a coarse-grained, poorly sorted, dark-grey fossil limestone with crinoids, micritic cement, some detrital mollusc shells. Isolated nests of coarse-grained dolomite crystals can be found, but no visible pore space. Despite some surface dissolution and leaching the weathering resistance is very good to good. Although the German name Baumberger Sandstein (Fig. 5d) uses the word sandstone, this stone is a yellow-grey–cream-coloured fine-grained limestone. The colour is due to the homogeneously distributed glauconite

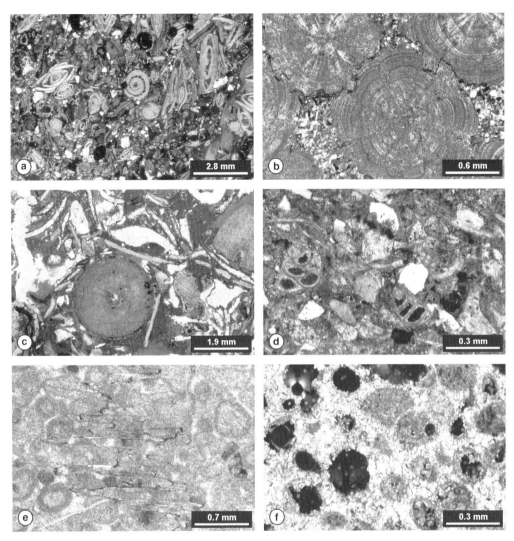

Fig. 5. Thin section microphotographs from (**a**) Enzenauer Nummulitenkalk, (**b**) Rogenstein, (**c**) Crailsheimer Muschelkalk, (**d**) Baumberger calcareous sandstone, (**e**) Harzer dolomite (Nüxeier dolomite) and (**f**) Elmkalk.

in the limestone, whereas the visible brown spots are from Fe-bearing compounds. The coarse-grained stone has no macroscopic visible layering. The small pores are inter- and intraparticle pores ranging from 0.2 mm or less in diameter, but mainly about 0.06 mm. The weathering resistance is moderate–poor, with dissolution, the break out of particles, exfoliation and gypsum efflorescence.

Grainstone

A grain-supported framework of carbonate grains with no observable fraction of lime mud characterizes a grainstone. The following stones are in this group: the Elmkalkstein (Fig. 5f) is a light-grey limestone with fine pores from the middle member of the German Triassic, the Lower Muschelkalk. It shows cross-bedding features and layers with medium-grained fossil detritus, where larger pores can be found. The layering is visible through the alignment of particles with their long axis, and by changes in the grain size. The macroscopically visible interparticle pores and molds are due to carbonate dissolution or incomplete cementation. The weathering resistance is good, mainly surface dissolution and bleaching. The Jurassic Hoheneggelsener Korallenoolith is a light brownish grey porous limestone, with alternating fine- to coarse-grained layers, and, in part, coarse-grained fossil detritus is also present. The layering is visible through the alignment of elongated particles and changes in grain size. The portion of ooides can be alternating. Visible pore space comprises interparticle pores and molds with an average size ranging from 0.05 to 0.30 mm. The stone has a moderate–poor weathering resistance. The Thüster Kalkstein, a limestone from the Thüster Berg area, shows a brownish grey colour, has a medium grain size and is generally homogeneous. Parallel layers of fossil detritus contain fine pores; however, the cementation is partly incomplete, resulting in interparticle pores and moldic porosity, with an average pore size ranging from 0.3 to 1.5 mm. This Lower Cretaceous stone has, in general, a good weathering resistance.

Rudstone

A rudstone is grain-supported with more than 10% of components larger than 2 mm. In this group two Eocene limestones investigated in this study fall into this classification. The Enzenauer Nummulitenkalk (Fig. 5a) is a red-brown, bulky and dense limestone mainly comprised of fossil detritus. White foraminifera, showing a diameter of 1–1.5 cm, are visible in a red-brown matrix, where the red colour is due to Fe-oxide compounds. No layering and pore spaces are macroscopically visible in this stone. The weathering resistance is good–moderate, but the polish shows accelerated deterioration. After a short exposure time, surface dissolution and leaching can occur. After a longer period a weathering of the matrix can be seen, where fossils are loosened and fall out. The Rosenheimer Kalkstein is grey-brown–blue-grey in colour, coarse-grained and mainly comprised of fossil detritus. The main fossils are Lithothamnien and Foraminifera, showing a random distribution in the stone. The overall colour of the stone can change owing to the colour of the components. No pore space is visible. In general, the weathering resistance of this stone is poor, with a strong leaching and loss of surface polish.

Boundstone, framestone and others

A boundstone and framestone are characterized by a grain-supported framework, where the components are organically bound together. The Holocene Gönninger Kalktuff is a white, crumbly, fine crystalline carbonate tuff with cell-like pores. The pore space, visible up to 50%, is inhomogeneously distributed with pore sizes ranging up to 5 mm (mega pores). In the 99% calcitic boundstone some pieces of plant material can be found. Surface dissolution can result in a disintegration of the stone, which shows in general a moderate weathering resistance. The Villmarer Kalkstein (Unika variety, Fig. 4f) shows a light-red colour, and has a dense, inhomogeneous appearance. The stone is mainly composed of colonial fossils and fossil detritus. Black stylolites in zigzag lines can be found, but no pore space is macroscopically visible. The Middle Devonian framestone has a very good–good weathering resistance, although outside exposure results in an accelerated degeneration of the polishing. The break out of particles occurs after a long period of exposure.

There are quite a number of limestone varieties from the Kirchheim area, near Würzburg, all from the Middle Triassic (Upper Muschelkalk). The Kirchheimer Muschelkalk (Kernstein variety) is a light-brown grey, dense-looking limestone, which is rich in shell and brachiopod fragments that are aligned in layers. The matrix shows a more grey-brownish colour. The floatstone shows no visible pore space, but has measurable porosity (see Table 1). The weathering resistance is good, especially when the stone is used outside because it also shows good preservation charactistics. The Kernstein variety can be further subdivided into several quarry varieties, like the Kuaker2 Kernstein (Fig. 4a), Mooser Muschelkalk and the Krensheimer Muschelkalk. Kirchheimer Muschelkalk itself has two varieties, Blaubank and Goldbank. The designated names relate to their colour, Blaubank (blue grey) or Goldbank (grey brown). The dense fossil limestone contains mollusc and brachiopod shell fragments. The matrix shows different colours, mainly from Fe-bearing compounds. The partly occurring auburn colour from Fe-oxide compounds is characteristic for the Goldbank variety. The floatstone shows no visible pore space. Although the stone has a good–moderate weathering resistance, the preferred use is for interior objects.

This suite of limestones also comprises two crystalline dolomites. The first is the Upper Permian Harzer Dolomit (Fig. 5e) from the Zechstein sequence. It is a brown-grey dense dolomite, indistinctly layered and slightly bituminous, with some fossil fragments, and with numerous stylolites. The stone shows a very good–good weathering resistance, although it can develop microkarst and weathering along the stylolites. The second is the Upper Jurassic Salzhemmendorfer Korallenoolith. The grey–brown-grey dolomite has sugar-like grains with small pores. It is a homogeneous stone that shows no layering as the original oolitic structure of the limestone was destroyed by the dolomitization. The intercrystalline pores between the idiomorphic dolomite crystals have an average grain size of 0.1–0.5 mm. In general, the dolomite exhibits a good weathering resistance.

Physical properties

A selection of the main technical values and parameters of the German carbonate dimension stone varieties are given in Table 1. Most of the data concerning bulk density, porosity, degree of saturation, water uptake, compressive strength, tensile strength, flexural strength abrasion and Young's modulus are taken from various sources in the literature. The data compilation aims to give an overview of the carbonate dimension stones in Germany, although it might not be representative in all cases. Most of the collected data were determined over the last several years; however, some data are much older. Therefore, the data quality might differ over time, which has to be taken into account. For a more detailed dataset and recent values, interested readers and possible buyers are referred to the certified data, which are available from the dimension stone companies. However, not all the varieties presented in this study have recently been mined or are currently being extracted.

Density and pore space properties

Carbonate rocks show a great variability in the composition and the pore space, and therefore also in the porosity and the bulk density. This mainly results from the variability in the sedimentation processes of the carbonate sediments, and the subsequent syn- and post-sedimentary diagenetic processes. The definition of the bulk density leds to a direct correlation with the porosity values. However, carbonate rocks show an exponential increase of the porosity with decreasing bulk density (Fig. 6) (Mosch & Siegesmund 2008). According to Quervain (1967), limestones cover a porosity range from less than 1% to more than 20%, from very compact to highly porous.

The selected limestones show significant differences in their porosity, density and pore radii distribution properties. The group from the Devonian limestones have grain density values between 2.71 and 2.72 g cm^{-3} and porosities of less than 1%. The pore radii distribution shows that 55–60% of the pores have a diameter of less than 0.0001 mm; sometimes it can be nearly 100%.

The Permian and Triassic limestones are a quite heterogeneous group. For example, the bulk density varies between 2.74 and 2.04 g cm^{-3}, with a corresponding porosity variation between 0.74 and 25.8%. In addition, the pattern of the pore radii distribution varies. Some rocks, where a narrowly spaced pore radii maximum was measured, can be described as equally porous (Fig. 6a). Other limestones are more unequally porous having a wider range of the pore radii distribution. The patterns of the pore radii distribution for these rocks are preferred unimodal and, rarely, bimodal (Fig. 6b, c). The overall heterogeneous impression of the post-Devonian limestones is amplified by a macroscopic and microscopic view. For example, in one rock less dense and denser parts can be found close together. Similar observations can be made for the Jurassic limestones – however, in a smaller range of variation – and for the Cretaceous and Tertiary rocks. Travertine rocks show an extreme heterogeneity; they are highly porous (see Fig. 4e) because of their sedimentary environment, and they are less compacted because of their young geological age (Clemens *et al.* 1990).

Water uptake properties

The very low porosities of the Palaeozoic Devonian limestones correspond to the low values of water uptake [in weight %] at an atmospheric pressure ranging from 0.02 to 1.82 and 0.05 to 4.65% when under a vacuum (see also Fig. 7). The degree of saturation (*S*) varies between 0.40 and 0.94. This value is quite important for the practical use of the limestones, as it gives first the evidence for the

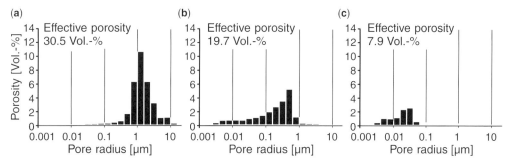

Fig. 6. Porosity and pore radii distribution of three selected limestones from (**a**) Elmkalkstein (ELKF), (**b**) Baumberger calcareous sandstone (BASF) and (**c**) Rogenstein (R2) showing the wide variation in porosity and the different pore radii distributions in the limestones.

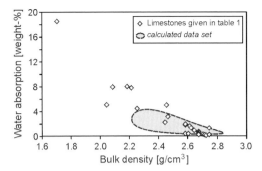

Fig. 7. Water absorption v. bulk density for the selected limestones given in Table 1 and in comparison with data illustrated by the grey field representing the confidential region of theoretically calculated populations based on an extended dataset from Germany (see Mosch & Siegesmund 2008).

frost resistance of the rock. When the saturation degree increases, the frost resistance decreases. With an S value above 0.9, the rock loses its frost resistance ability. The Permian and Triassic rocks show a large variability in these values. The water uptake ranges between 0.21 and 2.68 weight% at atmospheric pressure and between 0.27 and 12.04 weight% under vacuum, with corresponding saturation degrees between 0.25 and 0.89. For the Jurassic limestones the values of the water saturation degree correspond well with those of the Permo-Triassic rocks, better than the pore space properties. Similar observations can be made for the Cretaceous and Tertiary limestones. The water content values of the travertine and carbonate tuff are much higher, with 2.10–14.83 weight% at atmospheric conditions. Under vacuum the water uptake is 5.08–20.67 weight%, with corresponding values for the saturation degree of 0.41–0.72. These values indicate a good frost resistance. The more sandstone-like limestones have average values of 4.52 (atmosphere) and 5.85 weight% (vacuum), with a mean saturation degree of 0.79.

The capillarity is the rise or depression of a liquid, here water, in the pore spaces of a porous material, here a rock, owing to interfacial forces between the water molecules and the solid minerals and grains. These attractive forces balance the gravitational forces at a characteristic level. The related coefficient of (capillary) water uptake coefficient, w, is shown in Figure 7. In the literature the value of the water uptake coefficient, w, for dimension stones is 0.85–110 kg m^{-2} h$^{0.5}$ (Meng 1993). Carbonate rocks, in general, have water uptake values below 1 kg m^{2} h$^{0.5}$, whereas the porous Schaumkalke of the Lower Muschelkalk, like the Elmkalkstein, have much higher values. The main parameters for this are the porosity and the effective pore space, which are the pores that are interconnected with each other. For the dense limestones, for example, the pore space for an effective capillarity is not sufficient. The capillary water uptake is part of the hygric properties, together with the saturation degree, the moisture and adhesive water content, and the permeability. These parameters are important for the frost and weathering resistance of a rock.

Mechanical properties: compressive, tensile and flexural strength

The different mechanical strength values show large ranges for all the different groups of carbonate dimension stones presented in this study (see Fig. 8). The uniaxial compressive strength values

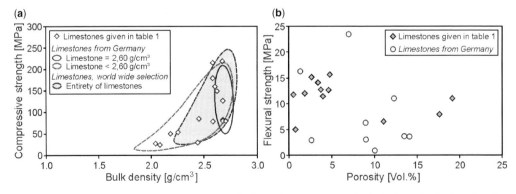

Fig. 8. (a) Compressive strength v. bulk density and (b) flexural strength v. porosity. In both graphs the data given in Table 1 are compared with a larger variety of limestones from Germany (the grey field) and with respect to the confidential region calculated for limestone with densities below and above 2.60 g cm^{-3} based on an international dataset (see Mosch & Siegesmund 2008) and those from Germany. The solid line ellipsoid gives the 80% probability density area.

vary between 36 and 219 MPa, with the lowest value for the travertine, and the highest value for the Nüxeier Dolomite. These upper and lower limit values correspond with the data presented by Mosch & Siegesmund (2008) and Mosch (2009). Tensile strength data are limited in the literature. Available data show a range between 4 and 15 MPa for carbonate rocks. Mosch & Siegesmund (2008) derived a clear relationship between the tensile strength, density and porosity. For the density the correlation is positive, whereas for the porosity it is negative. For dense limestones with porosities of less than 1% (Quervain 1967), the compressive strength values are between 50 and 220 MPa, and the flexural strength values between 5 and 35 MPa. The values of the splitting tensile strength (Brazil test) are lower, which is expected from its relationship to the flexural strength. An increasing occurrence of stylolites has a negative impact on the overall rock strength values, as they exhibit weak zones during the application of compressional and tensile forces. In general, it is obvious that in the lower porosity and upper density range no significant trend between the mechanical and basic physical properties can be expected. In the travertine group, however, no correlation between the strength values and the bulk density and porosity could be found, although the density and porosity values show a relatively wide range.

Salt crystallization test

According to the capillary pressure model of Wellmann & Wilson (1965), stones with mainly capillary pores, which are surrounded by a greater volume of micropores, are more sensitive and therefore more susceptible to salt crystallization processes. Ruedrich et al. (2005) and Ruedrich & Siegesmund (2007) have shown that with a decrease in the tensile strength and an increase in the porosity the samples investigated exhibited an increase in the sensibility towards the sodium sulphate crystallization test (Fig. 9). However, samples with a narrow distribution of the pore radii within the capillary pore size region are an exception, as they are more resistant to salt exposure than samples that have a higher percentage of micropores lower than the capillary pore size.

Hygric expanison

For most of the limestones the hygric expansion is a subordinate factor. The Thüste limestone (Table 1, Fig. 10) shows a hygric expansion of 0.02 mm m^{-1} parallel to the bedding, while the value perpendicular to the bedding is around 0.07 mm m^{-1} (Weiß et al. 2004). For the Travertine of Bad Langensalza the value is around 0.02 mm m^{-1} (Weiß et al. 2004), whereas Katzschman et al. (2006) reported values of less then 0.01 mm m^{-1}.

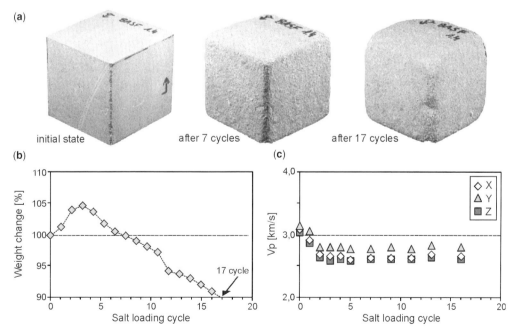

Fig. 9. (a) Salt weathering tests on Bamberg calcareous sandstones as a function of deterioration cycles, (b) weight loss after 17 loading cycles and (c) change in V_p velocities and its directional dependence during the weathering tests.

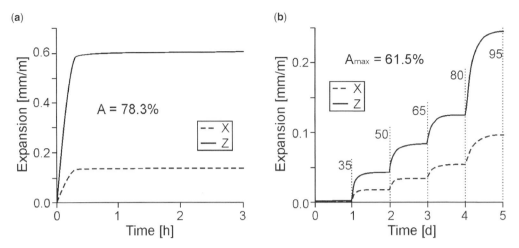

Fig. 10. Hygric expansion v. time relationship for the Baumberger calcareous limestone: (**a**) direct capillary water uptake and (**b**) expansion controlled by relative humidity levels (35%, etc.).

Comparable low values were reported for the limestones of the Unterer Muschelkalk (Terebratelkalk and Schaumkalk), where values of between 0.03 and -0.07 mm m^{-1} are given by Katzschmann et al. (2006). In contrast, the Baumberger calcareous sandstone, as well as the calcareous sandstone from Anröchte, exhibits quite different hygric behaviour. The Baumberger 'sandstone' shows a remarkable swelling of 0.6 mm m^{-1} parallel to the bedding and around 0.15 mm m^{-1} perpendicular to it. The time until maximum expansion is reached (see Table 1 & Fig. 10) can vary between 8 and 12 min depending on the direction, but for identical sample dimensions. The Anröchte 'sandstone' exhibits a smaller expansion, with about 0.08 mm m^{-1} parallel to the bedding and 0.16 mm m^{-1} perpendicular to it.

Thermal expansion

The thermal expansion coefficients of limestones are around 5.5×10^{-6} K^{-1}. Systematic investigations for limestones are still missing. In one case investigated here (Kuacker), residual expansions of 0.07 mm m^{-1} were observed, which indicates a strong anisotropy (Fig. 11, compare X-, Y- and Z- direction). This sample had a sparitic fabric type; however, further investigations are still necessary. A rare exception is the Baumberger 'sandstone'. Even if the directional dependence in this rock is weak, its magnitude is directly controlled by the mineral composition, that is, the relative frequency of highly expanding quartz and the lesser expansion in feldspar and calcite. This can be clearly

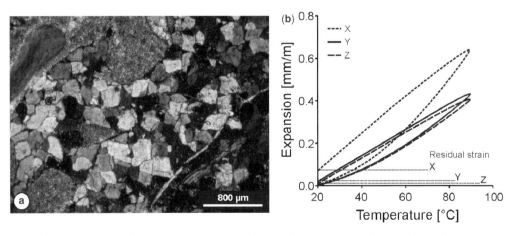

Fig. 11. Thermal expansion for the Kuacker Muschelkalk (see Table 1): (**a**) microfabrics showing a clear sparitic texture and (**b**) thermal expansion and its anisotropy as a function of temperature. Note the residual strain values after the heating cycle corresponding to a permanent length change.

documented by comparing the values from the Anröchte and Baumberger samples with those from Thüste and Bad Langensalza.

Fire damage

For carbonate rocks, heating leds to the decomposition of calcite and dolomite between 650 and 900 °C depending on the partial pressure of CO_2 (Sippel et al. 2007). The samples Anröchte and Thüste, which contain quartz and calcite, exhibit dicalciumsilicate (β-$2CaOSiO_2$) after heat treatment, a reaction product of CaO and SiO_2 at higher temperatures. The limestone from Eibelstadt was chosen for thermal expansion tests. This rock shows a directionally dependent thermal expansion. After decarbonatization and cooling down, an expansion is observed perpendicular to bedding, whereas a contraction occurs parallel to bedding. A phase transformation and an accompanied volume increase is observable for all carbonate rocks (including Anröchte) after the exposure to the heat impact followed by a few days at room temperature. Evidence of the reaction of CaO with atmospheric water to form portlandite $Ca(OH)_2$ can be discerned by X-ray diffraction (XRD). After 20 days of this exposure the new metastable phase vaterite (γ-$CaCO_3$) has already formed at the expense of portlandite owing to the exchange of OH-groups by atmospheric CO_2.

Anröchte shows a strong anisotropy of the thermal expansion up to 950 °C and a very high residual strain. Small-scale fire tests up to about 950 °C were also perfomed on the limestone. The main fracture plane of the Eibelstadt limestone was oriented parallel to a clay layer documenting the importance of the bedding plane as a pre-existing discontinuity and for fracture propagation. Furthermore, the atmospheric humidity supports the formation of portlandite at room temperature, which results in a total collapse of the rock structure in the outermost 4 mm. This indicates that the temperatures exceeded the critical value for the decarbonatization at this fire-exposed surface. The response of carbonate rocks to varying temperatures is decisively related to the behaviour of calcite and dolomite and their physical properties. At lower temperatures, the strongly anisotropic thermal expansion might control the stresses along the grain boundaries, although there might be a clear difference between micritic and sparitic limestones. At temperatures above 600 °C, however, the disintegration of carbonate minerals is accompanied by the release of CO_2 and should, therefore, be responsible for the intense shrinking of most carbonate rocks up to the final formation of CaO and MgO, respectively. Even for the calcareous sandstones, the release of CO_2 results in a very large positive residual strain indicating that volume effects similar to that of released water can be ascribed to CO_2. The formation of portlandite as a result of the reaction of CaO with water below 600 °C gives rise to a strong volume increase and further decay processes.

Decay history of limestones

Rocks are subjected to weathering because the geological processes of exhumation or uplift have brought them closer to the Earth's surface (Fig. 12a). During the process of uplift, the state of the stress changes as the gravity-driven vertical lithospheric pressure decreases in comparison to the horizontal tectonic stresses. This can result in fractures and joints in the near-surface rocks. Through these open pathways, rainwater can run deeper into the rockmass. The rainwater and the atmospheric CO_2 react together resulting in carbonic acid, a weak acid compound.

$$H_2O(l) + CO_2(g) \rightarrow H_2CO_3 (aq).$$

The major constituents of limestones are calcite and dolomite, which are quite resistant to freshwater. They are quite sensitive against acid compounds, like the natural carbonic acid (a low acid), with calcite being more sensitive than dolomite.

$$H_2CO_3 (aq) + CaCO_3 (s) \rightarrow 2HCO_3^- (aq) + Ca^{2+} (aq)$$

The low acid solutions can easily react with the calcium carbonate resulting in the dissolution of the rock, with caves and open pathways. Stronger acids can react much faster, resulting in an accelerated dissolution. Similar processes occur near the rock's surface, with organic humic acid from plants and micro-organisms covering the rocks. This can result in microkarst features near the surface. The acid solutions can also penetrate much deeper into the rock and dissolve the minerals there. The dissolved solids can then be transported to the surface, where they can precipitate, leaving behind a crust on the surface.

These natural weathering processes continue until human activities led to the excavation of the carbonate rock masses for the use as dimension stones (Fig. 12b). Primary processes involve cutting, sawing and the use of massive force to separate the rock mass into blocks of manageable sizes, which can result in (micro) crack formation in these blocks. Later polishing and further treatment can also alter the excavated rock mass, especially at or near the surface, which represents the contact or boundary to the environment. The reworking of the rock mass in order to produce ornamental stone

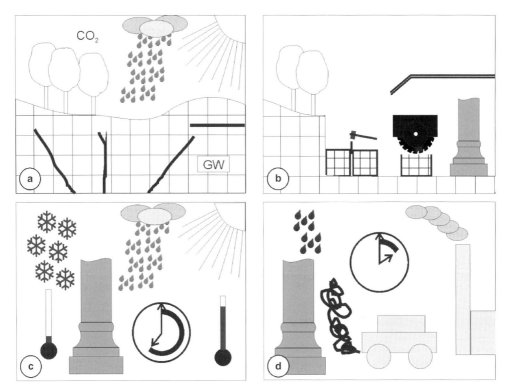

Fig. 12. Simplified sketch showing the different factors influencing the weathering of carbonate dimension stones. (**a**) Weathering of the carbonate rock mass in the natural environment near the surface: gravitational stress release results in naturally occurring fractures and cracks; dissolution processes related to naturally occurring acids, like carbonic acid formed from CO$_2$ and rainwater. The dissolved calcium carbonate is transported by groundwater (GW) flow. (**b**) Quarrying of the carbonate rocks and manufacturing processes can produce man-made fractures and cracks. The near-surface mass can be altered by cutting and polishing materials and procedures. (**c**) Weathering of the carbonate rocks after outdoor emplacement as a building stone owing to natural environmental factors, mainly seasonal and daily temperature change, and rainwater, frost and sun exposure over time. (**d**) Anthropogenic air pollution adds further weathering factors to (c) due to the release of sulphur and nitrogen compounds from transportation and industrial activities over time.

pieces also involves different processes, including using a hammer, cutting and polishing. All this changes or can change the near-surface part of the rock mass. After the dimension stone is brought to its final (outdoor) position it will be subjected to the natural weathering conditions again; however, with a larger contact surface area and no soil cover (Fig. 12c). The main parameters are the exposure to weak acid solutions through rain and different humidity levels, the seasonal (winter–summer) and daily (day–night) temperature changes, and the exposure to the sun. Depending on the geology and properties of the original rock mass, these factors can result in different weathering features on the dimension stones, whereas the exposure time is critical. Since industrial processes started to play a significant role worldwide and also German society, man-made factors are now added to the list of significant weathering parameters (Fig. 12d). The emission of sulphur and nitrogen compounds by high industrial smokestacks leds to chemical reactions in the atmosphere resulting in acids that are responsible for the creation of acid rain according to following chemical reactions:

$$S + O_2 \rightarrow SO_2$$
$$SO_2 + O_3^\bullet \rightarrow SO_3^\bullet + O_2$$
$$SO_3^\bullet + H_2O \rightarrow H_2SO_4 \text{ (sulphuric acid)}$$
$$SO_2 + H_2O \rightarrow H_2SO_3 \text{ (sulphurous acid)}$$
$$2\,NO + O_2 \rightarrow 2\,NO_2$$
$$2\,NO_2 + H_2O \rightarrow HNO_3 \text{ (nitric acid)}$$
$$+ HNO_2 \text{ (nitrous acid)}$$
$$2\,HNO_2 + O_2 \rightarrow 2\,HNO_3.$$

Because of the higher acidity, the acid rain is a much more effective medium than the natural acids described above. However, the chemical reactions are quite similar. Gypsum crusts can be found on the limestones as the result of the sulphuric acid reaction. In addition to the acid rain, dust and soot also contributes to the man-made weathering factors. They mainly come from the emission of coal-fired power plants, households, and from the emissions of car and truck engines powered by hydrocarbons, especially diesel fuel. The soot can contribute to the black crust often found on carbonate dimension stones; however, not all the processes involved are fully understood. Exposure time is also a main factor when looking at man-made weathering processes. Dimension stones used for construction in historical times have often experienced significant changes in the atmosphere that have occurred until today.

Decay characteristics of carbonate dimension stones

Most typical decay features of limestones from Germany are illustrated in Figure 13. Crusts on carbonate dimension stones can be observed relatively often, mostly with the precipitation of gypsum or a carbonate mineral. The chemical interaction between carbonate minerals and carbonic and sulphuric acids results in the dissolution of the minerals. The dissolved solids are then transported to the surface of the dimension stone and may precipitate new minerals. The formation of the carbonate crust is the result of dissolution and precipitation, whereas, for gypsum crusts, the carbonate provides the cation and the atmosphere brings in the anion SO_x. The destructive effect of these crusts or similar near-surface precipitations is mainly the result of the difference in the mechanical properties with respect to the original stone. The crust is often denser and therefore less elastic. This can result in shear stress at the contact during environmental changes in temperature and humidity. In addition, the gypsum crusts exhibit a bulge-like structure during further growth, which leds to the destruction of the surface of the dimension stones underneath. Furthermore, near-surface dissolution of carbonate material at the local scale often leds to a weakening of the overall dimension stone.

Limestones exhibit a moderate–very good salt and frost resistance. In most cases it is the result of the poor capillary water intake and the storage capability of aqueous solutions, which can be linked to two main structural features (see Fig. 3). Most of the limestones are quite dense with less or, even, no porosity. Another group of limestones exhibit cavernous–large pores, but the interconnections of these pores are relatively low. As a result the aqueous solutions may not easily be transported inwards by capillary forces or they may be expelled from the larger macro pores owing to the exceeding gravitational force. Conversely, limestones with smaller pores distributed throughout the stone, like the highly porous Miocene limestones from the Mediterranean area, are highly sensitive to inner crystallization processes.

Soft limestones, for example the sandstone-like limestones (Kalksandsteine) and the highly porous limestones, can show quite high hygric expansions. Dense limestones and those with a cavernous porosity do not exhibit this, or, if they do, then with much lower values. The hygric expansion in natural building stones is seen as the result of three main processes. Many authors mainly attribute the swelling process between the parallel layers in certain clay minerals to intracrystalline swelling. Others favour the disjoining pressure of the water as a mechanism, which can also explain hygric volume changes in stones lacking clay minerals. During this process thin water films on the opposite mineral surfaces oppose each other because they have the same electric charge. Of subordinate importance is the osmotic (intercrystalline) swelling; in those cases where the salinity of the pore liquid is low.

Thermal dilatation seems to be of only a minor relevance for the weathering of limestones, in contrast to marble as the metamorphic equivalent. The sensitivity of calcite and dolomite marble towards thermal dilatation is based on the extreme anisotropic single crystal thermal properties of dolomite and calcite. Calcite has a linear expansion coefficient of α_{11} $26 \times 10^{-6} K^{-1}$ parallel to the c-axis and of $\alpha_{22} = \alpha_{33} - 6 \times 10^{-6} K^{-1}$ parallel to the a-axis, whereas dolomite has an α_{11} of $26 \times 10^{-6} K^{-1}$ parallel to the c-axis and $\alpha_{22} = \alpha_{33} = 6 \times 10^{-6} K^{-1}$ parallel to the a-axis (Kleber 1959). For calcite these values mean that in the case of heating the single crystal expands in one direction and contracts in the other. This behaviour leds to a stress increase in the crystal during temperature changes, so that grain boundaries and intracrystalline planes, like cleavage planes and twin lamellas, can break up and diverge. The main important difference between marbles and limestones seems to be the grain size, as limestones in general have a much smaller grain size than marbles. Exceptions are the sparitic limestones (Fig. 11). They have larger grain sizes, and they can exhibit a higher thermal sensibility. Furthermore, some limestones, like the Nagelfluh, show the impressive phenomena of the bending of cladding plates. However, the intensity observed here is far less than what can be seen with marble plates.

In Germany the Devonian limestones in general have a good–very good weathering resistance. However, when exposed to the outdoor environment the surface polishing of these stones shows an

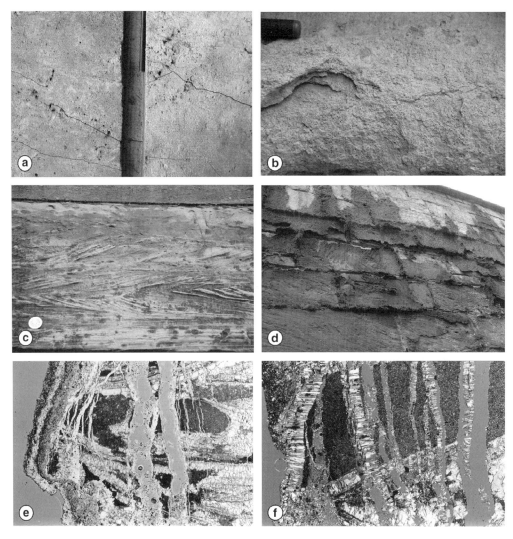

Fig. 13. Decay features of limestones: (**a**) crack formation, (**b**) scaling and flaking, (**c**) fabric-controlled back-weathering, (**d**) formation of black gypsum crusts, (**e**) flaking of a limestone due to the formation of gypsum and (**f**) counterscaling of surface-parallel cracks induced by the gypsum sealing phenomena (e and f are from Schlütter & Juling 2002).

accelerated disappearance. Bleaching is quite prevalent. Fossil fragments are often carved out by surface dissolution processes. Only after a very long exposure time is the breaking out of stone parts observable. Near stylolites, an increase of the surface roughness over time and further weathering is quite normal. The Harzer Dolomite also shows a moderate–very good weathering resistance. Bleaching, microkarst as the result of near-surface dissolution processes and back-weathering along the stylolites is readily apparent. Rogensteine has a good weathering resistance. Sometimes the break out of ooides can be seen, as well as a rare contour scaling. Fire damage can occasionally result in a significant loss in rock strength.

For the group of Triassic limestones weathering resistance is good–very good. The stone surface is often slightly dissolved resulting in the loss of the polishing by bleaching and microkarst. During these processes fossils can become loosened and fall out. An exception is the Füssener Steinbruchkalk, where surface weathering with an increase in surface roughness, granular disintegration and contour scaling may occur. Parts showing bioturbation exhibit a greater weathering than parts without owing to the different porosity and the related capillary water intake. The areas where stylolites occur are usually problematic zones when considering the weathering. The lower frost resistance of some limestone varieties can also be a problem. However, these weathering features can also be found within the stratigraphically younger varieties. Another type of weathering is associated with the sandstone-like limestones, such as the Anröchter Grünsandstein and the Baumberger Sandstein. They show a moderate–good weathering resistance, with dissolution, particle break outs, granular disintegration and seldom exfoliation of thicker parts, and gypsum efflorescence.

The dominant weathering feature in the limestones are the formation of microkarst and crusts (including biological actions), as they occur over larger areas. This is the result of extreme high levels of air pollution, and the high sensibility of limestones towards SO_2 and SO_3 at dry and humid environmental conditions. Various limestones show a white coloured patina at the side directed towards the main weather influence. These are places where the rainwater flows and the areas are exposed to splash water. The patina forms as a result of the chemical reaction from calcium carbonate to gypsum. Concurrently, the gypsum is washed down with the rainwater. At these rain-exposed areas, dust or soot cannot adhere long to the limestone surface. However, on surfaces that are not, or are less, exposed to rain these particles can stick to the surface, and thus form black crusts on the limestone surfaces. These crusts are mainly composed of dust, dirt, microbiological impact and soot that are primarily cemented by gypsum.

Most of the limestones exhibit a high weathering resistance. One of the main problems are the stylolites because their presence results in a faster surface recession for cladding plates in an outdoor environment. Another factor is the pore radii distribution, as it can lead to an increase in the frost and thaw sensibility, for example the Schaumkalk. All the limestones react quite sensitively to the extreme air pollution, which leads to crust formation and relief features in combination with granular disintegration. Relief formation is the result of weathering parallel to the original layering or the carving out of fossil parts (schill). Furthermore, scales, flakes, cracks, break outs and dirt can be observed. Crusts covering larger areas are preferred on the east and north side of buildings; for example, opposite the direction of the main weather influence. However, on the sides facing towards the direct weather impact, the west and south side, relief formation is the main weathering feature.

Summary

Limestones, which were deposited in various sedimentary environments throughout geological time, have been important stones used in building since early antiquity. The importance of limestones changed over time, reflected either in a local or regional use. Some limestone varieties received worldwide recognition. The favoured taste of a particular time led to greater export and trading activities at that time. Most of the limestones are good building and dimension stones, which in only a few cases show stronger weathering effects. Generally, the weathering patterns follow certain orders and trends, which can also be found in the German limestones. The portion and amount of non-carbonate minerals and components in limestones have a significant effect on dissolution processes, as well as on the crystal size, as limestones in general have a smaller grain size when having a higher portion of non-carbonate components. In grainstones the amount of non-carbonate components is less than 1%, whereas in mudstones and wackestones the amount is about 5%. These properties should have an effect on the intensity of the back-weathering. Over a period of years or tens of years, a polished smooth stone facade can turn into a very rough surface, where the larger grain size and weathering-resistant particles form a strong relief. Among the problems experienced in practice and often controversially discussed are the influences of stylolites. These zigzag-like structures with different amplitudes crossing the stones as irregular planes may act as preferential pathways for weathering processes, resulting in the often-observed back-weathered stylolites. The enrichment in clay or other minerals characterizes the stylolite planes, which is the result of the stylolite formation itself. As clay minerals in general are quite sensitive to changes in the water content, their abundant presence on stylolite planes can explain the preferred weathering there with the associated problems. The mechanical instability under stress or weight is a general problem of dimension stones containing stylolites. Occasionally, there are reports about the limited frost resistance of some limestones. In this case, the ratio between the pores with a larger diameter (>4 μm) and pores with a smaller diameter (<4 μm) is very significant and important. When limestones with mainly larger pore size diameters are water saturated, they can exhibit unfavourable behaviour during frost–thaw cycles. However, under these conditions a more favourable behaviour in terms of weathering can be expected during salt exposure.

Further important weathering phenomena are the formation of scales and shell-like structures. These are related to the processes of salt efflorescences and increasing salt crystallization pressures, and to the build up of thin crusts, as well as to local delaminating of thin layers parallel to the layering, for example in clay minerals. Gypsum crusts are the most important here. Dissolution and precipitation are the main physico-chemical processes involved in the crust formation. Dust particles in the air and small particles from air pollution and microbiological actions play a critical role during these processes and are important for this type of weathering in limestones. Consequently, limestones are often highly affected by these anthropogenic factors because the formation of scales and shell-like structures, at different scales, is associated with material loss of the original stone. With the society's recognition of the air pollution problems, political and technical efforts led to a reduction in the pollutants, especially in sulphur dioxide. Subsequently, some obvious and distinct weathering features and problems of limestones were also reduced to noncritical levels. However, for the sandy limestones the problems remain because as they also react quite sensitively to anthropogenic air pollution.

We are grateful for the unreserved help of many colleagues: F. Schlütter (Bremen), U. Kalisch (Halle), S. Mosch (Göttingen), M. Auras (Mainz), A. Wehinger (Mainz), J. Lepper (Hannover) and K. J. Stein (Waldsee). W.-D. Grimm would like to thank all his co-workers for their contributions to the book *Denkmalgesteine*: N. Ballerstädt, K. Clemens, K. Poschlod, G. Weiß, U. Schwarz, F. Niehaus, R. Lukas, R. Schürmeister, M. Simper and E. Erfle. Furthermore, special thanks to R. Snethlage for his constructive review.

References

ARCHIE, G. E. 1952. Classification of carbonate reservoir rocks and petrophysical considerations. *AAPG Bulletin*, **36**, 278–298.

CHOQUETTE, P. W. & PRAY, L. C. 1970. Geologic nomenclature and classification of porosity in sedimentary carbonates. *AAPG Bulletin*, **54**, 207–250.

CLEMENS, K., GRIMM, W.-D. & POSCHLOD, K. 1990. Zur Kennzeichnung des Korngefüges und des Porenraumes der Naturwerksteine. *In*: GRIMM, W.-D. (ed.) *Bildatlas wichtiger Denkmalgesteine der Bundesrepublik Deutschland*. Arbeitsheft 50 Bayerisches Landesamt für Denkmalpflege, München, 65–94.

COLLINSON, J. D. & THOMPSON, D. B. 1989. *Sedimentary Structures*. Chapman & Hall, London, 1–207.

DUNHAM, R. J. 1962. Classification of carbonate rocks according to depositional texture. *AAPG Memoir*, **11**, 108–121.

EMBRY, A. F. & KLOVAN, J. E. 1972. Absolute water depths limits of Late Devonian paleoecological zones. *Geologische Rundschau*, **61**, 672–686.

FOLK, R. L. 1959. Practical petrographic classification of limestones. *AAPG Bulletin*, **43**, 1–38.

FOLK, R. L. 1962. Spectral subdivision of limestone types. *AAPG Memoir*, **1**, 62–84.

FÜCHTBAUER, H. 1959. Zur Nomenklatur von Sedimentgesteinen. *Erdöl und Kohle*, **12**, 605–613.

GRIMM, W.-D. 1990. *Bildatlas wichtiger Denkmalgesteine der Bundesrepublik Deutschland*. Arbeitsheft 50 Bayerisches Landesamt für Denkmalpflege, München.

KATZSCHMANN, L. & LEPPER, J. 1999. Naturwerksteine der Germanischen Trias. *In*: HAUSCHKE, N. & WILDE, V. (eds) *Trias, eine ganz andere Welt: Mitteleuropa im frühen Erdmittelalter*. Dr Friedrich Pfeil, München, 429–448.

KATZSCHMANN, L., ASELMEYER, G. U. & AURAS, M. 2006. *Natursteinkataster Thüringen*. Institut für Steinkonservierung (IFS), Mainz, **23**.

KIM, D.-C., MANGHNANI, M. H. & SCHLANGER, S. O. 1985. The role of diagenesis in the development of physical properties of deep-sea carbonate sediments. *Marine Geology*, **69**, 69–91.

KLEBER, W. 1959. *Einführung in die Kristallographie*. VEB Verlag Technik, Berlin.

LOGAN, B. W. & SEMENIUK, V. 1976. *Dynamic metamorphism, Processes and Products in Devonian Carbonate Rocks, Canning Basin, Western Australia*. Geological Society of Australia Special Publications, **6**, 1–183.

LUCIA, F. J. 1983. Petrophysical parameters estimated from visual description of carbonate rocks: a field classification of carbonate pore space. *Journal of Petroleum Technology*, **35**, 626–637.

LUCIA, F. J. 1995. Rock fabric/petrophysical classification of carbonate pore space for reservoir characterization. *AAPG Bulletin*, **79**, 1275–1300.

LUCIA, F. J. 1999. *Carbonate Reservoir Characterization*. Springer, Berlin, 1–226.

MAZULLO, S. J., CHILINGARIAN, G. V. & BISSELL, H. J. 1992. Carbonate rock classification. *In*: CHILINGARIAN, G. V., MAZULLO, S. J. & RIEKE, H. H. (eds) *Carbonate Reservoir Characterization: A Geologic–Engineering Analysis, Part I*. Elsevier, Amsterdam, 59–108.

MENG, B. 1993. *Ein neues Verfahren zur Ermittlung von Porenstrukturkennwerten mit dem Ziel der Kalkulation von Transportkoeffizienten*. Jahresberichte Steinzerfall-Steinkonservierung. Ernst & Sohn, Berlin, 3–18.

MOSCH, S. 2009. *Optimierung der Exploration, Gewinnung und Materialcharakterisierung von Naturwerksteinen*. Dissertation, University of Göttingen, Germany. http://webdoc.sub.gwdg.de/diss/2009/mosch/mosch.pdf

MOSCH, S. & SIEGESMUND, S. 2008. Statistisches Verhalten petrophysikalischer und technischer Eigenschaften von Naturwerksteinen. *Zeitschrift der Deutschen Gessellschaft für Geowissenschaften*, **158**, 821–868.

MURRAY, R. C. & PRAY, L. C. 1965. Dolomitization and limestone diagenesis – an introduction. *In*: PRAY, L. C. & MURRAY, C. (eds) *Dolomitization and Limestone Diagenesis*. SEPM, Special Publications, **13**, 1–2.

NELSON, R. A. 1985. *Geologic Analysis of Naturally Fractured Reservoirs*. Contributions in Petroleum Geology and Engineering, **1**. Gulf Publishing, Houston, T.X.

O'BRIEN, D. K., MANGHNANI, M. H., TRIBBLE, J. S. & WENK, H.-R. 1993. Preferred orientation and velocity anisotropy in marine clay-bearing calcareous sediments. *In*: REZAK, R. & LAVOIE, D. L. (eds) *Carbonate Microfabrics*. Springer, New York, 1–313.

OLSEN, H. 1988. The architecture of a sandy braided–meandering river system: an example from the Lower Triassic Solling Formation (M. BUNTSANDSTEIN) in W-Germany. *Geologische Rundschau*, **77**, 797–814.

PARK, W. C. & SCHOT, E. H. 1968. Stylolite: their nature and origin. *Journal of Sedimentary Petrology*, **38**, 175–191.

PAUL, J. 1982. Der Untere Buntsandstein im Germanischen Becken. *Geologische Rundschau*, **71**, 795–811.

PAUL, J. 1999. Fazies und Sedimentstrukturen des Buntsandsteins. *In*: HAUSCHKE, N. & WILDE, V. (Hrsg.) (eds) *Trias, eine ganz andere Welt: Mitteleuropa im frühen Erdmittelalter*. Dr Friedrich Pfeil, München, 105–114.

QUERVAIN F. DE. 1967. *Technische Gesteinskunde*. Birkhäuser, Basel.

RATSCHBACHER, L., WETZEL, A. & BROCKMEIER, H.-G. 1994. A neutron texture goniometer study of the preferred orientation of calcite in fine-grained deep-sea carbonate. *Sedimentary Geology*, **89**, 315–324.

ROEHL, P. O. & CHOQUETTE, P. W. 1985. Introduction. *In*: ROEHL, P. O. & CHOQUETTE, P. W. (eds) *Carbonate Petroleum Reservoirs*. Springer, New York, 1–15.

RUEDRICH, J. & SIEGESMUND, S. 2007. Salt-induced weathering: an experimental approach. *Environmental Geology*, **52**, 225–249.

RUEDRICH, J., KIRCHNER, D., SEIDEL, M. & SIEGESMUND, S. 2005. Beanspruchungen von Naturwerksteinen durch Salz- und Eiskristallisation im Porenraum sowie hygrische Dehnungsvorgänge. *Zeitschrift der Deutschen Gesellschaft für Geowissenschaften*, **156**, 59–74.

SCHAFTENAAR, C. H. & CARLSON, R. L. 1984. Calcite fabric and acoustic anisotropy in deep-sea carbonates. *Journal of Geophysical Research*, **89**, 503–510.

SCHLANGER, S. O. & DOUGLAS, R. G. 1974. The pelagic ooze–chalk–limestone transition and its implication for marine stratigraphy. *In*: HSU, K. J. & JENKYNS, H. C. (eds) *Pelagic Sediments: On Land and Under the Sea*. International Association of Sedimentologists, **1**, 117–148.

SCHLÜTTER, F. & JULING, H. 2002. Mikroskopische Untersuchungen an Testflächen. *In*: SCHMUHL, B. (ed.) *Kalksteinkonservierung am Westportal des Halberstädter Domes St. Stephan und St. Sixtus*. Domstiftung, Leitzkau, 129–162.

SCHÖN, J. H. 1996. *Physical Properties of Rocks: Fundamentals and Principles of Petrophysics*. Pergamon, Oxford, 1–583.

SIEGESMUND, S. 1996. The significance of rock fabrics for the geological interpretation of geophysical anisotropies. *Geotektonische Forschung*, **85**, 1–123.

SINCLAIR, S. W. 1980. *Analysis of macroscopic fractures on Teton Anticline, Northwestern Montana*. MS thesis, Texas A&M University, College Station, TX.

SIPPEL, J., SIEGESMUND, S., WEIß, T., NITSCH, K. H. & KORZEN, M. 2007. Decay of natural stones caused by fire damage. *In*: PRIKRYL, R. & SMITH, B. J. (eds) *Building Stone Decay: From Diagnosis to Conservation*. Geological Society, London, Special Publications, **271**, 139–151.

STEARNS, D. W. 1967. Certain aspects of fracture in naturally deformed rocks. *In*: RICKER, R. E. (ed.) *NSF Advanced Sciences Seminar in Rock Mechanics*. National Science Foundation, Bedford, MA, 97–118.

VAN GOLF-RACHT, T. D. 1996. Naturally-fractured carbonate reservoirs. *In*: CHILINGARIAN, G. V., MAZULLO, S. J. & RIEKE, H. H. (eds) *Carbonate Reservoir Characterization: A Geologic–Engineering Analysis, Part II*. Elsevier, Amsterdam, 683–771.

WARDLAW, N. C. 1965. Pore geometry of carbonate rocks as revealed by pore casts and capillary pressure. *AAPG Bulletin*, **60**, 245–257.

WEBER, K. J. 1986. How heterogeneity affects oil recovery. *In*: LAKE, L. W. & CARROLL, H. B. JR (eds) *Reservoir Characterization*. Academic Press, Orlando, FL, 487–544.

WEIß, T., SIEGESMUND, S., KIRCHNER, D. & SIPPEL, J. 2004. Insolation weathering and hygric dilatation as a control on building stone degradation. *Environmental Geology*, **46**(3/4), 402–413.

WELLMANN, H. W. & WILSON, A. T. 1965. Salt weathering, neglected geological erosive agent in coastal and arid environments. *Nature*, **205**, 1097–1098.

ZIEGLER, P. 1990. *Geological Atlas of Western and Central Europe*, 2nd edn. Shell International Petroleum Maatappij B.V. Geological Society of London, Elsevier, Amsterdam.

Mechanisms of damage by salt

R. M. ESPINOSA-MARZAL & G. W. SCHERER*

Civil and Environmental Engineering, Princeton Materials Institute, Princeton University, Princeton, NJ 08544, USA

*Corresponding author (e-mail: scherer@princeton.com)

Abstract: Limestone is very susceptible to the aggressive action of salts. This paper describes the current understanding of the mechanisms by which salt crystallization causes damage to limestone. Crystallization pressure increases with the supersaturation of the solution, which may result from rapid drying and/or decrease in temperature. Salts with a tendency to achieve higher supersaturation owing to a high nucleation barrier are potentially able to induce more severe damage. In the presence of small pores (<100 nm), equilibrium thermodynamics indicates that crystallization pressure can result from the curvature dependence of the solubility of a salt crystal. Under non-equilibrium conditions, high transient stresses can occur even in larger pores. In the field, the complexity of salt weathering results from heat, moisture and ion transport coupled with in-pore crystallization during changing climatic conditions. This paper describes how progress in the modelling and numerical simulation of these coupled processes can contribute to a better understanding of the influencing factors and assessment of critical conditions.

Classically, tests such as the bursting test and the capillary rise experiment with simultaneous evaporation have been applied to evaluate qualitatively stone deterioration induced by salt crystallization. More recently our group has introduced other experimental methods to the field of salt weathering that provide quantitative information about nucleation and crystallization kinetics in porous materials (by differential scanning calorimetry), induced deformation and stress (by dynamic mechanical analysis and a novel warping test), and pore clogging caused by in-pore crystallization.

The final part of this paper is dedicated to a discussion of methods to prevent damage that may alter one of the crystallization steps, such as nucleation, crystal growth, disjoining pressure between mineral and crystal surfaces, or solution properties. Indeed, efficient treatments have been found for particular scenarios in the laboratory; however, the consequences of these treatments in the field, such as the behaviour at other temperatures and concentrations as well as the durability of the treatments, are not known yet.

Indeed, a lack of knowledge still exists in understanding the pore-level crystallization, such as the processes in the thin film between mineral surfaces and salt crystals that determine the disjoining pressure, or the dynamics of crystallization within the pore network that influence the salt distribution and stress in the stone. Atomic force microscopy, surface force measurements, nuclear magnetic resonance and simulations using molecular dynamics are promising methods to elucidate these points. By understanding these remaining questions a more reliable protection of stone against salt weathering will be achieved.

Limestone is very common in architecture, especially in North America and Europe. It was a very popular building block in the Middle Ages because it is hard, durable and easily accessible at surface exposures. Many medieval churches and castles in Europe are made of limestone, as are landmarks across the world, including the pyramids in Egypt. The longevity of such monuments is threatened by the same processes that sculpt natural landforms (Goudie & Viles 1997), prominent among which is the stress exerted by salts crystallizing in the pores of the stone (Chatterji & Jensen 1989; Flatt 2002).

Salts that are found in the pores of limestone are mainly absorbed with groundwater, result from reaction with atmospheric pollutants (such as sulphates in industrial regions), are transported with the air (such as chloride salts in coastal areas) or are supplied to the building by de-icing salts.

Salts originating from air pollutants affect many monuments, such as the Al-hambra, an immense and valuable archaeological site in Granada, and churches in northern parts of the UK made from dolomitic limestone and suffering the consequences of the coal-burning power plants formerly located in this area. The photographs in Figure 1 show severe efflorescence of magnesium sulphate salts (left) and flaking resulting from subflorescence of the same salt in the dolomitic limestone (right) used in Howden Minster, Yorkshire, UK.

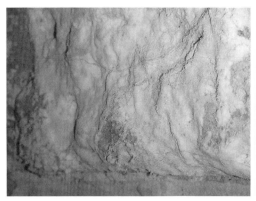

Fig. 1. Efflorescences of magnesium sulphate salts (left) and flaking (right) of the dolomitic limestone in Howden Minister, UK.

Salt-induced deterioration of architectural heritage and of geological, or geo-archeological, sites is accelerated drastically in marine environments, such as the Mediterranean Basin. Rothert *et al.* (2007) reported on the intense damage caused by the action of salt crystallization in monuments and buildings constructed of the local Globigerina Limestone in the Maltese Islands. Similar observations are presented about sites in Alexandria, mainly built from oolitic limestone blocks (Ibrahim & Kahm 2005), sites on the SW coast of France (Cardell *et al.* 2003) and Roman monuments in Tarragona (Vendrell-Saz *et al.* 1996).

A great effort has been made in recent decades to understand the damage mechanisms responsible for salt weathering (e.g. Chatterji & Jensen 1989; Rodriguez-Navarro & Doehne 1999; Scherer 1999, 2004; Rodriguez-Navarro *et al.* 2000*a*; Flatt 2002; Steiger 2005*a, b*; Espinosa-Marzal & Scherer 2008*a*; Steiger & Asmussen 2008), as well as to develop methods to prevent damage (e.g. Rodriguez-Navarro *et al.* 2000*b*, 2002; Houck & Scherer 2006; Lubelli & van Heess 2007; Ruiz-Agudo & Rodriguez-Navarro 2010). Indeed, the application of new protective treatments and consolidants to protect stone against weathering is an important issue worldwide (Delgado Rodrigues & Manuel Mimoso 2008).

There have been several reviews on salt damage in recent years concerning the understanding of salt crystallization in porous materials, the damage mechanisms and the major influencing factors, such as the supersaturation and the porosity (Goudie & Viles 1997; Charola 2000; Doehne 2002; Charola *et al.* 2006). This paper complements those reviews with recently introduced experimental methods, advances in the modelling and numerical simulation of salt weathering, and new directions in the prevention of salt damage, and points out possible directions for future research.

Salt crystallization and damage mechanisms

In this section we give an overview of the main factors influencing salt crystallization and damage of stone. Supersaturation is the driving force for crystallization, which leads to crystallization pressure and damage, even in the absence of small pores. The significance of the interaction between in-pore crystallization and moisture/salt transport, including pore clogging, is emphasized. The consequences of these interactions (in particular for favouring or preventing damage) are not well understood, so further study is necessary. The currently available computational power permits simulations of the coupled processes on both laboratory and field scales, which promises to advance our understanding of complex interactions, including changing climatic conditions, the behaviour of salt mixtures and the effectiveness of poultices, among others. The end of this section deals with the stress induced by the crystallization pressure, which is a decisive topic in terms of predicting durability of stone. Very recently, there have been a few attempts to predict damage.

Supersaturation – nucleation and crystal growth

A good background on the theory of heterogeneous nucleation and crystal growth can be found in Christian (1975), Nielsen (1984) and Mullin (1993), and, more recently, applied to salt crystallization in porous materials in Espinosa-Marzal & Scherer (2010*a*).

According to Gibbs' theory, supersaturation of the salt solution is thermodynamically necessary for nucleation (Mullin 1993). The driving force for crystallization is the difference between the chemical potentials of solution and nuclei, $\Delta\mu$, which is related to the supersaturation, β, of the salt in the solution:

$$\frac{\Delta\mu}{RT} = \ln(\beta). \quad (1)$$

The supersaturation ratio is

$$\beta = \left(\frac{a_\pm}{a_\pm^*}\right)^\nu \cdot \left(\frac{a_w}{a_w^*}\right)^{\nu_0} \quad (2)$$

where $a_\pm^*$ is the average ion activity of the saturated solution, ν the ion mole number per mole of salt, R the gas constant, T the temperature, a_w the water activity, a_w^* the water activity in equilibrium, and ν_0 the number of water molecules per mole of salt (Steiger 2005a).

If the supersaturation ratio β is larger than 1, the solution is metastable and crystallization may begin or not. But when β is larger than a threshold value, $\beta^* > 1$, nucleation starts abruptly (Mullin 1993).

An energetic barrier for nucleation results from the fact that the formation of a critical nucleus (which grows spontaneously) requires a high surface energy related to the formation of the crystal–solution interface. Thus, the energetic barrier is found from the total free energy of the embryo, ΔF_i^* (Christian 1975):

$$\Delta F_i^* = i^* kT \ln(\beta) + 4(i^*)^{2/3} \gamma_{cl} \Omega^{2/3} \quad (3)$$

with i^* the number of molecules in a critical nucleus obtained at the maximum of ΔF_i^*:

$$i^* = \left(\frac{8\gamma_{cl}\Omega^{2/3}}{3kT \ln(\beta)}\right)^3 \quad (4)$$

where k is the Boltzmann constant, T the temperature, β the supersaturation of the solution according to equation (2), Ω the molar volume of the solid and γ_{cl} the crystal-liquid interfacial energy.

Figure 2 shows the relationship between the supersaturation necessary for nucleation, β, and the number of molecules in a critical nucleus, i^*, for different salts at a temperature of 10 °C according to equation (4). The number of molecules to form a critical nucleus of mirabilite or epsomite is much larger at each supersaturation (mainly owing to the effect of its larger crystal–liquid interfacial energy), which means that the nucleation of these salts is energetically more difficult and requires higher supersaturation of the solution. Thus, mirabilite and epsomite are potentially more damaging salts, according to Figure 2.

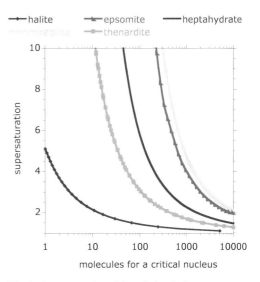

Fig. 2. Supersaturation of the solution for homogeneous nucleation of halite (NaCl), sodium sulphate heptahydrate ($Na_2SO_4 \cdot 7H_2O$), thenardite (Na_2SO_4), mirabilite ($Na_2SO_4 \cdot 10H_2O$) and epsomite ($MgSO_4 \cdot 7H_2O$) v. the number of molecules in a critical nucleus at 10 °C.

The presence of nucleation sites on the pore wall (impurities, defects or certain orientations of the mineral grains) decreases the energetic barrier and, consequently, the threshold supersaturation. In Espinosa-Marzal & Scherer (2008a), the threshold supersaturation for the nucleation of sodium sulphate heptahydrate and mirabilite in two different limestones is obtained by using differential scanning calorimetry (see the section on 'Methods of investigation'), and it confirms the tendency, predicted in Figure 2, of sodium sulphate heptahydrate to form at a lower supersaturation than mirabilite.

Drying – efflorescences, subflorescence and pore clogging

Salts may be carried into the stone with groundwater by capillary rise, or may be dissolved from the mortar joints, or result from chemical reaction between atmospheric pollutants (especially SO_2) and minerals [such as calcite ($CaCO_3$) in limestone]. As the water evaporates, the supersaturation of the pore solution increases until salt precipitates. A change in temperature, generally a decrease, leads also to a supersaturation of the solution resulting in the crystallization of salts. If evaporation occurs on the surface of the stone, then the crystals form a harmless (but unattractive) deposit on the surface called 'efflorescence'. However, if salts precipitate beneath the material surface (a phenomenon called subflorescence or cryptoflorescence), severe damage can be induced.

Although efflorescence does not generally affect the coherence and endurance of building materials, it impairs the surface appearance, which can be significant in case of historical buildings. Indeed, there are some efforts to prevent the formation of efflorescence. For example, a penetrating sealer can help prevent or lessen the occurrence of efflorescence by soaking in and blocking the pores below the surface, thus preventing the water from moving to the surface and bringing the salts with it. However, this process might lead to the crystallization of salts some millimetres beneath the surface (as subflorescence) or even deeper, possibly causing more severe damage. Therefore it is important to understand the interaction between in-pore crystallization and moisture and salt transport, as well as how it may affect damage of the stone.

Whether efflorescence or subflorescence forms depends on the drying rate, which is strongly affected by the presence of salts. Initially, the evaporation takes place on the exterior surface, driven by the gradient of vapour pressure. Air penetrates first into the large pores. The liquid flow from the large into the small pores keeps the small pores filled with water, such that evaporation continues on the surface of the material; in the absence of salts, drying takes place at a constant rate (see Scherer 1995 for the general theory of drying).

The drying rate, m_d, is directly proportional to the gradient of the vapour pressure of the surrounding air, p_a, and of the boundary layer of vapour on the drying surface, p_s:

$$m_d = k_v(p_s - p_a). \qquad (5)$$

The dissolved salts decrease the water activity, which decreases the pressure gradient in the vapour phase and, thus, the evaporation rate. This effect depends on the particular salt and is more significant with increasing supersaturation of the solution. For example, for Na_2SO_4, at a concentration of 3 mol kg^{-1} and 20 °C, the water activity is $a_w = 0.898$ and therefore p_s is smaller than that of pure water ($a_w = 1$ in absence of salts). Thus, at 50% relative humidity and neglecting the temperature difference between the air and the surface, the estimated decrease of the evaporation rate is 20%. This effect and the decrease in the area of the drying surface by efflorescence (Sghaier & Prat 2009) are the reasons for the lack of a period of constant evaporation rate in the presence of salts.

During the initial period of decreasing drying rate, the meniscus penetrates progressively into smaller pores. As long as a continuous liquid film covers the pore walls from the meniscus up to the exterior surface, then liquid transport can proceed by flow of the film, causing evaporation to take place on the surface and forming efflorescence. At the same time some liquid evaporates within the unsaturated pores, which causes salt to precipitate as subflorescence.

The decrease in the moisture content, the pore clogging with crystals and the increase of the solution viscosity with concentration may retard the liquid (capillary) flow. If the capillary flow becomes slower than the evaporation rate then the film becomes discontinuous and the flow to the surface is partly interrupted. Thereafter, evaporation takes place only inside of the material, indicating the start of the second (more accentuated) decreasing rate period of drying. As shown by Espinosa-Marzal & Scherer (2008b), at 21 °C and 30% relative humidity (RH) for Indiana limestone the first decreasing drying period is of short duration in the presence of salt; it vanishes completely in the case of $MgSO_4$, mainly because of the slow advective flow due to the high viscosity of the solution. This also explains the absence of efflorescence of magnesium sulphate salts in this experiment.

The vapour transfer from the stone surface to the surrounding air is affected by efflorescence. In addition, liquid transport through the efflorescence is affected by its porosity or by the presence of a film. One can imagine that if efflorescence forms and impedes the capillary flow to the surface, but the drying rate remains high because the efflorescence permits diffusion of vapour, the drying front may move deeper into the material.

Recently, the effect of salts on the drying rate has been studied in limestone (Espinosa-Marzal & Scherer 2008b, 2010b). Salts lead to a significant decrease of the drying rate. The intensity of pore clogging was investigated through the uptake of decane by salt-contaminated samples. Very effective pore clogging is caused by sodium chloride (halite) and magnesium sulphate (epsomite/hexahydrite). Experiments show that the drying rate is strongly dependent on the type and amount of salt, and on the porosity of the substrate.

It is not understood yet whether pore clogging can enhance or retard damage. Thus, it can be expected that pore clogging close to the surface may reduce the evaporation rate and, thereby, prevent the solution from achieving high supersaturation within the stone. However, if pore clogging occurs within the drying front, evaporation might lead to supersaturation of the residual solution, which is present as a thin film between the crystals, and between the salt and the pore wall (Scherer 2004). As crystal growth cannot continue owing to spatial confinement, an increase of the crystallization pressure might be induced. The same effect could result from a decrease in the temperature. Moreover, the hydraulic pressure caused by the freezing of water in pores whose entrances are clogged with salt crystals might cause significant damage. Thus, there are still important unanswered questions regarding pore clogging by salt crystals that should be the subject of future research.

In summary, the formation of either efflorescence or subflorescence, as well as the effectiveness of the pore clogging, depends on porosity of the stone, solution properties, nucleation and crystal growth kinetics, as well as on the environmental conditions. The only way to consider the interactions between all of these factors and to make a rough estimation of their effects is by means of numerical simulation.

Numerical models for salt crystallization coupled with transport in porous stone

Prediction of the position of the crystallization front under real climatic conditions requires the coupled heat, moisture and salt transport through the pore network to be analysed, together with the processes of phase changes of salts in the pores and chemical reactions (e.g. dissolution of carbonate rocks). From the point of view of comparing the behaviour of different salts, porous materials or the effects of diverse climatic conditions, the development of such numerical models is of interest for practical applications. Currently, there are several efforts to simulate these phenomena in porous building materials using models based on physical principles or empirical formulas (Černy et al. 2007; Franke et al. 2007; Nicolai 2007; Derluyn et al. 2008).

The coupled advective–diffusive partial differential equations for the transport of moisture and dissolved substances in porous media (Bear 1972), or its extended version including migration in an electric field based on the Nernst–Plank model (Samson & Marchand 1999), are well known.

The thermodynamics of salt solutions are also well known for the salts of interest in salt weathering. With the help of the Pitzer parameterization (Pitzer 1991; Steiger et al. 2008) it is possible to determine solution density, activity coefficients, supersaturation ratio (also of metastable salts: Steiger & Asmussen 2008), enthalpy of dissolution and specific heat capacity of solutions very accurately. This thermodynamic model permits prediction of which salts can precipitate from a solution consisting of an ion mixture at a given temperature, relative humidity and concentration; evidently, this may be of interest for assessing the cause of damage of particular sites.

In contrast, the kinetics of in-pore crystallization and the interaction between salt and substrate (pore surface) are not completely understood. For example, it is unclear whether the interfacial energy might influence where crystals form (at the solid–liquid or liquid–air interface) according to Rodriguez-Navarro & Doehne (1999) and Shahidzadeh-Bonn et al. (2008) or in which pores the crystals will form preferentially in non-equilibrium conditions.

Recent publications accentuate the significant role of the kinetics of crystallization on the damage mechanism (Scherer 2004; Steiger 2005a, b; Espinosa et al. 2008a; Espinosa-Marzal & Scherer 2008a, 2010a; Espinosa-Marzal et al. 2010) and especially emphasize the role of the supersaturation of the solution. Rodriguez-Navarro & Doehne (1999) carried out evaporation experiments at 60 and 35% RH, respectively, with an environmental scanning electron microscope (ESEM). Depending on the evaporation rate, they observed several different phases including mirabilite, both anhydrous forms (Rodriguez-Navarro et al. 2000a) and, eventually, the heptahydrate (Rodriguez-Navarro & Doehne 1999), apparently dependent on the supersaturation ratio achieved.

Espinosa et al. (2008b) investigated the phase change of salts (crystallization, dissolution, hydration, dehydration and deliquescence) in capillary porous materials and used a diffusion-reaction model (Nielsen 1984) to calculate the average crystallization rate of salts in porous materials, S, according to:

$$S = K(\beta - 1)^g \quad \text{with } \beta > \beta^* \qquad (6)$$

where K and g are kinetic parameters, and β^* is the necessary supersaturation required for crystallization to start. This equation is based on the fact that the driving force for crystallization is the supersaturation ratio of the solution with respect to a salt, β (Nielsen 1984). Three kinetic parameters are necessary to calculate the crystallization rate (K, g, β^*), which generally depend on both pore surface and salt, and must be obtained experimentally for each salt–substrate combination. The starting supersaturation ratio gives the threshold condition for the phase change. If salts are already present in a pore, the threshold supersaturation for further crystallization becomes $\beta = 1$.

The following example shows how numerical simulation can help the damage mechanism of salt crystallization to be better understood. A capillary rise experiment of sodium sulphate in two bricks (A and B) of different porosity was performed in the laboratory at 50% RH and 23 °C (Espinosa-Marzal et al. 2008), and showed completely different behaviour for the two materials. While severe efflorescence formed in brick A, it was negligible in brick B even after 2 weeks. In contrast damage (flaking) was only observed in brick B.

A significant difference between the materials is the presence of large pores (c. 10 μm) in brick A, while most of the pores in brick B are smaller than 2 μm. Thus, the measured sorptivity of brick A is approximately 3 times larger than that of brick B: $S_A = 1.71\ g^2\ s^{-1}$ and $S_B = 0.57\ g^2\ s^{-1}$. For the crystallization rate of mirabilite equation (6) was applied. The kinetic parameters for the crystallization of mirabilite in this brick according to equation (6) were determined experimentally elsewhere (Espinosa et al. 2008b).

The numerical simulation of this laboratory experiment shows that the higher permeability and larger pores of brick A are responsible for low pore blocking and rapid transport of the solution to the surface forming efflorescence. Figure 3 shows the distribution of mirabilite in brick B computed with the engineering tool ASTra (Franke et al. 2007). Owing to the axisymmetric geometry, the simulation of the coupled transport with crystallization was performed in only half of the sample (cross-hatched surface in inset). Thus, the x-axis goes from 0 to 3 cm, the y-axis goes along the height of the sample (14 cm) and the computed mirabilite content is depicted on the z-axis. The bottom 4 cm of the sample is placed inside the pan, which contains the unsaturated solution that keeps the relative humidity constant at 95%. Therefore, the evaporation inside of the container is negligible and no crystallization takes place. Above this 4 cm, mirabilite precipitates. The predicted width of the crystallization front is about 0.8 cm and no appreciable efflorescence forms, according to the simulation. The pores on the sample surface (at $x = 0.03$ m) contain salt crystals (c. 15 wt%) but there is no growth out of the pores (see arrow). Thus, the simulation with ASTra predicts just the formation of subflorescence, which is in agreement with the experimental results.

Espinosa et al. (2008b) show how the crystallization rate affects the results of the simulation. Thus, if a slower crystallization of mirabilite in brick B is assumed for the simulation, both the total amount of salt and the evaporation rate are larger, and so the damage more intense. The reason for this is simply the reduction in the number of crystals on the surface during the first stage of drying as a result of the lower crystallization rate, which leads to a smaller reduction in the evaporation rate as pore clogging is less effective. Because the evaporation rate is raised relative to the capillary transport rate, the crystallization front moves into the interior

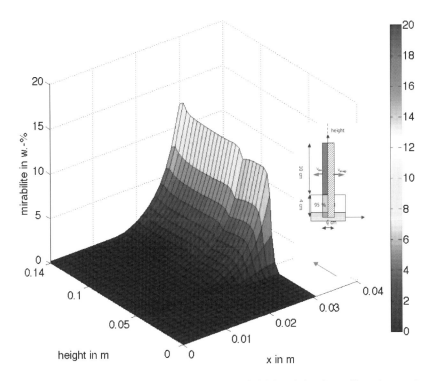

Fig. 3. Computed distribution of the content of mirabilite (in wt%) in brick B during the capillary-rise experiment after 196 h. Owing to the axisymmetric geometry, the simulation was performed on half of the sample (cross-hatched surface in inset). Simulations were performed with the engineering tool AStra (Franke et al. 2007).

of the sample. As a result of the lower crystallization rate, the width of the crystallization zone reaches approximately 1.8 cm beneath the surface.

This example shows how a numerical simulation can help in the understanding of the interaction between mechanisms under particular boundary conditions. However, the applicability of the computational models is still restricted to specific scenarios.

Crystallization pressure

Crystallization pressure is the main reason for the damage caused by the crystallization of salts in a supersaturated solution (Chatterji & Jensen 1989; Flatt 2002; Rodriguez-Navarro et al. 2002; Scherer 2004; Steiger 2005a, b). Correns (1949) (see also the annotated translation by Flatt et al. 2007) measured the growth pressure of a crystal against a load, and derived an expression for the crystallization pressure as a function of the supersaturation ratio by assuming that a thin layer of aqueous solution always remains between the crystal and the internal solid walls of the porous network, providing for the diffusion of the ions to the growing surface.

This concept can also be applied to a crystal growing in a pore: the growth of the crystal in a pore is impeded by the pore wall, but the crystal remains in contact with a supersaturated solution, at least temporarily. The reason for the formation of the thin film is the action of repulsive forces (i.e. disjoining pressure) between the two approaching solid surfaces (viz. the growing crystal and the pore wall). If this thin layer did not exist the crystal would come into contact with the pore wall, the growth would stop and no crystallization pressure would be exerted. Therefore, the existence of this thin layer is a necessary condition for crystallization pressure.

From thermodynamics it follows that a crystallization pressure, Δp, must be exerted to maintain a crystal in equilibrium in a supersaturated solution. According to Flatt (2002), Scherer (2004) and Steiger (2005a, b) the crystallization pressure is given by:

$$\Delta p = \frac{RT}{V_c} \ln \beta - \gamma_{cl} \kappa_{cl} + \frac{\Delta V}{V_c} \gamma_{lv} \kappa_{lv} \qquad (7)$$

where $\Delta V = V_L - V_c$, $V_L = \sum V_i$ is the sum of molar volumes, V_c the molar volume of crystal, κ_{cl} the curvature of the interface between crystal and solution, κ_{lv} the curvature between liquid and vapour, γ_{cl} the crystal–solution surface energy and γ_{lv} the vapour–solution surface energy. The influence of the capillary pressure (the third term in equation 7) on the solubility and on the crystallization pressure is relevant only in unsaturated porous materials.

The influence of the crystal curvature (given by the second term in equation 7) on the crystallization pressure must be considered in small pores. For example, if the threshold supersaturation for the nucleation of a salt is equal to 2, then the influence of the curvature is only relevant in pores with entries smaller than 12 nm. If the material contains such small pores, then a crystallization pressure can be expected in equilibrium (Scherer 2004; Steiger 2005b). Thus, for a cylindrical crystal with hemispherical ends in a cylindrical pore, if the solution is in equilibrium with the hemispherical surface of the crystal it is supersaturated with respect to the cylindrical surface and equation (7) reduces to:

$$\Delta p^\infty = \gamma_{cl}(\kappa_1 - \kappa_2) \qquad (8)$$

where κ_1 and κ_2 are the curvatures of the cylindrical side $(1/R)$ and of the hemispherical ends $(2/R)$, respectively, and R is the pore radius. According to this, in a material with a bimodal pore size distribution, with $R_2 = 1$ μm and $R_1 = 10$ nm, confined mirabilite would exert 7.96 MPa (assuming $\gamma_{cl} = 0.04$ N m^{-1}, $\kappa_2 = 1/R_2$ and $\kappa_1 = 2/R_1$).

In a non-equilibrium state, even in the absence of small pores, high mechanical stresses may arise owing to high supersaturation ratios, according to the first term in equation (7). That is, crystals come into contact with the pore walls while the solution is highly supersaturated, and they exert force on the wall until equilibrium is established (Scherer 2004; Espinosa et al. 2008a; Steiger & Anmussen 2008). At equilibrium, the large pores are filled with a saturated solution and stress-free crystals. A local increase in supersaturation ratio in the pores must be reduced by ion diffusion or solution transport; if these processes are slow, mechanical stress can be exerted over a considerable period of time and even cause material damage. This was proved experimentally for Indiana Limestone and Cordova Cream Limestone (which do not contain small pores) during the crystallization of sodium sulphate salts (Espinosa-Marzal & Scherer 2008a).

Indeed, the magnitude of the local supersaturation during the crystallization process depends on the dynamic interaction between nucleation and growth kinetics of each single crystal and the ion transport rates between connected pores, as well as on temperature. This is relevant for the prediction and prevention of damage, and therefore it should be an important field of future research.

Hydrostatic pressure from the change in volume has also been considered to be a reason for damage caused by salt crystallization (Correns 1949).

Certainly there is an increase in the total volume consisting of solution and salt crystals compared to the volume of the supersaturated solution before crystallization occurs. For example, when cooling a sodium sulphate solution with an initial concentration of 3.34 mol kg^{-1} from 27 °C down to 0 °C, a maximal volume increase of 3% is measured (Haynes 2006). However, the pore system is only partially filled with solution in practice and, consequently, the hydrostatic pressure can be simply released by pushing the solution into the empty pores without generating stress in the material. However, it cannot be excluded that hydraulic pressure becomes relevant in the presence of pore clogging or when ice and salt crystals (cryohydrates) form in the pore system simultaneously.

Stress

The crystallization pressure causes a compressive radial stress on the pore wall and this is accompanied by a tensile hoop stress in the material (Scherer 1999). The net tensile stress, which is reflected in the expansion of the body during crystallization, can cause damage.

There are several comparative studies of salt damage in porous materials in the literature (e.g. Scherer 2004; Coussy 2006; Lubelli & van Hees 2007; Espinosa et al. 2008a; Espinosa-Marzal et al. 2008). According to equation (8), materials with smaller pores are more susceptible to salt damage, which has been confirmed experimentally (e.g. in Espinosa et al. 2008a; Espinosa-Marzal & Scherer 2008a; Espinosa-Marzal et al. 2009). The stress induced in the material depends on both pore size and salt distribution within the pore network, since these factors control the area over which the crystallization pressure is exerted.

According to equilibrium thermodynamics salt precipitates preferentially in large pores (>0.1 μm), which constitute most of the pore volume of limestone. However, Espinosa-Marzal & Scherer (2008a, 2009a) give evidence that a non-equilibrium state is responsible for the high stress and damage induced by salt crystallization in stone with large pores. Since the salt distribution depends on various kinetic factors, there is no universal salt distribution. Efforts to characterize the salt distribution have been made; for example, by using mercury intrusion porosimetry. This method requires a severe drying of the sample, which may modify the original salt distribution by inducing solution transport or dehydration of the salt. Research is needed to examine the salt distribution using methods that do not require evaporation, such as freeze-drying, or in situ methods, such as nuclear magnetic resonance (NMR) (Pel et al. 2000) or radiography(Ketelaars 1995).

The stress exerted in a pore by a single crystal is not likely to generate cracking and failure of the body since the volume under stress is too small. Thus, for the stress generated by crystallization to act on the largest strength-limiting flaw in the material, the crystals must first propagate through a substantial volume of the pore space until its stress field reaches the flaw (Scherer 1999). A few attempts have been made to predict the average stress in the material. A good estimation can be obtained by multiplying the crystallization pressure by the fraction of the pore volume occupied by the crystals (Hamilton et al. 2008).

Poromechanics has been classically used to determine the deformation and the stress caused by pore pressure in saturated and unsaturated materials. The average stress results from a thermodynamically consistent overall elastic energy induced in the matrix by the pore pressure; damage is predicted when the average stress exceeds the tensile strength of the material. Recently it has been applied to problems of freezing in concrete (Coussy 2005; Sun & Scherer 2009) and of salt crystallization (Coussy 2006; Espinosa-Marzal & Scherer 2009a), and the estimations are very reasonable (see Cooling experiments in the section on 'Methods of investigation').

Another approach to predict damage is based on the existence of strength-limiting flaws in the stone. Thus, the tensile strength required to extend the flaw and cause failure can be determined according to fracture mechanics (Scherer 1999; Zehnder 2008). The application of finite-element method (FEM) simulations might help in detecting critical conditions for crack growth induced by salt crystallization.

Methods of investigation

In the last two decades salt crystallization and capillary-rise tests have been widely applied to evaluate the resistance of stones to the damaging action of salt crystallization. We start this section with a short overview of the classical tests and describe briefly recent modifications that provide additional information. The second part of this section deals with experimental methods [such as pore-clogging and warping tests, differential scanning calorimetry (DSC) and differential mechanical analyser (DMA)] that provide *quantitative* information on different aspects discussed in the previous section, such as supersaturation for nucleation, kinetics of in-pore crystallization, pore clogging, crystallization pressure and resulting deformation.

Several other analytical techniques have been used to understand the nature of the phase changes of salts in solution or in the porous material, such as environmental scanning electron microscopy

(SEM) (Rodriguez-Navarro & Doehne 1999; Ruiz-Agudo *et al.* 2007), environmental X-ray diffraction (XRD) (Genkinger & Putnis 2007; Steiger & Linnow 2008), nuclear magnetic resonance (NMR) (Rijniers *et al.* 2005), and, more recently, synchrotron measurements (Hamilton *et al.* 2008; Espinosa-Marzal *et al.* 2009) and atomic force microscopy (A. Hamilton pers. comm.). We do not describe these methods specifically in this section, but include the information obtained from them.

Bursting test or salt crystallization test

The salt crystallization, or bursting, test (RILEM 1980) has been widely used for assessing stone durability against salt weathering, even if the validity of this accelerated method is controversial due to the extremely aggressive crystallization conditions (Price 1996; Tsui *et al.* 2003). The test consists of impregnation–drying cycles using sodium sulphate. During the impregnation, an unsaturated sodium sulphate solution is adsorbed by cubes of stone at constant room temperature. By drying at high temperature (>60 °C), thenardite precipitates in the pores of the stone. Environmental scanning electron microscopy (ESEM) observations by Rodriguez-Navarro *et al.* (2000*a*) show that, during the wetting of thenardite, thenardite first dissolves and then mirabilite precipitates from the solution.

As the dissolution of thenardite leads to a solution highly supersaturated with respect to mirabilite below 32 °C, a high crystallization stress exerted by mirabilite can be expected. Tsui *et al.* (2003) showed that damage occurs during the impregnation at 20 °C (where mirabilite is stable), but not at 50 °C (where thenardite is stable). This demonstrates that during the bursting test any damage is linked to mirabilite precipitation that occurs during the wetting step.

More recently, Steiger & Asmussen (2008) performed a similar test with the difference that the impregnation took place only by capillary uptake of the solution through the bottom surface of the samples, and at 27 °C to avoid the possible formation of heptahydrate. Their modified set-up allowed the length change of the sample to be monitored using a dilatometer and confirmed that, indeed, an expansion was caused during the rewetting by the crystallization of mirabilite. Thus, the bursting test confirms the influence of the supersaturation on the damage mechanism by salt crystallization.

Capillary-rise experiment with simultaneous drying

It is well known, from evaporation experiments under conditions of continuous capillary absorption

Fig. 4. Damage of Cordova Cream Limestone caused by the capillary-rise of an unsaturated sodium sulphate solution. The drying conditions are 34% RH and 21 °C.

of an unsaturated solution, that crystal growth during evaporation can generate substantial stress leading to severe damage. This is illustrated in Figure 4 for Cordova Cream Limestone adsorbing sodium sulphate solution (12 wt%) and drying at 22 °C and 34% RH (Scherer 2004). This experiment, introduced by Lewin (1982), simulates the penetration of salts with groundwater into stone or masonry in the field, while evaporation takes place on the surface. Thus, it includes the interaction between moisture and ion transport, environmental conditions and crystallization kinetics, which determine whether efflorescence or subflorescence forms in the field, as discussed in the previous section. However, the experiment fails to reproduce natural conditions in some important respects. First of all, the concentration of the salt solution is much greater in the laboratory than anything found in nature in order to accelerate the process. In addition, the samples are not subjected to the fluctuations in temperature and humidity that they would experience in nature, but this is necessary in order to have proper control of the experiment.

Sodium sulphate is most often used in the capillary-rise experiments, but sodium chloride was also used in Rodriguez-Navarro & Doehne (1999) and Espinosa-Marzal *et al.* (2008) and magnesium sulphate in Ruiz-Agudo *et al.* (2007). Both sulphate salts are very often linked to severe damage under the conditions of this experiment.

Although this experiment does not allow a quantitative analysis of the damage, it offers the possibility of comparing the action of different salts, substrates and even of different preventive treatments qualitatively. Moreover, it is an ideal experiment for numerical simulation, as it allows analysis of the influence of factors affecting the crystallization and damage pattern, as explained in the previous section (see Fig. 3).

In Kiencke (2005) a very interesting variation of the capillary-rise experiment was used with the aim of determining the salt content required for damage owing to crystallization of potassium chloride, sodium sulphate and gypsum in different bricks. Here, the evaporation surface (upper surface) was treated with a hydrophobic coating to avoid the capillary transport of the solution to the surface and thereby preventing efflorescence. While solution uptake and evaporation take place, the length change of the sample was measured using a dilatometer gauge placed on the upper surface (see Fig. 5). The salt distribution was measured by X-ray fluorescence (XRF) and showed an enrichment of the precipitated salt just below the $c.$ 2 mm coating depth, concentrated within a crystallization front approximately 1 mm thick. Thus, with this method it was possible to link damage with the salt content within the crystallization front or *critical salt content*.

Pore-clogging test

Pore clogging results from the interaction between in-pore crystallization and transport processes. A first attempt to evaluate the pore clogging induced by different salts in limestone has been made by Espinosa-Marzal & Scherer (2008b). The test consists of an initial impregnation of prismatic specimens ($5 \times 2.5 \times 2.5$ cm) with an unsaturated solution under atmospheric pressure. In this work NaCl (3.2 and 5.8 mol kg^{-1}), Na$_2$SO$_4$ (0.85 mol kg^{-1}) and MgSO$_4$ (1.68 mol kg^{-1}) were used. After impregnation five surfaces were coated with a special epoxy that seals wet surfaces to allow drying to occur through only one surface. Drying at $31 \pm 2\%$ RH and 21 ± 1 °C took place over months. While reference samples impregnated with water were dry after 2 weeks, the presence of salt induced a significant slowing of the drying. After the samples reached constant weight, a sorptivity test was performed in which decane was absorbed through the only uncoated surface. Decane does not dissolve the precipitated salt and, consequently, comparison of the decane uptake before and after drying allows the resistance to the uptake caused by pore clogging with salt crystals to be compared. In Indiana Limestone, halite and magnesium sulphate salts lead to the highest pore clogging, while less intensive pore clogging was obtained with sodium sulphate salts, mainly due to the enhanced formation of efflorescence of thenardite. To determine the depth of the crystallization front and the pore filling within the crystallization front, the method developed by Hall & Hoff (2002) to calculate the sorptivity of a bilayer composite can be applied (work in progress).

Cooling experiments

The objective of the cooling experiments was to study nucleation and crystallization kinetics of salts in the porous material, as well as the induced stress. Crystallization of salts can be caused by cooling (or, in a few cases, heating) the solution, if the solubility is strongly dependent on temperature. For such salts the nucleation temperature (i.e. where the supersaturation threshold is reached) can be determined by differential scanning calorimetry, as crystallization is an exothermic (or endothermic) process. The threshold supersaturation gives an indication of the damaging nature of the salt because it is directly related to the crystallization pressure. The crystallization rate in the porous material can also be obtained, provided that the crystallization enthalpy is known (Espinosa-Marzal & Scherer 2008a). This technique permits comparison of the catalytic influence of different porous materials or the efficiency of nucleation inhibitors, among other things.

Some salts (notably including sodium sulphate and magnesium sulphate) can precipitate as different metastable or stable hydrated phases. For example, the concentration of the sodium sulphate solution (*in vitro* and in the pores of limestone), measured using NMR (Rijniers et al. 2005) as well as synchrotron X-ray analysis by Hamilton et al. (2008), indicates that the metastable sodium

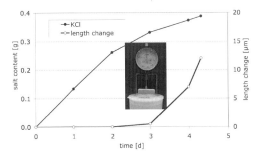

Fig. 5. Salt content in a brick ($4 \times 4 \times 4$ cm) and measured length change during the capillary uptake and simultaneous evaporation of a potassium chloride solution over 5 days.

sulphate heptahydrate always forms prior to mirabilite when cooling down the solution. This has consequences from the point of view of stress because the crystallization of heptahydrate reduces the concentration and, consequently, the crystallization pressure that mirabilite can exert.

Espinosa-Marzal & Scherer (2008a) reported that the supersaturation required for the nucleation of mirabilite is much higher than that for heptahydrate, which suggests that mirabilite will be more damaging. This was confirmed with cooling experiments performed with the dynamic mechanical analyser (Espinosa-Marzal & Scherer 2008a, 2010a). Here, the expansion of different limestones saturated with sodium sulphate were measured during cooling-heating cycles. Damage was very often found to be related to the crystallization of mirabilite. Figure 6 shows the measured strain owing to the crystallization of both salts.

The expansion caused by heptahydrate starts at approximately 4 °C and remains during the isothermal hold (c. 60 μm m^{-1}). After 600 min the rapid crystallization of mirabilite is responsible for the abrupt expansion, followed by a rapid relaxation and a much slower relaxation. The residual strain when heating up to 50 °C reveals damage of the limestone.

Using the theory of thermoporoelasticity (Coussy 2006), the crystallization pressure exerted by the precipitated salt, as well as the stress, can be estimated from the measured strain. Under the conditions of the cooling experiments, the crystallization pressure exerted by sodium sulphate heptahydrate is much smaller than that of mirabilite (at 0 °C, $\Delta p_{hep} \approx 11$ MPa and $\Delta p_{mir} \approx 22$ MPa). Moreover, the resulting stress induced by the crystallization of heptahydrate is not damaging, but it exceeds the tensile strength (3 MPa) of the stone during the crystallization of mirabilite (Espinosa-Marzal & Scherer 2010a). Similar values for the stress are estimated by multiplying the crystallization pressure (equation 7) by the fraction of the pore volume occupied by the crystals (Hamilton et al. 2008).

Warping experiment

Warping of a stone–glass composite has been used to determine the deformation caused by the rewetting of samples containing thenardite, which induces crystallization of mirabilite (Espinosa-Marzal et al. 2010). A plate of glass is glued to a plate of thenardite-bearing stone; water or a salt solution is then allowed to wick into the stone, and the composite bends as crystallization pressure causes the stone to expand. Using a linear variable differential transformer (LVDT), the deflection of a limestone–glass composite (Indiana Limestone or Cordova Cream Limestone) was measured. Synchrotron radiation, producing hard X-rays, was used to examine the thenardite distribution in the initially dry sample and the kinetics of transformation to mirabilite on wetting. The warping results indicate that drying-induced crystallization of thenardite (in the oven at 105 °C) puts the stone into tension without causing any damage under the conditions of this experiment. The stress is relieved during the process of rewetting, when thenardite dissolves, as indicated by the initial positive deflection in Figure 7. Expansion is caused by the subsequent crystallization of mirabilite, which puts the stone plate again into tension and leads to the increasingly negative deflection in Figure 7.

To determine the stress related to the measured deflection of the glass–limestone composite during rewetting, the mechanical problem of bending was solved. This requires knowledge of the location of the advancing crystallization front of mirabilite, which is estimated by numerically modelling the solution transport coupled with sodium sulphate dissolution–recrystallization. Finally, the crystallization pressure exerted by the mirabilite crystals was obtained by applying poroelasticity theory. Thus, the estimated crystallization pressure exerted by mirabilite crystals is approximately 14 MPa at 22 ± 1 °C. According to equation (7), the expected crystallization pressure under the conditions of this experiment is 12.8 MPa at 21 °C, considering that the maximal concentration of the solution is given by the solubility of thenardite at this temperature (3.69 mol kg^{-1}). The resulting stress depends on the amount of salt and on the pore-size distribution

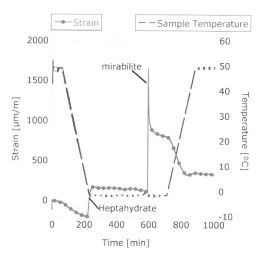

Fig. 6. Measured strain owing to crystallization and to thermal expansion/contraction of a sample of Cordova Cream Limestone impregnated with a sodium sulphate solution (2.9 mol kg^{-1}).

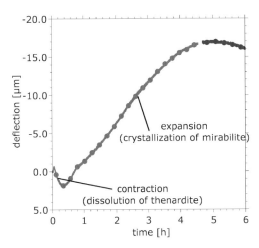

Fig. 7. Measured deflection during the rewetting of a sample of Indiana Limestone with sodium sulphate solution (20 wt%) at 20 °C in the warping experiment. A positive deflection indicates contraction of the stone plate, while a negative deflection corresponds to expansion.

of the material. Thus, the crystallization pressure is slightly overestimated by the model, but given the complexity of the coupled phenomena (transport, dissolution and crystallization) this result is reasonable.

In the previous sections it was shown that progress has been made in understanding the in-pore crystallization and the coupling with the transport processes; however, from a macroscopic point of view, there is still a lack of knowledge in a few aspects related to the *pore-level crystallization*. Thus, still an open question is the direct measurement of the crystallization pressure exerted by a crystal as it has not been possible to reproduce the Correns experiment so far (Flatt *et al.* 2007), except for ice (Buil & Aguirre-Puente 1981). There is indirect evidence that crystallization pressure can only be exerted if a thin film forms between crystal and pore surface, but the thin film has not been measured directly for salt, although it has been measured for ice (Wilen & Dash 1995). The interaction forces between the minerals of the substrate and the crystal, as well as the transport processes in the thin film, have not been quantified. The disjoining pressure could be addressed using atomic force microscopy or the surface force apparatus, and the mobility in the film could, perhaps, be measured using NMR, and these phenomena could all be simulated using molecular dynamics.

ESEM has provided important information by allowing direct observation of crystallization in a pore (see, for example, Rodriguez-Navarro & Doehne 1999). Determining the pore-size-dependent salt distribution and dynamic redistribution in connected pores exceeds the current resolution of the ESEM. In contrast, NMR might help to visualize the pore-size-dependent crystal distribution (L. Pel pers. comm.), but it has not been carried out to date.

The propagation of cracks induced by the tensile stress involved in the crystallization process has not been studied in detail. Rapidly increasing computational power, however, may permit FEM models to simulate cracking induced by salt crystallization at the pore level.

Another important subject of investigation and debate deals with the control of nucleation, crystal growth and crystallization pressure to prevent damage from salt. This is discussed in detail in the next section.

Methods for prevention

Methods that have been used to prevent or retard stone decay caused by salt crystallization include cleaning, desalination (poultices and methods based on electromigration), a variety of consolidants and surface coatings, such as water repellents, polymers to passivate the mineral surface and crystallization inhibitors (Price 1996). However, the effectiveness of these methods for the prevention of salt decay in porous materials is still controversial.

In the following discussion we focus on the additives that directly modify a step of the in-pore crystallization process (Füredi-Milhofer & Sarig 1996): nucleation (acting as inhibitor or promoter); crystal growth (as inhibitor of growth or modifier of the crystal habit); aging process, salt transport through the pore network (by modifying properties of the solution); and crystallization pressure (by changing the disjoining pressure between crystal and mineral surfaces). Indeed, there is no unique classification for additives as they usually affect rates and/or mechanisms of more than one of the crystallization steps.

Examples of crystallization additives with technological and industrial uses are phosphates and polyphosphates, carboxylic acid derivatives, polyelectrolytes, ferrocyanides, surfactants and phosphonates. In the present century studies have been carried out on the effect of some of them on the crystallization of salts associated with salt weathering of building materials.

Modifier of nucleation

A crystallization or nucleation inhibitor causes salt to nucleate at a higher supersaturation. The purpose is to avoid crystallization by maintaining the salt dissolved in the solution until it is carried to the surface, where it precipitates as harmless efflorescence instead of damaging subflorescence. However, there are several problems associated with nucleation or crystallization inhibitors.

- If crystallization takes place inside the stone, higher crystallization pressure will be exerted as a consequence of the higher supersaturation, and therefore more severe damage might be caused.
- The expected transport of the solution to the surface might be affected by several factors that are not accounted for. Thus, blocking of pores in the surface or a much more rapid evaporation rate (e.g. when submitted to extreme conditions of wind) might move the crystallization front into the stone.
- Currently, the capillary-rise experiment with simultaneous evaporation is used to determine the efficiency of inhibitors (and also of nucleation promoters). However, it cannot be excluded that the action of additives is dependent on the crystallization conditions. Rodriguez-Navarro *et al.* (2002) reported that one surfactant, a nucleation promoter, reduced damage during the capillary-rise experiment. However, after submerging the sample in water, the recrystallization of thenardite to mirabilite caused more severe damage in the treated stone than in the untreated stone.

Nucleation promoters catalyse the nucleation of salt so that it precipitates at a lower supersaturation. The effect of small quantities of borax as a nucleation promoter of mirabilite and of epsomite in biocalcarenite has recently been reported by Ruiz-Agudo *et al.* (2008). Here, the capillary-rise experiments clearly showed that the presence of borax in the absorbed solution is associated with a reduction in damage caused by salt crystallization and, therefore, it seems to be a promising method to prevent damage. It has to be remarked, though, that it might enhance pore clogging in some materials, the consequences of which are not yet predictable.

Modifiers of crystal growth

Some additives can either suppress or promote the growth of a crystal by adsorption on all its faces, whereas others act only on certain faces and thus also change the morphology of the crystal. The combination of ESEM and molecular modelling by Ruiz Agudo *et al.* (2006) shows that the most efficient growth inhibitors of mirabilite crystals among the studied phosphonates are those that display the best stereochemical matching with mirabilite surfaces. Classically, successful retardation of crystal growth was attributed to good matching between inhibitor and crystal interionic distance; however, the type, number and strength of the chemical bonds with the crystal surface seem to be the determining factors (Füredi-Milhofer & Sarig 1996; Veintemillas-Verdaguer 1996). A different crystal habit leads to a different tensile stress owing to the different surface contact between the crystals and the pore.

Modifiers of aging process

There are multitudes of studies concerning the action of additives changing the aging process for industrial applications (Füredi-Milhofer & Sarig 1996). It is not obvious how an agglomeration process could cause more or less damage to stone. However, additives that could retard crystal aging by recrystallization from a metastable to a stable phase might be potentially interesting. Metastable salts are more soluble than stable salts and, consequently, they exert a lower crystallization pressure (Espinosa-Marzal & Scherer 2008*a*, 2010*a*). This has not been subjected to investigation in the field of salt weathering, but in other fields of material science, such as alloy casting (Magnin & Kurz 1988) or the pharmaceutical industry (Zhang *et al.* 2004), methods to promote the formation of metastable phases have been investigated to improve certain properties of the product. For example, we have shown that sodium sulphate heptahydrate exerts less stress on stone than mirabilite, so encouraging the formation of heptahydrate would appear to be beneficial; however, it is not clear that transformation of heptahydrate to mirabilite can be completely suppressed.

Modifiers of solution properties

Surfactants contain a hydrophobic part and a hydrophilic ionic group. They can adsorb on a crystalline surface through electrostatic interaction with the polar part (as the polyelectrolytes do), but further adsorption can take place at higher concentration due to hydrophobic interactions between their organic parts. Surfactants that lower the mineral–solution contact angle will enhance the capillary transport; this favours the formation of harmless efflorescence, preventing the formation of subflorescence. However, surfactants may also influence nucleation, and may induce significant morphological changes in crystallizing salts mostly due to preferential adsorption on particular crystal faces (Rodriguez-Navarro *et al.* 2000*b*). Therefore, it is necessary to study the effect of each surfactant on each particular salt before a reliable evaluation can be performed.

Modifiers of disjoining pressure

When a crystal grows from a supersaturated solution in a pore, a repulsive force (disjoining pressure) acts between the two approaching surfaces, primarily resulting from electrostatic and solvation forces; in the case of ice, the van der Waals forces are also repulsive (Scherer 1999). As a consequence of the disjoining pressure, a gap between the crystal and the pore surface of approximately a few nanometres is filled by the solution. The existence of a supersaturated thin film is necessary for crystals to exert a pressure on the pore wall, as discussed before. Moreover, the disjoining pressure gives the maximal crystallization pressure that a salt crystal can exert on the confining surface. Atomic force microscopy allows measurement of the disjoining pressure in diverse research fields (e.g. in biomechanics) and is currently also applied to mineral–crystal interactions (Hamilton 2008 pers. comm.).

Indeed, if the crystal–stone interaction force was attractive, instead of repulsive, crystal and pore surface would come into contact and no thin film would form. The growth would stop and no pressure would be exerted. Based on this idea, additives capable of reducing the disjoining pressure or, ideally, converting repulsive into attractive forces might be very interesting from the point of view of preventing salt weathering, as proposed by Scherer et al. (2001). Houck & Scherer (2006) studied the behaviour of several polymers on the crystallization pressure. They argued that an effective polymer should be strongly adsorbed on the pore wall, and their ligands should be attracted to the surface of the salt crystal, allowing the pore wall to attract, rather than to repel, the salt. The result of the salt crystallization test on Indiana Limestone showed that polyacrylic acid, with a low molecular weight, offered the most protection against damage without changing the crystallization pattern and, consequently, its efficiency was associated with a decrease of the disjoining pressure. However, one of the problems associated with polacrylic acid, and carboxylic acids in general, is its susceptibility to be washed out due to its weak binding to calcite. In addition, it is not known whether these polymers could also act as nucleation or crystallization inhibitors; if so, they might retard nucleation, but once crystallization starts the damage might be worse. Research is currently being carried out to clarify these questions.

Part of the complexity of the crystallization–additive interaction is the dependence of nucleation and crystal growth on the concentration of the additive, as well as on the additive–crystal mass ratio, as reported in Füredi-Milhofer & Sarig (1996). By increasing the additive concentration, an enhancement of nucleation is first observed; it reaches a maximum and then the trend reverses. The question is how to control the concentration in the field under changing climatic conditions. Moreover, temperature and pressure also alter the effect of additives (Veintemillas-Verdaguer 1996). The effectiveness of these additives is also salt- and substrate-dependent (e.g. ferrocyanide in Rodriguez-Navarro et al. 2000b; Selwitz & Doehne 2002 and Lubelli & van Hees 2007; or diethylenetriaminepentakis methylphosphonic acid in Lubelli & van Hees 2007; Ruiz-Agudo et al. 2006). In the field we have to expect the presence of salt mixtures rather than of single salts. Therefore, laboratory work is required for each particular salt–substrate combination to assess the efficiency of the additive.

The application technique of such additives in the field is still in debate. Sprayed additives did not lead to satisfactory results in Lubelli & van Hees (2007). The application of additives by a poulticing technique might have bad consequences if salts dissolved and were transported deeper into the wall. Conversely, putting additives into the poultice might be favourable because it would allow salt to be sucked into the poultice without precipitating. Thus, the efficiency of this technique depends on the transport properties of the poultice, substrate and interface, and this has not yet been studied. Generally, desalination prior to the application of the additive seems advisable.

Currently a lot of effort is being invested to find proper additives, and progress is being made for specific applications in the laboratory. Of course, there will never be one universal treatment or one universal test to assess the reliability of each additive. Each site is subjected to particular conditions (environment, substrate porosity, salt mixture, presence of biofilm and swelling clays), so each site will require a skilled diagnosis before an appropriate treatment can be designed.

Conclusions

This paper summarizes the main issues related to crystallization of salts in pores and the mechanisms of damage. The dynamic nature of the salt weathering process is emphasized through the accentuation of the influence of moisture and salt transport in the porous stone, crystallization kinetics, and the change in the pore structure by the clogging of pores with salts. Salt crystallization tests, capillary rise with simultaneous evaporation, cooling-induced crystallization and warping experiments are used to evaluate the damaging action of salts in stone, and to get quantitative information about crystallization kinetics and the induced deformation of the porous material.

Finally, an overview of the use of additives to reduce salt weathering is given. The philosophy of these additives is to decrease damage by altering a specific crystallization step (e.g. nucleation, growth), the solution properties or the disjoining pressure. Laboratory experiments show that additives work successfully for specific scenarios, and therefore an accurate diagnosis and laboratory tests are required prior to the application of any treatment in the field. The capillary-rise test with evaporation and the salt crystallization test are used to determine the effectiveness of a treatment. Their efficiency under other crystallization conditions in the field is still questionable and therefore more effort is still necessary to understand better the dynamics of salt crystallization within the pore network, as well as the interaction with the additives.

This paper reviews the current state of salt weathering from a scientific point of view and indicates the questions that remain in debate. How does crystallization take place within a pore network? How do the processes in the thin film determine the crystallization pressure? Does pore clogging enhance damage? Can we modify the disjoining pressure? Can we control nucleation and crystal growth? Under what conditions does salt crystallization lead to crack propagation and failure of the material? Can we predict damage when salt mixtures or the combined action of salt and swelling clays are involved? By clarifying these questions, it will be possible to develop useful and practicable methods that can be applied to control salt damage in practice. Fortunately, research on these topics is progressing in many laboratories around the world, so the scientific foundation for protection of stone against salt crystallization is developing rapidly.

The authors thank the Deutsche Forschungsgemeinschaft, the Getty Conservation Institute and the National Centre for Preservation Technology & Training (NCPTT) for financial support.

References

BEAR, J. 1972. *Dynamics of Fluids in Porous Materials.* Dover Publications, New York.

BUIL, M. & AGUIRRE-PUENTE, J. 1981. Thermodynamic and experimental study of the crystal growth of ice. *Proceedings of the ASME Winter Annual Meeting,* 81-WA/HT-69, 1–7.

CARDELL, C., DELALIEUX, F., ROUMPOPOULOS, K., MOROPOULOU, A., AUGER, F. & VAN GRIEKEN, R. 2003. Salt-induced decay in calcareous stone monuments and buildings in a marine environment in SW France. *Construction and Building Materials,* **17**(3), 165–179.

ČERNY, R., FIALA, L., PAVLÍK, Z. & PAVLIKOVA, M. 2007. *Mathematical modelling of water and salt transport in porous materials.* Transactions on Modelling and Simulation, Vol 1, WIT Press, Southampton, 339–348.

CHAROLA, A. E. 2000. Salts in the deterioration of porous materials: an overview. *Journal of the American Institute for Conservation,* **39**(3), 327–343.

CHAROLA, A. E., PÜHRINGER, J. & STEIGER, M. 2006. Gypsum: a review of its role in the deterioration of building materials. *Environmental Geology,* **52**, 339–352.

CHATTERJI, S. & JENSEN, A. D. 1989. Efflorescence and breakdown of building materials. *Nordic Concrete Research,* **8**, 56–61.

CHRISTIAN, J. W. 1975. *The theory of Transformation in Metals and Alloys, Part I: Equilibrium and General Kinetic Theory.* 2nd edn, Pergamon Press, Oxford.

CORRENS, C. W. 1949. Growth and dissolution of crystals under linear pressure, Discuss. *Faraday Society,* **5**, 267–271.

COUSSY, O. 2005. Poromechanics of freezing materials. *Journal of the Mechanics and Physics of Solids,* **53**, 1689–1718.

COUSSY, O. 2006. Deformation and stress from in-pore drying-induced crystallization of salt. *Journal of the Mechanics and Physics of Solids,* **54**, 1517–1547.

DELGADO RODRIGUES, J. & MANUEL MIMOSO, J. 2008. *Proceedings of the International Symposium on Stone Consolidation in Cultural Heritage.* Laboratório Nacional de Engenharia Civil, Lisbon.

DERLUYN, H., MOONEN, P. & CARMELIET, J. 2008. Modelling of moisture and salt transport incorporating salt crystallization in porous media. *In*: SCHLANGEN, E. (ed.) *International RILEM Symposium on Concrete Modelling-CONMOD'08.* Delft, The Netherlands, 28–30 May. RILEM Publications S.A.R.L., Bagneux, France.

DOEHNE, E. 2002. Salt weathering: a selective review. *In*: SIEGESMUND, S., VOLLBRECHT, S. A. & WEISS, T. (eds) *Natural Stone, Weathering Phenomena, Conservation Strategies and Case Studies.* Geological Society, London, Special Publications, **205**, 51–64.

ESPINOSA, R. M., FRANKE, L. & DECKELMANN, G. 2008a. Model for the mechanical stress due to the salt crystallization in porous materials. *Construction and Building Materials,* **22**, 1350–1367.

ESPINOSA, R. M., FRANKE, L. & DECKELMANN, G. 2008b. Phase changes of salts in porous materials. *Construction and Building Materials,* **22**, 1758–1773.

ESPINOSA-MARZAL, R. M. & SCHERER, G. W. 2008a. Study of sodium sulphate salts crystallization in limestone. *Environmental Geology,* **50**, 605–621; doi: 10.1007/s00254-008-1441-7.

ESPINOSA-MARZAL, R. M. & SCHERER, G. W. 2008b. Study of the pore clogging induced by salt crystallization in Indiana limestone. *In: Proceedings of the 11th International Congress on Deterioration and Conservation of Stone.* Volume I. Nicolaus Copernicus University Press, Torun, Poland, 81–88.

ESPINOSA-MARZAL, R. M. & SCHERER, G. W. 2010a. Crystallization kinetics of sodium sulfate salts in limestone and resulting stress. In preparation.

ESPINOSA-MARZAL, R. M. & SCHERER, G. W. 2010b. Determining pore clogging caused by salt crystallization in limestone. In preparation.

ESPINOSA-MARZAL, R. M., FRANKE, L. & DECKELMANN, G. 2008. *Predicting Efflorescence and Subflorescences of Salts.* Materials Research Society Symposium Proceedings, **1047**.

ESPINOSA-MARZAL, R. M., HAMILTON, A., SCHERER, G. W., MCNALL, M. & WHITAKER, K. 2010. The chemomechanics of sodium sulfate crystallization in thenardite impregnated limestones during re-wetting. *Journal of Geophysical Research,* submitted.

FLATT, R. J. 2002. Salt damage in porous materials: how high supersaturations are generated. *Journal of Crystal Growth,* **242**, 435–454.

FLATT, R. J., STEIGER, M. & SCHERER, G. W. 2007. A commented translation of the paper by C. W. Correns and W. Steinborn on crystallization pressure. *Environmental Geology,* **52**, 187–203.

FRANKE, L., KIEKBUSCH, J., ESPINOSA, R. & GUNSTMANN, C. 2007. CESA AND ASTRA – two program systems for cement and salt chemistry and the prediction of corrosion processes in concrete. *In*: International Conference on Durability of HPC and Final Workshop of CONLIFE. Aedificatio, Essen, 501.

FÜREDI-MILHOFER, H. & SARIG, H. 1996. Interactions between polyelectrolytes and sparingly soluble salts. *Progress in Crystal Growth and Characterization of Materials,* **32**, 45–74.

GENKINGER, S. & PUTNIS, A. 2007. Crystallization of sodium sulfate: supersaturation and metastable phases. *Environmental Geology,* **52**, 295–303.

GOUDIE, A. & VILES, H. 1997. *Salt Weathering Hazards.* Wiley, Chichester.

HALL, C. & HOFF, W. 2002. *Water Transport in Brick, Stone and Concrete.* Spon Press, London.

HAYNES, H. 2006. A mechanism of distress to concrete during crystallization of mirabilite. *In*: MALHOTRA, V. M. (ed.) *Proceedings of the Seventh CANMET/ACI International Conference on Durability of Concrete, Supplemental Volume,* Montreal, June 1–16, ACI, Michigan, USA.

HAMILTON, A., HALL, C. & PEL, L. 2008. Salt damage and the forgotten metastable sodium sulfate heptahydrate: direct observation of crystallization in a porous material. *Journal of Physics D: Applied Physics,* **41**, 212002.

HOUCK, J. & SCHERER, G. W. 2006. Controlling stress from salt crystallization. *In*: KOURKOULIS, S. K. (ed.) *Fracture and Failure of Natural Building Stones.* Springer, Dordrecht, chap. 5; doi: 10.1007/978-1-4020-5077-0_19.

IBRAHIM, H. & KAMH, G. 2005. The negative effect of environmental geological conditions of some geo-archaeological sites of North Coast and Alexandria. *Environmental Geology,* **49**, 179–187.

KETELAARS, A. A. J., PEL, L., COUMANS, W. J. & KERKHOF, P. J. A. M. 1995. Drying kinetics: a comparison of diffusion coefficients from moisture concentration profiles and drying curves. *Chemical Engineering Science,* **50**, 1187–1191.

KIENCKE, S. 2005. *Beanspruchungsmechanismen und kritischer Gehalt von Salzen in porösen Materialien.* Dissertation TU-Hamburg-Harburg. GCA Verlag (only in German).

LEWIN, S. Z. 1982. The mechanism of masonry decay through crystallization. *In: Conservation of Historic Stone Buildings and Monuments.* National Academies Press, Washington, DC, 120–144.

LUBELLI, B. & VAN HEES, R. 2007. Effectiveness of crystallization inhibitors in preventing salt damage in building materials. *Journal of Cultural Heritage,* **8**, 223–234.

MAGNIN, P. & KURZ, W. 1988. Competitive growth of stable and metastable Fe–C–X eutectics: Part I. experiments. *Metallurgical and Materials Transactions A,* **19**, 1955–1963.

MULLIN, J. W. 1993. *Crystallisation.* 3rd edn. Butterworths, London.

NICOLAI, A. 2007. *Modeling and Numerical Simulation of Salt Transport and Phase Transitions in Unsaturated Porous Building Materials.* PhD thesis, Syracuse University.

NIELSEN, A. E. 1984. Electrolyte crystal growth mechanisms. *Journal of Crystal Growth,* **67**, 289–310.

PEL, L., KOPINGA, K. & KAASSCHIETER, E. F. 2000. Saline absorption in calcium silicate brick observed by NMR scanning. *Journal of Physics D: Applied Physics,* **33**, 1380–1385.

PITZER, K. S. 1991. *Activity Coefficients in Electrolyte Solutions.* 2nd edn. CRC Press, Boca Raton, FL.

PRICE, C. A. 1996. *Stone Conservation – An Overview of Current Research.* The Getty Conservation Institute, Los Angeles, CA.

RILEM. 1980. *Recommended Tests to Measure the Deterioration of Stones and Assess the Effectiveness of Treatment Methods.* Commission 25-PEM. Protection et Erosion des Monuments, 175–253.

RIJNIERS, L. A., HUININK, H. P., PEL, L. & KOPINGA, K. 2005. Experimental evidence of crystallization pressure inside porous media. *Physical Review Letter* **94**, 075503.

RODRIGUEZ-NAVARRO, C. & DOEHNE, E. 1999. Salt weathering: influence of evaporation rate, supersaturation and crystallization pattern. *Earth Surface Processes and Landforms,* **24**, 191–209.

RODRIGUEZ-NAVARRO, C., DOEHNE, E. & SEBASTIAN, E. 2000a. How does sodium sulfate crystallize? Implications for the decay and testing of building materials. *Cement and Concrete Research,* **30**, 1527–1534.

RODRIGUEZ-NAVARRO, C., DOEHNE, E. & SEBASTIAN, E. 2000b. Control crystal growth influencing crystallization damage in porous materials through the use of surfactants: experimental results using sodium dodecyl sulfate and cetyldimethylbenzylammonium chloride. *Langmuir,* **16**, 947–954.

RODRIGUEZ-NAVARRO, C., LINARES-FERNANDEZ, L., DOEHNE, E. & SEBASTIAN, E. 2002. Effects of ferrocyanide ions on NaCl crystallization in porous stone. *Journal of Crystal Growth,* **243**, 503–513.

ROTHERT, E., EGGERS, T., CASSAR, J., RUEDRICH, J., FITZNER, B. & SIEGESMUND, S. 2007. Stone properties and weathering induced by salt crystallization of Maltese Globigerina Limestone. *In*: PRIKRYL, R. & SMITH, B. J. (eds) *Building Stone Decay: From Diagnosis to Conservation.* Geological Society, London, Special Publications, **271**, 189–198.

RUIZ-AGUDO, E., MEES, F., JACOBS, P. & RODRIGUEZ-NAVARRO, C. 2007. The role of saline solution properties on porous limestone salt weathering by magnesium and sodium sulfates. *Environmental Geology,* **52**, 269–281; doi 10.1007/s00254-006-0476-x.

RUIZ-AGUDO, E., RODRIGUEZ-NAVARRO, C. & SEBASTIAN-PARDO, E. 2006. Sodium Sulfate Crystallization in the Presence of Phosphonates: Implications in ornamental stone conservation. *Crystal Growth and Design*, **6**, 1575–1583.

RUIZ-AGUDO, E. & RODRIGUEZ-NAVARRO, C. 2010. Suppression of salt weathering of porous limestone by borax-induced promotion of sodium and magnesium sulfate crystallization. *Journal of the Geological Society*, in press.

SAMSON, E. & MARCHAND, J. 1999. Numeric solution of the extended Nernst–Planck model. *Journal of Colloid and Interface Science*, **215**, 1–8.

SCHERER, G. W. 1995. Fundamentals of drying and shrinkage. *In*: HENKES, V. E., ONODA, G. Y. & CARTY, W. M. (eds) *Science of Whitewares*. American Ceramic Society, Westerville, OH, 199–211.

SCHERER, G. W. 1999. Crystallization in pores. *Cement and Concrete Research*, **29**, 1347–1358.

SCHERER, G. W. 2004. Stress from crystallization of salt. *Cement and Concrete Research*, **34**, 1613–1624.

SCHERER, G. W., FLATT, R. & WHEELER, G. 2001. Materials Science Research for Conservation of Sculpture and Monuments. *MRS Bulletin*, Jan. 2001, 44–50.

SELWITZ, C. & DOEHNE, E. 2002. The evaluation of crystallization modifiers for controlling salt damage to limestone. *Journal of Cultural Heritage*, **3**, 205–216.

SGHAIER, N. & PRAT, M. 2009. Effect of efflorescence formation on drying kinetics of porous media. *Transport in Porous Media*, **12**, 549–559; doi: 10.1007/s11242-009-9373-6.

SHAHIDZADEH-BONN, N., RAFAÏ, S., BONN, D. & WEGDAM, G. 2008. Salt crystallization during evaporation: impact of interfacial properties. *Langmuir*, **24**, 8599–8605.

STEIGER, M. 2005a. Crystal growth in porous materials I–: The crystallization pressure of large crystals. *Journal of Crystal Growth*, **282**, 455–469.

STEIGER, M. 2005b. Crystal growth in porous materials II–: Influence of crystal size on the crystallization pressure. *Journal of Crystal Growth*, **282**, 470–481.

STEIGER, M., KIEKBUSCH, J. & NICOLAI, A. 2008. An improved model incorporating Pitzer's equations for calculation of thermodynamic properties of pore solutions implemented into an efficient program code. *Construction of Building Materials*, **22**, 1841–1850.

STEIGER, M. & ASMUSSEN, S. 2008. Crystallization of sodium sulfate phases in porous materials: the phase diagram $Na_2SO_4-H_2O$ and the generation of stress. *Geochimica et Cosmochimica Acta*, **72**, 4291–4306.

STEIGER, M. & LINNOW, K. 2008. Hydration of $MgSO_4 \cdot H_2O$ and generation of stress in porous materials. *Crystal Growth and Design*, **8**(1), 336–343.

SUN, Z. & SCHERER, G. W. 2009. Effect of air voids on salt scaling and internal freezing. *Cement and Concrete Research*. in press, doi: 10.1016/j.cemconres.2009.09.027.

VEINTEMILLAS-VERDAGUER, S. 1996. Chemical aspect of the effect of impurities in crystal growth. *Progress in Crystal Growth and Characterization*, **2**, 75–109.

VENDRELL-SAZ, M., GARCIA-VALLES, M., ALARCON, S. & MOLERA, J. 1996. Environmental impact on the Roman monuments of Tarragona, Spain. *Environmental Geology*, **27**, 263–269.

TSUI, N., FLATT, R. J. & SCHERER, G. W. 2003. Crystallization damage by sodium sulfate. *Journal of Cultural Heritage*, **4**, 109–115.

WILEN, L. A. & DASH, J. G. 1995. Frost heave dynamics at a single crystal interface. *Physics Review Letters*, **74**(25), 5076–5079.

ZEHNDER, A. 2008. *Lecture Notes on Fracture Mechanics*. http://ecommons.library.cornell.edu/handle/1813/3075.

ZHANG, G., LAW, D., SCHMITT, E. A. & QIU, Y. 2004. Phase transformation considerations during process development and manufacture of solid oral dosage forms. *Advanced Drug Delivery Reviews*, **56**(3), 23, 371–390.

Primary bioreceptivity of limestones used in southern European monuments

ANA Z. MILLER[1]*, NUNO LEAL[2], LEONILA LAIZ[3], MIGUEL
A. ROGERIO-CANDELERA[3], RUI J. C. SILVA[4], AMÉLIA DIONÍSIO[5],
MARIA F. MACEDO[1] & CESAREO SAIZ-JIMENEZ[3]

[1]*Departamento de Conservação e Restauro, Faculdade de Ciências e Tecnologia, Universidade Nova de Lisboa, Monte de Caparica, 2829-516 Caparica, Portugal*

[2]*Centro de Investigação em Ciência e Engenharia Geológica (CICEGe), Faculdade de Ciências e Tecnologia, Universidade Nova de Lisboa, Monte de Caparica, 2829-516 Caparica, Portugal*

[3]*Instituto de Recursos Naturales y Agrobiologia, IRNAS-CSIC, Av. Reina Mercedes 10, 41012 Sevilla, Spain*

[4]*CENIMAT/I3N and DCM, Faculdade de Ciências e Tecnologia, Universidade Nova de Lisboa, Quinta da Torre, 2829-516 Monte de Caparica, Portugal*

[5]*Centro de Petrologia e Geoquímica, Instituto Superior Técnico, Av. Rovisco Pais, 1049-001, Lisboa, Portugal*

*Corresponding author (e-mail: azm@fct.unl.pt)

Abstract: Different Mediterranean Basin limestones, like *Calcário Ançã* (Portugal), *Calcário Lioz* (Portugal), *Piedra San Cristobal* (Spain), *Piedra Escúzar* (Spain) and *Pietra di Lecce* (Italy), have been widely used as building materials in the European architecture. The aim of this study was focused on biodeterioration, mainly on evaluation of the primary bioreceptivity of those materials.

A set of samples was inoculated with a cultured photosynthetic biofilm under laboratory conditions. Several assessment tools were applied to monitor the colonization overtime of the different lithotypes. After 3 months of incubation the colonization occurred endolithically in some lithotypes, namely *Piedra San Cristobal* and *Piedra Escúzar*. Spectrophotometric determination of chlorophyll *a* was a useful analytical technique to achieve the total amount of photosynthetic biomass on rock substrates, demonstrating that *Piedra Escúzar* and *Calcário Lioz* were the highest and lowest bioreceptive lithotypes, respectively. Microscopic and image analyses were essential to understand the stone colonization process and its pattern of distribution. Physical stone parameters and exposure conditions were shown to play an important role in the establishment and development of photosynthetic colonization.

Stone monuments, statues and historic buildings are exposed to the effects of physical, chemical and biological weathering factors. Several pieces of evidence reveal that many types of deterioration on limestone are due to biological activity through a process referred as biodeterioration. Cyanobacteria and microalgae have a global significance in stone biodeterioration process emphasized by their photoautotrophic nature, and depending on light, carbon dioxide and a few other elements. They colonize stone surfaces forming biofilms composed of cells immobilized on the stone surface and embedded in a hydrated polysaccharide matrix produced by polymerases affixed to the lipopolysaccharide component of the cell wall (Morton *et al.* 1998; Warscheid & Braams 2000). These extracellular polymeric substances (EPS), secreted by the cells, allow the adhesion of micro-organisms to surfaces. In addition, the intrinsic properties of the lithic substrate (e.g. surface roughness, porosity, permeability, chemical composition), and its decay degree, determine the adhesion and spreading of the micro-organisms on and within the stone material, providing a protective niche on which micro-organisms can develop. The totality of material properties that contribute to biological colonization has been defined as bioreceptivity by Guillitte (1995), who further defined different types of it, such as primary, secondary and tertiary bioreceptivity. Laboratory experiments focused on stone primary bioreceptivity, that is, the initial potential of unaltered stone material for microbial colonization, revealed that carbonate rocks have a high risk of colonization by micro-organisms and are extremely susceptible to cyanobacteria and microalgae (Guillitte & Dreesen 1995; Tiano *et al.* 1995; Tomaselli *et al.* 2000; Miller *et al.* 2006). Moreover, these experiments suggested that mineralogy, porosity, surface roughness and water uptake, as well as chemical composition, of the substrate are significant parameters promoting microbial colonization. It is important to consider that specific microbial genera or species may specialize in particular types of stones and require specific environmental conditions (Koestler *et al.* 1996; Tomaselli *et al.* 2000).

In the present study, special emphasis was placed on the primary bioreceptivity of five sedimentary carbonate rocks used as construction materials in Portuguese, Spanish and Italian buildings and monuments. This study was based on the application of laboratory analyses to evaluate the colonization of these stone materials by photosynthetic micro-organisms, and to assess the relationship between the intrinsic properties of the stones and microbial colonization.

Materials and methods

Studied lithotypes

Five types of limestone with different petrographical, chemical, physical and mechanical characteristics were studied to investigate their primary bioreceptivity. The lithotypes used were *Calcário Ançã* (Portugal), *Calcário Lioz* (Portugal), *Piedra San Cristobal* (Spain), *Piedra Escúzar* (Spain) and *Pietra di Lecce* (Italy).

Calcário Ançã (CA) is a Bajocian–Bathonian limestone that occurs in a large outcrop area, located in the Beira Litoral region, in the Coimbra District (Dionísio 1997). It is extensively used in monuments, buildings and sculptures in the central zone of Portugal, such as Santa Cruz Church and the *Porta Especiosa* of Coimbra Old Cathedral, as well as in most parts of Spain, and reaching other places in Europe.

Calcário Lioz (CL) is a limestone (with several varieties) present in many Portuguese emblematic monuments, such as the Jerónimos Monastery and the Tower of Belém, both in Lisbon. During the seventeenth, eighteenth and nineteenth centuries huge blocks of this rock were taken to Brazil to build churches, mainly in the Bahia region (Silva 2007). This rock is mainly extracted from the Pêro Pinheiro region (Lisbon's surroundings), where it widely occurs and where many quarries are worked. From a geological point of view, it is a Middle Turonian (Middle Cretaceous) limestone.

Piedra San Cristobal (SC) is an Upper Miocene calcarenite present in Salvador's Church and in Seville Cathedral, both in Seville (Sapin). Samples of this lithotype were collected from the El Puerto de Santa María quarry (Cádiz), from where a large amount of material was extracted for the construction of Seville Cathedral.

Piedra Escúzar (PF), also known as *Piedra Franca*, is a very soft and porous biocalcarenite from the Tortonian age. It has been widely used in historic monuments of Granada (southern Spain), such as the cathedral, the Royal Hospital and the Alhambra walls.

Pietra di Lecce (PL) comes from the most important city of the Salento region, in SE Italy. It is a limestone dating from the Miocene period – Langhian age – widely used in the Late Baroque. It is present in the *Basilica di Santa Croce*, the most important baroque monument in the Salento region, as well as in many other temples.

Characterization of lithotypes. Samples of the selected lithotypes were collected from quarries located in the vicinities of the sites in which monuments and buildings were raised, and cut into cylinders with dimensions ϕ 4.4 × 2 cm. The stone probes were described by visualization with the naked eye and under a microscope. Petrographical characterization was made by means of 30 μm epoxy-impregnated thin sections studied under a polarizing microscope. Four samples of each lithotype were analysed. From a physical point of view, stones were characterized by determining water absorption capillarity and open porosity according to European Standards EN 1925:1999 and 1936:1999, respectively. These physical assays were carried out on 27 probes of each lithotype, oriented perpendicularly to rock stratification.

Stone bioreceptivity experiment

Microbial enrichment. In order to study the primary bioreceptivity of the selected stone materials, the stone probes were inoculated with a natural photosynthetic enrichment previously identified and cultured. The natural green biofilm developed over *Calcário Ançã* was collected at 50 cm above the ground from columns of the north façade of the Santa Clara-a-Velha Monastery in Coimbra (central Portugal). Sampling was performed by scraping the biofilm into sterile tubes and it was then stored at 4 °C until being processed. This microbial community was identified by molecular techniques and cultured in liquid BG11 medium for 3 months (Miller *et al.* 2008). Denaturing gradient gel electrophoresis (DGGE) analysis was used to evaluate the major microbial components of the green biofilm. Cyanobacteria and green microalgae were identified by DNA-based molecular analysis targeting the 16S and 18S ribosomal RNA genes, as described by Miller et al. (2008). The genomic DNAs were extracted using the Nucleospin Food DNA Extraction Kit (Macherey-Nagel, Düren, Germany). Amplification of DNA was carried out by Polymerase chain reaction (PCR) in a BioRad iCycler iQ thermal cycler (BioRad, Hercules, CA) using the primer pair Cya106F and Cya781R, and PCR conditions as described by Nübel et al. (1997). Eukaryotic 18S rRNA genes were amplified with the primer pair EukA and EukB (Diez et al. 2001). ExTaq (Takara, Shiga, Japan) was the DNA polymerase used for PCR, following the manufacturer's recommendations.

Prokaryotic community fingerprints were obtained by DGGE. A nested-PCR reaction was performed using the primer pair 341F-GC and 518R, following the method described by Muyzer et al. (1993) and Gonzalez & Saiz-Jimenez (2004). In order to obtain 16S and 18S rRNA gene clone libraries used for sequencing and identification, PCR products were purified with the JetQuick PCR Purification Spin Kit (Genomed, Löhne, Germany) and cloned with the TOPO TA Cloning Kit (Invitrogen, Carlsbad, CA). Clone screening was carried out by DGGE as previously described (Gonzalez et al. 2003). Migration markers (16S rRNA gene fragments from *Pseudomonas* sp., *Escherichia coli*, *Paenibacillus* sp. and *Streptomyces* sp.) were used throughout this study as reference for locating the position of cloned fragments in the microbial community fingerprints obtained by DGGE. A restriction digestion, using *Hin*6I and *Msp*I enzymes, combined with electrophoresis in agarose was used for selecting algal clones. Plasmids were purified using the JetQuick Plasmid Purification Spin Kit (Genomed, Löhne, Germany) and sequenced by SECUGEN Sequencing Services (Madrid, Spain). Sequence data were edited using the software Chromas, version 1.45 (Technelysium, Tewantin, Australia). Homology searches with those sequences were performed using the Blast algorithm (Altschul et al. 1990) on the NCBI database (http://www.ncbi.nlm.nih.org/blast/).

Inoculation and incubation of stone probes. Twenty-four stone probes of each lithotype were washed with distilled water and sterilized at 120 °C and 1 atm for 20 min. After cooling 0.750 ml of the cultured photosynthetic community was inoculated with a pipette on the surface of the stone samples, in a laminar flow cabinet, and placed in a non-commercial incubator system – equipped with fluorescent lamps for photosynthesis (FLUORA, Osram) – containing sterile water at the bottom (20 l). The samples were incubated under constant conditions at 20 ± 2 °C (laboratory temperature) and 12 h dark–light (1200 lux) cycles for 3 months inside the incubation chamber. Moisture inside the chamber was maintained by water circulation provided by a water pump that favoured condensation. Sterile water was periodically added.

Monitoring lithotype colonization. The development of photosynthetic biofilms on the lithic probes was monitored by taking six replicate stone probes of each lithotype out of the incubator chamber after the first, second and third months of incubation, in order to determine the amount of chlorophyll *a* and *in vivo* chlorophyll *a* fluorescence.

Chlorophyll *a* is a photosynthetic pigment present in all photoautotrophic micro-organisms, including cyanobacteria and microalgae, and thus it is reliable and commonly used to estimate the amount of photosynthetic biomass present in liquid media, in soil and also on rock substrates. According to the Vollenweider et al. (1974) periphyton pigment extraction protocol, chlorophyll *a* was extracted by crushing each stone sample into fragments (0.20–0.50 cm^3), which were then added to 50 ml of dimethyl-sulphoxide (DMSO) and heated to 65 °C for 1 h. The samples were filtered to remove stone particles, and absorbancies of the extracts were measured at 664 and 750 nm before and after acidification with 1 N HCl, in a PerkinElmer Lambda 35 UV/Vis Spectrophotometer. Vollenweider et al.'s (1974) equation was used to calculate chlorophyll *a* concentrations.

As this method is based on the extraction of chlorophyll *a* from disintegrated cells in an organic solvent, a non-destructive technique for chlorophyll *a* quantification was also applied in this work. It consists of a spectrofluorometric method employing a fibre-optic accessory, which allows measurements of *in vivo* chlorophyll *a* fluorescence on solid substrates without sampling procedure. This technique allows an estimation to be made of photosynthetic biomass colonizing stone substrates according to fluorescence intensities measured on the surface materials (Cecchi et al. 2000; Miller et al. 2006). For *in vivo* chlorophyll *a* fluorescence analysis, pigment emission spectra were determined immediately after inoculation and after the first, second and third months of incubation, using a spectrofluorometer (SPEX Fluorolog-3 FL3-22) fitted with a fibre-optic accessory (Horiba Jobin Yvon F-3000). This allowed measurement when the fibre-optic end-piece was held steady facing the sample surface at a distance of 2 mm. Five spectrofluorometric measurements were made on each stone probe at an excitation wavelength of 430 nm, the optimum for chlorophyll *a* molecules (APHA/AWWA/WEF 1992). It used an integration time of 0.3 s, an increment of 1.0 nm and slits of 4.5 nm.

After an incubation of 3 months stone samples were examined under a binocular stereo microscope (Zeiss Discover V8 with phototube) and photographically recorded (Canon Powershot A630). In order to appreciate the penetration depth of the micro-organisms inside the stone profiles, image analysis was performed by means of principal component analysis (PCA), a technique that allows the detection of minority elements inside a digital image. The objective of PCA is to simplify images, avoiding the redundant data present in the different bands of the image. This approach has been used to improve the visualization of rock art motifs in highly correlated images (Rogerio-Candelera 2008), and to record separately different elements (of different nature and composition) present in mural paintings (Rogerio-Candelera *et al.* 2008). PCA was performed using HyperCube 9.5 software (US Army Topographic Engineering Centre, Alexandria, VA, USA). From the resulting bands, the first and third principal components (PC1 and PC3) were selected to elaborate false-colour images in a RGB colour space (three bands, each one corresponding to the red, green and blue values of a trichrome image). PC1, which plots around 95% of the total information included in the three bands of each image, was used as the Red band, in order to create a homogeneous reddish background in which the minority PC3 plots (around 0.5% of the total information of the images), which was used twice for representing the spatial dimension of the endolithic biofilm (Green and Blue bands).

Electronic microscopy analysis was also applied in order to assess penetration depth of photosynthetic micro-organisms inside the stone probes and to estimate potential physical damage in the stones after 3 months of incubation. Fragments of each lithotype were cut perpendicularly to the colonized surfaces. The selected samples were studied under a scanning electron microscope (SEM), Zeiss model DSM 962, with a secondary electrons detector (SE) and a back-scattered electrons detector (BSE). The samples were previously coated with a high conductance thin layer (gold film).

The experimental data were subject to analysis of variance (ANOVA) using Statistica 6.0 software for Windows, and the averages were compared using the Tukey HDS Test at the 5% level of significance.

Results and discussion

Micro-organisms identification

DNA-based molecular analysis of the green biofilm collected from the Santa Clara-a-Velha Monastery revealed a complex microbial community. 16S and 18S rRNA genes libraries were used to identify the microbial components of the community. The phototrophic micro-organisms detected in this study were Chlorophyta and Cyanobacteria. Within Chlorophyta, *Chlorella*, *Stichococcus*, *Trebouxia* and *Myrmecia* were the genera identified. Among the Cyanobacteria, two genera were detected, *Leptolyngbya* and *Pleurocapsa*. Most of these micro-organisms are very widespread on building stones according to several studies carried out by other authors in samples taken from stone monuments located in Europe (Ortega-Calvo *et al.* 1993; Tomaselli *et al.* 2000; Bellinzoni *et al.* 2003; Ascaso *et al.* 2004).

The culture of this natural green biofilm was chosen as a standardized phototrophic community adequate for the inoculation of stone probes. The use of a complex microbial community gives the advantage of simulating the existence of competition and/or synergy between colonizing micro-organisms. Micro-organisms act singly or in co-association with other micro-organisms, or with physical and chemical factors, to deteriorate stones (Koestler *et al.* 1996).

Lithotypes characterization

All of the studied lithotypes are light-coloured rocks that have a homogeneous appearance at a macroscopic scale, with exception of *Calcário Lioz*, which exhibits a heterogeneous texture mainly conditioned by the presence of fossil debris. All of the studied materials are almost exclusively composed of carbonate material, except for *Piedra San Cristobal*, which has significant prescence of other materials, mainly quartz grains.

Calcário Ançã (CA) is a homogeneous, very-fine-grained and oolitic-tendency limestone, with colours ranging from white to grey, with some yellowish variations. Low compressive strength and hardness are significant features of this rock. This carbonate rock is composed of a micrite matrix, in which abundant bioclasts (formed by fibrous calcite) are present, less than 500 μm long, containing very little spathized micritic cement (Fig. 1). *Calcário Ançã* is a pure limestone, with a $CaCO_3$ relative weight proportion greater than 96.5% (Dionísio 1997).

Calcário Lioz (CL) is a light-coloured crystalline medium-grained carbonate rock, almost exclusively composed of sparite carbonate. It has a heterogeneous texture, mainly conditioned by the presence of fossil debris, most of them of rudists, and by the occurrence of compositional veins with different colours. These veins seem to be a result of incipient metamorphic physical conditions. In some areas, strongly recrystallized bioclasts are present (Fig. 2). Although this rock has not been

Fig. 1. General view of CA lithotype under the petrographic microscope (crossed polars).

Fig. 3. General view of PF lithotype under the petrographic microscope (crossed polars).

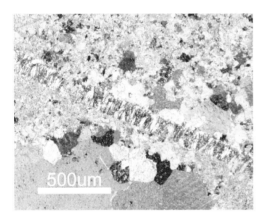

Fig. 2. General view of CL lithotype under the petrographic microscope (crossed polars).

classified as a marble, the prominent sparite carbonate crystal development suggests that the physical conditions reached during the recrystalization process went beyond the diagenesis limits, entering a low-temperature metamorphic facies. So, in terms of mechanical and chemical behaviour, this rock must be considered as a carbonate low-temperature metamorphic rock.

Piedra San Cristobal (SC), in hand specimen, can be described as a yellowish quartz-bearing fossiliferous calcarenite, with a bioclastic character. This lithotype exhibits an inequigranular heterogeneous texture, with very scarce matrix and low cementation level, causing a fair development of intergranular porosity. Pitting is observed on the stone surface, which could, in turn, enhance granular disaggregation processes. Under petrographic microscope this rock appears as medium grained (about 1 mm, attaining 2 mm), mainly composed of micrite carbonate clasts and less than 50% quartz clasts, involved in a micritic carbonate matrix (Fig. 3). Most of the carbonate clasts are of biological origin, featuring as microfossils. The quartz grains exhibit subeuedral (sometimes rounded) forms. In many places, the micrite carbonate matrix has turned into sparite, sometimes with white mica development associated with the micrite to sparite transformation process. Also the micrite carbonate clasts have been recrystallized to sparite, mainly in the inner parts of the fossilized organisms, thus revealing their internal structure. This stone presents a $CaCO_3$ and silica content that oscilates between 50–80 and 48–20%, respectively (Ortega *et al.* 1988).

Piedra Escúzar (PF) can be described as a light-coloured fine- to medium-grained biocalcarenite, in which only carbonate materials seem to be present. It is a biomicritic carbonate rock, almost completely composed of bioclasts up to 5 mm long. Some of the bioclast micrites have been transformed into sparite, apparently because they probably have a different geological origin. Although the rock has almost no matrix, which is a well-crystallized sparite, there are plenty of open spaces between the bioclasts, apparently reflecting the existence of a former matrix that has probably has been dissolved. Sometimes empty spaces can be observed inside the bioclasts (Fig. 4).

Pietra di Lecce (PL) is a 'honey'-coloured, homogeneous, highly porous, fine-grained limestone. It is almost exclusively composed of sparite bioclasts, in general less than 100 μm long, included in a very-fine micrite matrix (Fig. 5). The scarce cementation allows a high porosity to exist,

Fig. 4. General view of PL lithotype under the petrographic microscope (crossed polars).

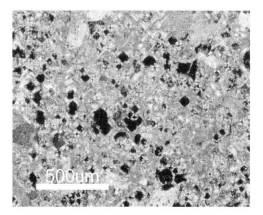

Fig. 5. General view of SC lithotype under the petrographic microscope (crossed polars).

mainly present as 1–10 μm-long cavities. It is soft and easy to cut, and can be carved with a penknife.

Concerning physical parameters, the differences in porosity values determined for the five lithotypes are show in Figure 6, ranging from about 0.5 to 43%. PL is the most prominent lithotype, attaining 43%

open porosity, followed by PF, SC and CA, with 35, 28 and 19%, respectively. The lowest open porosity is observed for CL lithotype, with only 0.5%. The results for this lithotype are in fair agreement with the petrographical characterization, as the strong recrystallization processes suffered by this rock caused a marked decrease in the volume of the pores. As described before, SC and PF present a huge number of large pores, and, consequently, their open porosity values reflect this textural feature. Although CA and PL lithotypes have high open porosity values, this characteristic is not visible under petrographic microscope observation, owing to the pore size dimension.

All lithotypes, with the exception of CL, absorb large quantities of water by capillarity suction (Fig. 7). They present regular curves obtained from capillary penetration that correspond to the filling of a unimodal and regular network. When the capillary suction phenomenon starts, the water suction rate is, in general, constant for all lithotypes. This lineal behaviour ends with a sharp change in suction rate. This evolution is clearly shown in Figure 7 for all lithotypes. However, their different hydraulic behaviour is clearly illustrated for their capillarity rates, and in Table 1 for their coefficients of capillarity. The pore space of these stones is characterized by conduits that give access to most of its open porosity (values higher than 92.6% were obtained from the ratio of water absorption by capillarity to free water saturation).

CL has the lowest uptake of water, showing that almost no water is absorbed by capillarity (Fig. 7, inset). CA and SC have similar maximum water uptake by capillarity; however, presenting different

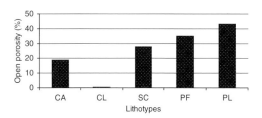

Fig. 6. Open porosity of the studied lithotypes: CA (*Calcário Ançã*), CL (*Calcário Lioz*), SC (*Piedra San Cristobal*), PF (*Piedra Escúzar*) and PL (*Pietra di Lecce*).

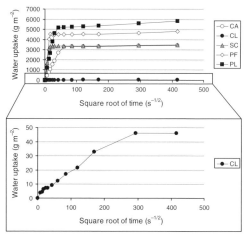

Fig. 7. Representative kinetics of capillarity water uptake of the studied lithotypes: CA (*Calcário Ançã*), CL (*Calcário Lioz*), SC (*Piedra San Cristobal*), PF (*Piedra Escúzar*) and PL (*Pietra di Lecce*).

Table 1. *ANOVA results obtained for the physic parameters analysed in this work*

Lithotype	Open porosity (%)	Capillary coefficient (g m^{-2} s$^{-1/2}$)	Water saturation capillarity (%)
a. CA	18.95 ± 0.27	57.25 ± 1.17	99.29 ± 0.04 bc
b. CL	0.53 ± 0.06	0.20 ± 0.06	99.91 ± 0.02 ac
c. SC	28.07 ± 1.88	199.30 ± 31.05	95.71 ± 0.23 abde
d. PF	35.08 ± 13.15	268.79 ± 77.54	93.04 ± 18.60 ce
e. PL	43.24 ± 0.68	128.76 ± 3.04	93.03 ± 0.23 cd

The values correspond to average ± SD ($n = 27$). Average followed by the same letters in a column are not significantly different by the Tukey HDS test at $p < 0.05$.

kinetics: SC absorbs water by capillarity more rapidly than CA. The highest capillary coefficient is observed for PF (269 g m^{-2} s$^{-1/2}$), closely followed by SC (199 g m^{-2} s$^{-1/2}$) and PL (129 g m^{-2} s$^{-1/2}$). CA has a lower value of 57 g m^{-2} s$^{-1/2}$ (Table 1). For PF, SC, PL and CL, these values reflect the microscopic observations; for the first three lithotypes abundant pores were observed, thus allowing effective water absorption; for the last one, extensive recrystallization occurred, reducing the existence of pores between grains. For CA, although almost no pores can be seen under microscope, their existence has already been referred to in terms of open porosity values. For this lithotype, porosity is mainly represented by a large number of very small pores with slow water absorption.

ANOVA revealed that the studied lithotypes are significantly different in terms of open porosity and capillary coefficient. With regard to water saturation by capillarity, ANOVA revealed significant differences between samples of the Portuguese lithotypes (CA and CL) and PF and PL lithotypes (Table 1).

As will be exposed and discussed later, both open porosity and capillarity strongly influence the bioreceptivity of the stones.

Stone bioreceptivity experiment

At the inoculation time, the absorption of the inoculum (photosynthetic liquid culture) clearly influenced the distribution pattern of the biofilms on the stone surfaces. This occurred by distinct ways in the five lithotypes, accordingly to their physical properties.

For CL liquid culture absorption and penetration was almost non-existent owing to its low porosity and water uptake coefficient values, 0.5% and 0.2 g m^{-2} s$^{-1/2}$, respectively. When the liquid culture was inoculated on the CA and PL surfaces a rapid absorption took place into the samples, which were enhanced by their high porosities combined with the high water uptake coefficient values. Owing to the presence of large pores, a faster process occurred for SC and PF lithotypes.

Figure 8 depicts one representative probe of each lithotype at the time of inoculation, and after 1, 2 and 3 months of incubation in the laboratory chamber. After the first month of exposure the colonization process led to increased green biofilms on the surfaces, which in the case of CL resulted in a brownish colour and an apparent cessation of colonization. This phenomenon seems to have occurred due to the evaporation of the inoculum that was not absorbed because of the very compact nature of this lithotype. The rapid growth of photosynthetic biomass on the CA, CL and PL surfaces during the first month was due to the presence of residual medium elements added with the inoculum, providing the necessary nutrients for the development of micro-organisms. With further incubation lithotypes surfaces were covered by brownish-green biofilms, in contrast to the vertically exposed surfaces of SC and PF, which showed a light green coloration. The decrease in this green colour could be attributed to the lack of nutrients provided by the total consumption of the inoculum and to a negative adaptation to the new type of nutrients supplied by the rock materials. The high water capillarity and porosity values of SC and PF seemed to positively influence the growth of the photosynthetic biomass on the vertically exposed surfaces of SC and PF. Although CA and PL (especially PL) have significant open porosity space, the growth of the photosynthetic biomass did not occur on these lithotypes probably because their porosity is characterized by the presence of very small pores inhospitable to a more significant colonization, as in the case of SC and PF or because pores/voids of these stones (their shape, size and connections) do not permit the movement of fluids. These results suggest that the penetration time and growth on stones may depend on stone pore size and permeability. Exposure conditions have also influenced development of the phototrophic community on the stone probes. As all of the stone surfaces had developed a brownish colour after 2 months of incubation, it could have been related to the light intensity inside the chamber. Therefore, it is evident that

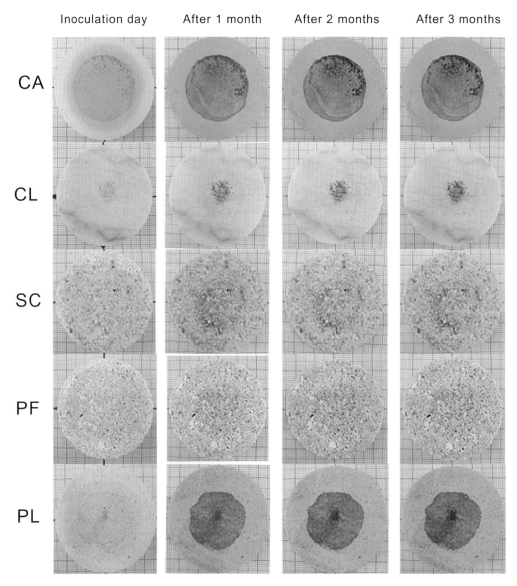

Fig. 8. Photosynthetic colonization along the incubation time of one representative sample of each lithotype.

the amount of biological growth varied with the intrinsic properties of the different lithotypes and exposure conditions.

The changes in colour of the biofilms on the stone probes can be attributed to a chromatic adaptation to optimize the photosynthetic process and adaptation as a consequence of the new exposure conditions. Cells of photoautotrophic microorganisms contain several pigments for photosynthetic activity; they are mainly chlorophylls and carotenoids that in the cyanobacteria are also associated with secondary pigments such as phycobiliproteins – a pigment prosthetic group (bilines) associated to proteins. These substances are inside the cellular structure on the thylakoid surface or close to them. The chromatic adaptation has genotypic and phenotypic bases. The cyanobacteria, for example, can take on a yellow-brown colour when the nitrogen is scarce, leading to a dramatic reduction in chlorophyll and phycocyanin, and an increase in carotenoids. In other cases, it has been shown that pigmentation changes can be an expression of different ecological stages and environmental adaptations, such as light intensity, temperature and cells age (Alakomi *et al.* 2004; Bartolini *et al.* 2004).

Scanning electron microscopy analysis

The visual examination performed in the present work was also complemented with microscopic observations of the colonized stone probes. SEM analysis on fragments transversally cut to the colonized surface of each lithotype enabled the presence of micro-organisms in the stones to be observed. In all samples micro-organisms adhered to the stone surfaces embedded in matrix of extracellular polymeric substances (EPS) were seen to be present, with the exception of CL samples. In CA samples this is particularly evident (Fig. 9). Algal cells were contained in a glue-like substance, allowing the adhesion of the microbial cells onto the stone substrate. Kaplan *et al.* (1987) described the chelating properties of the *Chlorella* exopolysaccharides, which are very important in the way in which these organisms deteriorate stone. The stone surface can lose cohesion owing to contraction and expansion of these biofilms because EPS incorporate large amounts of water into their structure, ensuring the maintenance of moisture by balancing changes in

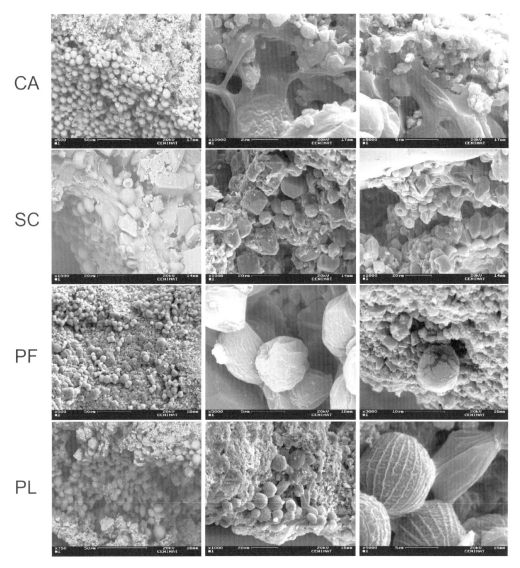

Fig. 9. SEM images of transversally cut samples of CA, SC, PF and PL lithotypes after incubation for 3 months.

humidity and temperature. Moreover, polysaccharides promote adhesion of stone fragments and film, eventually detaching from the original material (Saiz-Jimenez 1999). This aggressive action, mainly of cyanobacteria and microalgae on building stones, is more relevant because of their penetration into the substrates. This analysis also allowed some details of cellular penetration into the substrates to be appreciated. In SC and PF samples algal cells can easily penetrate into the large pores of these lithotypes. In these substrates micro-organisms used structural irregularities of the stones for their endolithic growth. Endolithic micro-organisms are generally described as slow-growing, stress-tolerant organisms as they can survive in inhospitable habitats, penetrating depths ranging from a few centimetres to almost 1 m (Koestler 2000; Pohl & Schneider 2002; Alakomi et al. 2004). The penetration of growing organisms into rock, and the diffusion of their excreted products into the intergranular fissures, are likely to lead to enhanced weathering reactions and decreased cohesion between grains (Alakomi et al. 2004). It can be observed in Figure 9, mainly for CA and PL lithotypes, how EPS contributes to the disaggregation of the rock grains.

Detection of endolithic growth, achieved by optical microscopy and image analysis of transversally cut samples, confirmed the presence of a thin green band under the surface that was attributed to endolithic development of the phototrophic biofilm. The resulting false-colour images show the phototrophic biofilm inside the stone samples (Fig. 10).

Spectrometric determinations of chlorophyll a

Photosynthetic biomass, presented in the five studied lithotypes, determined by the spectrophotometric method showed differences between CL and the other lithotypes (Fig. 11). Practically no change in photosynthetic biomass was observed in CL over 3 months of incubation (average value 0.25 ± 0.09 $\mu g\ cm^{-2}$). After 1 month the average amount of chlorophyll a determined by the spectrophotometric method was not significantly different for the studied lithotypes, except for CL, which presented significantly lowers values (Table 2).

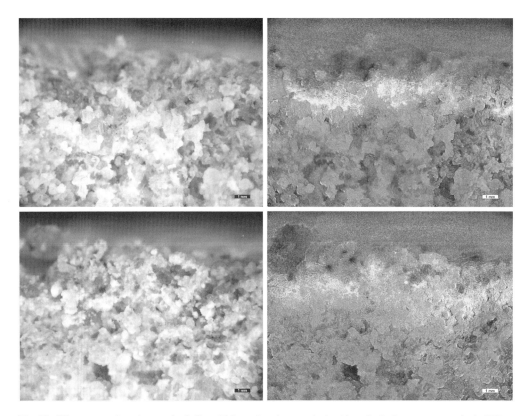

Fig. 10. SC stone samples micrographs (left) and false-colour images obtained by principal component analysis (PC1, PC3, PC3) (right) in order to enhance the extent of the endolithic phototrophic biofilm.

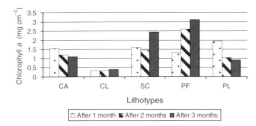

Fig. 11. Maximum values of chlorophyll *a* by the spectrophotometric method.

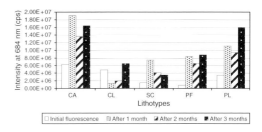

Fig. 12. Mean values of the chlorophyll *a* fluorescence at 684 nm (excitation wavelength: 430 nm). Each column is an average of 15 measurements.

Over 3 months PF was the only lithotype that presented an increase in photosynthetic biomass, indicating a successful colonization; whereas in CA and PL the quantity of chlorophyll *a* tended to decrease over the 3 months of incubation. This observation is in agreement with macroscopic examination; which showed that the surfaces of these lithotypes developed a brown coloration.

Regarding the application of the *in vivo* chlorophyll *a* fluorescence technique measured on surface materials (Fig. 12), the results were quite different to those obtained with the determination of the chlorophyll *a* concentrations on the stone probes. According to this non-destructive technique, PF was the only lithotype where the mean values of chlorophyll *a* fluorescence remained approximately the same over the 3 months of experimentation. After 1 month of incubation the values of *in vivo* chlorophyll *a* fluorescence for CA, SC, PF and PL lithotypes were about three times higher than those immediately after inoculation. These values revealed the development of micro-organisms. For CA, CL and SC the mean values of chlorophyll *a* were significantly different between the first and second month (Fig. 12 and Table 3), that is, for CA and SC there was a significant decrease in the intensity of chlorophyll *a*. This tendency was maintained for SC during the next month, whereas for CL the quantity of chlorophyll *a* increased. *In vivo* chlorophyll *a* in SC decreased over 3 months.

The quantity of chlorophyll *a* estimated by the spectrophotometric method reflected a total amount of this pigment in the samples, while the *in vivo* chlorophyll *a* fluorescence only estimated the photosynthetic biomass present at the sample surface. Consequently, if a ratio of the two variables is calculated, an estimation of the relative importance of the chlorophyll *a* present at the surface can be achieved, being also a way of determining the relative importance of chlorophyll inside the samples. Table 4 shows the relative contribution of chlorophyll *a* present at sample surfaces to total chlorophyll *a*. From the first to the second month these values presented almost no change, except for CL lithotype. From the second to the third month CA, CL and PL lithotypes showed a rise of these values, indicating that the relative contribution of chlorophyll *a* present at the surface increased at the same time the total chlorophyll *a* had also increased. SC and PF behaved in a different way, maintaining constant values during the 3 months. The fact that *in vivo* chlorophyll *a* fluorescence did not raise its values with time may look like a drawback in the experimental scenario. However, this information, together with an increase in chlorophyll *a* concentrations, indicates that microbial

Table 2. *ANOVA results obtained for the amount of chlorophyll* a *present on the studied lithotypes at 1, 2 and 3 months of incubation*

Lithotype	Amount of chlorophyll *a* ($\mu g\ cm^{-2}$)		
	At 1 month	At 2 months	At 3 months
a. CA	1.11 ± 0.38 cde	0.78 ± 0.58 bce	1.01 ± 0.11 bce
b. CL	0.27 ± 0.05	0.25 ± 0.07 ace	0.22 ± 0.14 ae
c. SC	1.38 ± 0.30 ade	1.12 ± 0.31 abe	1.89 ± 0.56 ade
d. PF	1.02 ± 0.28 ace	2.20 ± 0.35	2.40 ± 0.75 c
e. PL	1.73 ± 0.20 acd	0.99 ± 0.06 abc	0.75 ± 0.16 abc

The values correspond to average $\pm$ SD ($n = 3$). Average followed by the same letters in a column are not significantly different by the Tukey HDS test at $p < 0.05$.

Table 3. *ANOVA results obtained for the* in vivo *chlorophyll* a *fluorescence measured on the studied lithotypes at 1, 2 and 3 months of incubation*

Lithotype	Intensity of chlorophyll *a* fluorescence (cps*)		
	At 1 month	At 2 months	At 3 months
a. CA	$1.93 \times 10^7 \pm 3.06 \times 10^5$	$1.36 \times 10^7 \pm 2.80 \times 10^6$	$1.64 \times 10^7 \pm 2.52 \times 10^6$ e
b. CL	$1.37 \times 10^7 \pm 9.54 \times 10^4$	$1.92 \times 10^6 \pm 8.00 \times 10^5$ c	$6.49 \times 10^6 \pm 1.78 \times 10^6$ cd
c. SC	$7.43 \times 10^6 \pm 1.76 \times 10^6$ d	$4.11 \times 10^6 \pm 2.61 \times 10^5$ bd	$3.49 \times 10^6 \pm 3.40 \times 10^5$ b
d. PF	$8.33 \times 10^6 \pm 2.26 \times 10^6$ ce	$6.47 \times 10^6 \pm 1.55 \times 10^6$ ce	$8.79 \times 10^6 \pm 1.32 \times 10^6$ b
e. PL	$1.11 \times 10^7 \pm 1.03 \times 10^6$ d	$9.41 \times 10^6 \pm 8.95 \times 10^5$ d	$1.60 \times 10^7 \pm 4.58 \times 10^5$ a

The values correspond to average $\pm$ SD ($n = 15$). Average followed by the same letters in a column are not significantly different by the Tukey HDS test at $p < 0.05$.
*cps, counts per second.

Table 4. *Ratio between the average values of the* in vivo *chlorophyll* a *fluorescence method and the chlorophyll* a *estimated by the spectrophotometric method* ($cps* \, \mu g^{-1} \, cm^2$)

Lithotype	At 1 month	At 2 months	At 3 months
CA	1.74×10^7	1.74×10^7	7.13×10^7
CL	5.07×10^7	7.68×10^6	2.95×10^7
SC	5.38×10^6	3.67×10^6	1.85×10^6
PF	8.17×10^6	2.94×10^6	3.66×10^6
PL	6.42×10^6	9.51×10^6	2.13×10^7

*cps, counts per second.

colonization occurred mainly inside the SC and PF samples, and not on their surfaces. The endolithic growth may be related to intrinsic properties of the studied lithotypes, excess of light intensity during the experiment, differences in nutrient availability, interspecific competition or a combination of these reasons.

Conclusions

The application of different techniques allowed the complexity of microbial colonization on stone to be approached. Chlorophyll *a* extraction and subsequent determination proved to be an important analytical technique in determining photoautotrophic micro-organisms on rock substrates. The other techniques permitted the stone colonization process and the bioreceptivity of the studied lithotypes to be corroborated and understood. As the environmental conditions of the stone colonization experiment were the same, the factor involved in the differences observed in terms of bioreceptivity must be related to the different intrinsic properties of stone materials. Our experimental data showed that PF can be considered the most bioreceptive lithotype to phototrophic colonization among those tested. The physical characteristics of this lithotype, high porosity and capillary absorption, were the keys for the development of the microbial communities. As previous reports, a high porosity linked to significant open capillarity allows a rapid development of colonizing organisms (Guillitte & Dreesen 1995). Our analyses indicate that the preferential colonization in PF and, also, SC lithotypes occurred endolithically, which was related to stone pore size and permeability. In contrast, CL lithotype showed the lowest bioreceptivity. The compact nature of this lithotype and its extremely low porosity constituted the impediment to microbial growth, as has been demonstrated in low porous rocks such as marbles and granites (Tiano *et al.* 1995; Guillitte & Dreesen 1995; Miller *et al.* 2006). Moreover, the endolithic growth of organisms into the stones, and the diffusion of excreted products into the porous system, may enhance their biodeteriorating potential, promoting the decreasing of grain cohesion.

Further work is required through laboratory experiments to fully assess the implications of permeability on the primary bioreceptivity of these stones. An increased knowledge of this inherited characteristic of the stone will provide valuable insights and greater understanding of the vulnerability of building stone to biodeterioration.

This study has been financed by the Ministério da Ciência, Tecnologia e Ensino Superior, Portugal, with a doctoral grant (SFRH/BD/21481/2005), by the 2007PT0041 (CSIC-FCT) project and by the CEPGIST FCT subproject DECASTONE. The work also received support from Programa de Financiamento Plurianual de Unidades de Investigação da FCT, financed by European Union FEDER and national budget of the Portuguese Republic. N. Leal acknowledges support from the 'Compromisso com a Ciência' project C2007-446-CEGeol. This is a TCP CSD2007-00058 paper.

References

ALAKOMI, H. L., ARRIEN, N. *ET AL*. 2004. Inhibitors of biofilm damage on mineral materials (Biodam). *In*: KWIATKOWSKI, D. & LÖFVENDAHL, R. (eds) *Proceedings of the 10th International Congress on*

Deterioration and Conservation of Stone. ICOMOS, Stockholm, 399–406.

ASCASO, C., GARCIÁ DEL CURA, M. A. & DE LOS RÍOS, A. 2004. Microbial biofilms on carbonate rocks from a quarry and monuments in Novelda (Alicante, Spain). *In*: CLAIR, S. & SEAWARD, M. (eds) *Biodeterioration of Stone Surfaces*. Kluwer Academic, Dordrecht, 79–98.

ALTSCHUL, S. F., GISH, W., MILLER, W., MYERS, E. W. & LIPMAN, D. J. 1990. Basic local alignment search tool. *Journal of Molecular Biology*, **215**, 403–410.

APHA/AWWA/WEF. 1992. *Standard Methods for the Examination of Water and Wastewater*. 18th edn. American Public Health Association (APHA), American Water Works Association (AWWA), Water Environmental Federation (WEF), Washington, DC.

BARTOLINI, M., RICCI, S. & DEL SIGNORE, G. 2004. Release of photosynthetic pigments from epilithic biocenosis alter biocida treatments. *In*: KWIATKOWSKI, D. & LÖFVENDAHL, R. (eds) *Proceedings of the 10th International Congress on Deterioration and Conservation of Stone*. ICOMOS, Stockholm, 519–526.

BELLINZONI, A. M., CANEVA, G. & RICCI, S. 2003. Ecological trends in travertine colonization by pioneer algae and plant communities. *International Biodeterioration and Biodegradation*, **51**, 203–210.

CECCHI, G., PANTANI, L., RAIMONDI, V., TOMASELLI, L., LAMENTI, G., TIANO, P. & CHIARI, R. 2000. Fluorescence lidar technique for the remote sensing of stone monuments. *Journal of Cultural Heritage*, **1**, 29–36.

DIEZ, B., PEDROS-ALIO, C., MARSH, T. L. & MASSANA, R. 2001. Application of denaturing gradient gel electrophoresis (DGGE) to study the diversity of marine picoeukaryotic assemblages and comparison of DGGE with other molecular techniques. *Applied and Environmental Microbiology*, **67**, 2942–2951.

DIONÍSIO, A. 1997. *Mineralogical, Petrographical and Petrophysical Characterization of Ançã Limestone (Portugal)*. MScE thesis, Universidade Técnica de Lisboa, Instituto Superior Técnico, Portugal (in Portuguese).

EN 1925:1999. *Natural Stone Test Methods – Determination of Water Absorption Coefficient by Capillarity*. European Committee for Standardization (CEN), Brussels.

EN 1936:1999. *Natural Stone Test Method – Determination of Real Density and Apparent Density, and of Total and Open Porosity*. European Committee for Standardization (CEN), Brussels.

GONZALEZ, J. M. & SAIZ-JIMENEZ, C. 2004. Microbial activity in biodeteriorated monuments as studied by denaturing gradient gel electrophoresis. *Journal of Separation Science*, **27**, 174–180.

GONZALEZ, J. M., ORTIZ-MARTINEZ, A., GONZALEZ-DEL VALLE, M. A., LAIZ, L. & SAIZ-JIMENEZ, C. 2003. An efficient strategy for screening large cloned libraries of amplified 16S rDNA sequences from complex environmental communities. *Journal of Microbiological Methods*, **55**, 459–463.

GUILLITTE, O. 1995. Bioreceptivity: a new concept for building ecology studies. *Science of the Total Environment*, **167**, 215–220.

GUILLITTE, O. & DREESEN, R. 1995. Laboratory chamber studies and petrographical analysis as bioreceptivity assessment tool of building materials. *Science of the Total Environment*, **167**, 365–374.

KAPLAN, D., CHRISTIAEN, D. & ARAD, S. 1987. Chelating properties of extracellular polysaccharides from *Chlorella* spp. *Applied and Environmental Microbiology*, **53**, 2953–2956.

KOESTLER, R. 2000. Polymers and resins as food for microbes. *In*: CIFERRI, O., TIANO, P. & MASTROMEI, G. (eds) *Of Microbes and Art – The Role of Microbial Communities in the Degradation and Protection of Cultural Heritage*. Kluwer Academic, New York, 153–167.

KOESTLER, R., WARSCHEID, T. & NIETO, F. 1996. Biodeterioration: risk factors and their management. *In*: BAER, N. S. & SNETHLAGE, R. (eds) *Saving Our Architectural Heritage – The Conservation of Historic Stone Structures*. Wiley, Chichester, 25–35.

MILLER, A., DIONÍSIO, A. & MACEDO, M. F. 2006. Primary bioreceptivity: a comparative study of different Portuguese lithotypes. *International Biodeterioration and Biodegradation*, **57**, 136–142.

MILLER, A. Z., LAIZ, L., GONZALEZ, J. M., DIONISIO, A., MACEDO, M. F. & SAIZ-JIMENEZ, C. 2008. Reproducing stone monument photosynthetic-based colonization under laboratory conditions. *Science of the Total Environment*, **405**, 278–285; doi: 10.1016/j.scitotenv.2008.06.066.

MORTON, L. H. G., GREENWAY, D. L. A., GAYLARDE, C. C. & SURMAN, S. B. 1998. Consideration of some implications of the resistance of biofilms to biocides. *International Biodeterioration and Biodegradation*, **41**, 247–259.

MUYZER, G., DE WAAL, E. C. & UITTERLINDEN, A. G. 1993. Profiling of complex microbial populations by denaturing gradient gel electrophoresis analysis of polymerase chain reaction-amplified genes coding for 16S rRNA. *Applied and Environmental Microbiology*, **59**, 695–700.

NÜBEL, U., GARCIA-PICHEL, F. & MUYZER, G. 1997. PCR primers to amplify 16S rRNA genes from cyanobacteria. *Applied and Environmental Microbiology*, **63**, 3327–3332.

ORTEGA, J., MARTIN, A., APARICIO, A. & GARCIA, J. 1988. Bioalteration of the Cathedral of Seville. *In*: *Proceedings of the 6th International Congress on Deterioration and Conservation of Stone*. Nicholas Copernicus University Press Department, Torun, 1–8.

ORTEGA-CALVO, J. J., SANCHEZ-CASTILLO, P. M., HERNANDEZ-MARINE, M. & SAIZ-JIMENEZ, C. 1993. Isolation and characterization of epilithic chlorophyta and cyanobacteria from two Spanish cathedrals (Salamanca and Toledo). *Nova Hedwigia*, **57**, 239–253.

POHL, W. & SCHNEIDER, J. 2002. Impact of endolithic biofilms on carbonate rock surfaces. *In*: SIEGESMUND, S., WEISS, T. & VOLLBRECHT, A. (eds) *Natural Stone, Weathering Phenomena, Conservation Strategies and Case Studies*. Geological Society, London, Special Publications, **205**, 177–194.

ROGERIO-CANDELERA, M. A. 2008. *Una propuesta no invasiva para la documentación integral del arte rupestre*. MScE thesis, Universidad de Sevilla, Spain.

Rogerio-Candelera, M. A., Laiz, L. & Saiz-Jimenez, C. 2008. Una experiencia de laboratorio para la separación de cubiertas en la documentación de pinturas rupestres y murales afectadas por biodeterioro. *In*: Saiz-Jimenez, C. & Rogerio-Candelera, M. A. (eds) *Novena Reunión de la Red Temática del CSIC de Patrimonio Histórico y Cultural 'Avances recientes en la investigación sobre Patrimonio', Sevilla*. RTPHC, Sevilla, 71–72.

Saiz-Jimenez, C. 1999. Biogeochemistry of weathering processes in monuments. *Geomicrobiology Journal*, **16**, 27–37.

Silva, Z. C. 2007. *O Lioz Português. De lastro de navio a arte na Bahia*. Porto, Edições Afrontamento, Porto.

Tiano, P., Accolla, P. & Tomaselli, L. 1995. Phototrophic biodeteriogens on lithoid surfaces: an ecological study. *Microbial Ecology*, **29**, 299–309.

Tomaselli, L., Lamenti, G., Bosco, M. & Tiano, P. 2000. Biodiversity of photosynthetic micro-organisms dwelling on stone monuments. *International Biodeterioration and Biodegradation*, **46**, 251–258.

Vollenweider, R. A., Talling, J. F. & Westlake, D. F. (eds) 1974. *A Manual on Methods for Measuring Primary Production in Aquatic Environments*. 2nd edn. Blackwell, Oxford.

Warscheid, T. & Braams, J. 2000. Biodeterioration of stone: a review. *International Biodeterioration and Biodegradation*, **46**, 343–368.

Suppression of salt weathering of porous limestone by borax-induced promotion of sodium and magnesium sulphate crystallization

E. RUIZ-AGUDO* & C. RODRIGUEZ-NAVARRO

Departamento de Mineralogía y Petrología, Universidad de Granada, Fuentenueva s/n, 18002 Granada, Spain

*Corresponding author (e-mail: encaruiz@ugr.es)

Abstract: The effects of borax on the crystallization of sodium and magnesium sulphate, two of the most damaging salts affecting porous stones, have been studied. Borax promotes the crystallization of mirabilite and inhibits epsomite crystallization in open glass beakers. The additive is preferentially adsorbed onto $\{140\}_{mirabilite}$ and $\{111\}_{epsomite}$ faces, thus acting as an effective habit modifier. In contrast, in the presence of a calcitic support (either Iceland spar single crystals or a porous limestone – a biocalcarenite) crystallization is promoted in the presence of borax, irrespective of the salt tested. Apparently, this is due to a high stereochemical affinity between borate molecules adsorbed (and/or co-precipitated) onto calcite, and mirabilite and epsomite crystals. Salt weathering tests using a biocalcarenite show a significant damage reduction upon borax addition to the saline solutions. Borax promotes the crystallization of both mirabilite and epsomite within the pores of the stone, reducing its porosity. Crystallization promotion favours nucleation at a low supersaturation, thereby resulting in very low crystallization pressure and minimal damage. Application of borax to porous limestones affected by mirabilite and/or epsomite crystallization could be a new means of suppressing salt weathering.

Salt crystallization within the void spaces of porous materials is a major cause of damage in historic buildings and statues, as well as in new constructions and civil engineering works (Goudie & Viles 1997; Rodriguez-Navarro & Doehne 1999; Rodriguez-Navarro *et al.* 2000*a*; Ruiz-Agudo *et al.* 2007). Salts also contribute to the weathering of rocks in a range of natural environments (Goudie & Viles 1997; Rodriguez-Navarro & Doehne 1999). Highly soluble sodium and magnesium sulphates are among the most damaging salts associated with such a weathering phenomenon (Ruiz-Agudo *et al.* 2007). In fact, owing to their deleterious effects, both Na_2SO_4 and $MgSO_4$ have been used for testing the resistance of building materials to salt damage; for example, the American Association for Testing and Materials (ASTM) aggregate soundness tests (ASTM 1997).

The ubiquity and virulence of salt weathering has prompted the search for methods to avoid its damaging effects. However, in the field of cultural heritage, traditional conservation treatments aimed at minimizing salt damage have generally failed and new methods are needed to solve this problem (Rodriguez-Navarro *et al.* 2000*b*). Recently, it has been suggested that the use of compounds (additives) that modify salt crystallization may represent a new means of reducing the salt weathering effects on porous materials (Rodriguez-Navarro *et al.* 2000*b*, 2002; Selwitz & Doehne 2002). Additives dosed in very small proportions (ppm) influence nucleation, agglomeration, crystal growth and shape of crystals (Al-Jibbouri & Ulrich 2004). Some additives can either suppress or promote the growth of a crystal by adsorption on all its faces, whereas others act only on certain faces and thus change the morphology of the crystal (Al Jibbouri *et al.* 2002). When the additive inhibits salt crystallization, the induction time, that is, the period between the establishment of supersaturation and the formation of a new phase, is increased (Tantayakom *et al.* 2005). This effect allows transport of the saline solution to the surface of the porous stone where salt crystallization occurs as efflorescence causing negligible damage (Rodriguez-Navarro *et al.* 2002). In the case of additives that promote crystallization, salt precipitation occurs at a low supersaturation within the pore network, resulting in low crystallization pressure. As a consequence, damage to the substrate in which salts crystallize is reduced (Rodriguez-Navarro *et al.* 2000*b*). The catalytic (i.e. promoter) effect of heterogeneous materials on crystallization from supersaturated solutions has been thoroughly studied (Sheikholeslami 2003). However, in the field of cultural heritage, this effect (i.e. crystallization promotion) has received almost no attention within the sparse literature on the use of crystallization additives to prevent salt damage.

Borax ($Na_2B_4O_7 \cdot 10H_2O$) acts both as a crystallization inhibitor or as a promoter depending on the saline system, and may also play this dual role for a single salt (Jiang et al. 2007). The inhibitory effect of borax on the crystallization of epsomite ($MgSO_4 \cdot 7H_2O$) has been studied by several authors (Rubbo et al. 1985; Al-Jibbouri & Ulrich 2004). The influence of borax on the crystallization of mirabilite ($Na_2SO_4 \cdot 10H_2O$) has also been studied (Telkes 1952). Sodium sulphate solutions are used for thermal energy storage (Biswas 1977). This is because of the high heat of transition (78.9 kJ mol^{-1}) between mirabilite and the anhydrous sodium sulphate (thenardite) + a saturated solution, and its convenient temperature (32.4 °C) (Brodale & Giauque 1958). For this application, nucleation has to occur invariably with the minimum of undercooling and without the formation of metastable phases (i.e. sodium sulphate heptahydrate). This is achieved by adding small amounts of borax that promote the nucleation of mirabilite. Apparently, the crystallographic structures of mirabilite and borax are similar enough to account for the catalytic effect on nucleation (Telkes 1952).

It is the aim of this paper to study the effects of borax on the crystallization of sulphates (mirabilite and epsomite) within a porous ornamental stone (in this case, a biocalcarenite). The influence of borax on the saline solution's migration and evaporation within the porous media has also been studied. This information may aid in the development of new methods to preserve ornamental stones affected by damage associated with the crystallization of Na and Mg sulphates. It may also contribute to a better understanding of this damage process.

Materials and methods

Saturated Na_2SO_4 and $MgSO_4$ solutions were prepared from crystalline solid (Fluka, puriss.) using deionized water and decanted to eliminate any undissolved crystal. Reagent-grade borax (Fluka) was added to the saline solutions in concentrations ranging from 10^{-4} to 10^{-1} M. Solution surface tension was measured using the Wilhelmy Plate method on a Krüss GmbH Tensiometer (model K11) coupled to a Clifton thermostatic bath (TR32T Tempatron thermostat). Measurements were taken over 17 min in 1 min intervals. In order to avoid initial fluctuations, the value of the surface tension was calculated as the average of the data collected over the final 10 min. Possible variations in the vapour pressure of saturated Na and Mg sulphate solutions associated with the presence of borax were determined by measuring the equilibrium relative humidity, RH_{eq}, in airtight containers where saline solutions were placed and allowed to equilibrate at a constant temperature, T (21 °C). RH_{eq} values were recorded on an Oregon Scientific DTH-ETHG889 thermo-hygrometer. These measurements were aimed at detecting any possible change in the solutions' physical properties associated with the additive that could modify the dynamics of solution flow and evaporation, and, thereby, how and where salts precipitate within the tested porous stone. The selected porous stone for salt crystallization tests was a biomicritic limestone (biocalcarenite) from Granada (Spain), profusely used in Granada's architectural and sculptural heritage, and with well-known problems connected with salt weathering. A detailed description of this stone type and its behaviour towards sodium and magnesium sulphate weathering has been presented elsewhere (Ruiz-Agudo et al. 2007).

Macroscale salt crystallization experiments

Saline solutions (with and without additive) were used for salt crystallization tests that were carried out in a room with a controlled environment (20 ± 1 °C, and 45 ± 5% relative humidity). Two sets of crystallization tests were carried out in (a) open glass beakers and in (b) 3 × 3 × 20 cm porous stone slabs, following the procedure outlined by Rodriguez-Navarro et al. (2002). Salt crystals were collected after crystallization in glass beakers and examined by X-ray diffraction (XRD) without further sample preparation (no grinding). A Philips PW-1710 X-ray diffractometer with automatic slit, CuKα radiation ($\lambda = 1.5405$ Å), 3–45 °2θ explored area, with steps of 0.0281 °2θ and 0.011 °2θ s^{-1} goniometer speed, was used. Bragg peak intensities were compared with standard JCPDF powder patterns in order to determine which crystal faces were preferentially developed. Crystal morphologies were studied by means of optical microscopy (Olympus) and environmental scanning electron microscope (ESEM, Philips Quanta 400). Following salt crystallization tests in porous stone, small stone pieces were collected and studied using the ESEM. Salt distribution within the stone pore system was inferred by means of mercury intrusion porosimetry (MIP, Micromeritics Autoscan 6500). Stone samples were collected before and after salt crystallization tests and dried overnight in an oven (110 °C) prior to MIP analysis.

In situ XRD and ESEM analyses

Salt weathering of porous limestones occurs when the saline solution crystallizes within the pores of the stone and the growing salt crystals interact with the calcite substrate. In order to gain an insight into the role of the calcite substrate in the process of salt crystallization and growth, drops of saturated saline solution were deposited and allowed to crystallize onto $\{10\bar{1}4\}$ surfaces of cleaved Iceland spar crystals (ca 10 × 10 × 2 mm in size). Supersaturation was achieved by evaporation at constant temperature and relative humidity within the XRD chamber. XRD patterns were collected at 1 min intervals while evaporation and the eventual crystallization of salt was taking place on the Iceland spar crystal. Analysis of the relative intensity of the main Bragg peaks was performed in order to identify which faces were preferentially developed (with a preferred crystallographic orientation) in the presence of the additive and the calcite substrate.

Changes in habit and size of sodium and magnesium sulphate crystals precipitated from drops of saline solution with and without additive were studied *in situ* and at high magnification using the ESEM. Condensation and evaporation of water on salt samples deposited onto an Al sample holder was achieved by modifying the gas pressure (water vapour) inside the ESEM chamber from 6.5 to 2.5 torr, at a rate of 0.01 torr s^{-1}, keeping constant the temperature of the sample holder (5 °C) using a Peltier stage, following the methodology outlined by Rodriguez-Navarro & Doehne (1999).

Morphology simulation

The equilibrium morphology of mirabilite and epsomite crystals was calculated using the Bravais–Friedel–Donnay–Harker (BFDH) algorithm (Donnay & Harker 1937) that is part of the MSI Cerius^2 package. Morphology changes due to preferential adsorption of the additive were computer simulated by reduction of the relative growth rate of the faces where the interaction with the additive is most likely to have taken place.

Results and discussion

Effect of borax on critical supersaturation and crystal morphology

Epsomite and mirabilite were the phases formed in both control and additive solutions (open beakers) and were identified by XRD. Significant changes in both the induction time and the critical supersaturation (i.e. relative supersaturation reached at the onset of crystallization, calculated using the formula $\sigma = 100 \times (C-C_0)/C_0$, where C and C_0 are the actual and the saturation concentrations, respectively) were observed following the addition of borax (Fig. 1). The additive has no significant effect on the evaporation rate of saline solutions in open beakers (1.26 × 10^{-6} and 1.37 × 10^{-6} g s^{-1} cm^{-2} for magnesium and sodium sulphate solutions, respectively). This is in agreement with RH$_{eq}$ measurements showing no changes in the

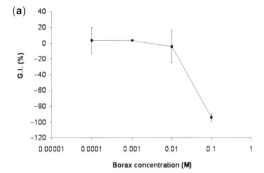

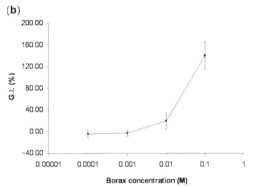

Fig. 1. Percentage of growth inhibition (% G.I.) v. borax concentration for saturated solutions of (**a**) sodium sulphate and (**b**) magnesium sulphate.

vapour pressure of the saturated saline solutions upon addition of borax (23 and 19.95 mbar for Mg and Na sulphate solutions, respectively). Therefore, it is assumed that changes in the induction time and critical supersaturation are due to the effect of borax on the nucleation of sulphates. Quantification of the inhibitory capability of the tested additive was performed by calculating the *percentage of growth inhibition, G.I.* (see Ruiz-Agudo et al. 2006). A clear (and opposite) trend in crystallization inhibition associated to additive concentration was observed. In the case of sodium sulphate, crystallization inhibition decreases with additive concentration or, in other words, crystallization promotion increases with borax concentration (Fig. 1a). This is in agreement with the well-known catalytic effect of borax on mirabilite nucleation (Telkes 1952). In contrast, the presence of borax inhibits the crystallization of epsomite, and this effect is more pronounced at higher borax concentrations (Fig. 1b). Adsorption of B$_4$O$_5$(OH)$_4^{2-}$ ions on {111} surfaces of epsomite may account for the latter effect (Rubbo et al. 1985 and references therein). Al-Jibbouri & Ulrich (2004) concluded that the adsorption of borax onto epsomite crystals is physical in nature and occurs at kinks in the ledges. However, no clear experimental evidence for this mechanism of interaction has been provided so far.

Mirabilite crystals grown from pure solutions have {100} as the dominant form, as shown by the increase in the relative intensity of the 200 Bragg peak. In the presence of borax, the increase in the relative intensity of the 140 Bragg peak indicates that {140} faces are morphologically more important and account for the largest proportion of the crystal surface (Fig. 2a). These results are in agreement with ESEM observations (see later). The pronounced structural similarity between Na–H_2O chains parallel to the [001] direction in both mirabilite and borax (Levy & Lisensky 1978) seems to be the basis for the utility of the latter as a nucleating agent for the crystallization of mirabilite. This explains why mirabilite preferentially grows along the (140) plane, which contains such a direction (i.e. [001] is the zone axis of {140} planes). Both the detected crystallization promotion effect and the observed morphological change may

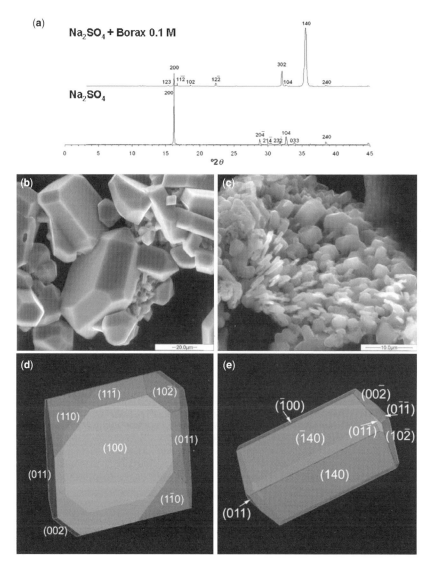

Fig. 2. (a) XRD patterns of mirabilite crystals precipitated in open-beaker crystallization experiments; (b) ESEM image of pure, *in situ* precipitated mirabilite crystals; (c) ESEM image of borax-doped mirabilite crystals; (d) morphology of mirabilite predicted using the BFDH algorithm; and (e) schematic of the overdeveloped {140} form.

be the result of mirabilite nucleation catalyzed by oriented growth on {140} planes of borax (i.e. epitaxia), as the structural misfit between $(140)_{mirabilite}$ and $(140)_{borax}$ is only about 5%. This effect was simulated by molecular modelling of the crystal morphology using the Cerius^2 software. An overdevelopment of {140} faces (simulated through a reduction in the distance to the centre of the crystal) leads to a change in mirabilite morphology from prismatic to plate-like (Fig. 2d, e). ESEM observations are in agreement with both XRD and morphology simulation results. *In situ* ESEM experiments (5 °C and 72% RH as environmental conditions in the chamber) show that the precipitation of sodium sulphate resulted in the direct formation of a few, large (>10 μm) mirabilite crystals showing an equilibrium morphology (Fig. 2b); that is, with {100} as the dominant form (Ruiz-Agudo *et al.* 2006). The formation of mirabilite was corroborated by the observation of its dehydration when the RH was reduced in the ESEM chamber (5 °C, 37% RH). Changes in mirabilite morphology were observed following the addition of borax. Such changes were most significant when borax was dosed at a concentration of 0.1 M. Mirabilite crystals of less than 5 μm in size grew from solution drops when borax was present. Crystals changed from equilibrium prismatic shape to plate-like (Fig. 2c), a morphological change similar to that obtained by molecular modelling (Fig. 2e). Again, these observations indicate that {140} are the most developed faces in the mirabilite crystals grown in the presence of borax. Oddly, the nucleation density increased significantly, an effect that would be expected in the case of a nucleation inhibitor, but not when crystallization is promoted.

XRD patterns of epsomite crystals precipitated in batch experiments show {110} as the most developed form in pure crystals, as is deduced from the high intensity of the 220 Bragg peak (Fig. 3a). In contrast, {111} seems to be the predominant form in epsomite crystals precipitated in the presence of borax, as shown by the relative increase in intensity of 111 and 222 Bragg peaks. This could be due to the incorporation of borate ions into active sites of {111} surfaces, hindering the growth of such faces that become more important in the final crystal morphology. Such an effect was modelled using Cerius^2. The reduction in the relative growth rate of {111} faces results in a morphology change from prismatic to wedge-like; this was confirmed by observations of epsomite crystals using optical microscopy (Fig. 3b, c). This effect has been known since the nineteenth century (Groth 1908), although the molecular mechanism responsible for this habit modification has not been clearly established. In the case of *in situ* precipitation of epsomite within the ESEM chamber, the crystal habit showed a similar modification in the presence of borax as that observed by optical microscopy.

Salt crystallization on Iceland spar and within the porous limestone: damage development

In situ XRD tests showed that in the absence of borax mirabilite randomly crystallized on the Iceland spar crystal, with no preferred orientation. Conversely, in the presence of borax a clear overdevelopment of the {114} faces was deduced from the XRD pattern (Fig. 4a). Apparently, borate ions adsorbed onto calcite crystals and formed a periodic two-dimensional (2D) network that acted as a template for mirabilite crystallization. Crystallization of mirabilite was promoted by such a template and took place in a crystallographically oriented manner at a low supersaturation. Note that formation of an organic or inorganic 2D template on a substrate is a very common phenomenon that contributes to a number of crystallization processes including controlled epitaxial growth and biomineralization (Addadi & Weiner 1992). Ichikuni & Kikuchi (1972) have shown that borate ions are easily adsorbed onto calcite surfaces via H-bonding with carbonate groups. Co-precipitation has also been reported to contribute to the incorporation of borate ions into calcite (Hobb & Reardon 1999). Therefore, the possible formation of a thin layer of calcium (and possibly sodium) borate–carbonate phase, which could not be detected by XRD, is not ruled out. In any case, such a layer should have the same template effect of the 2D network as described above. Overall, this crystallization promotion effect might result in a significant reduction in the damage to the stone support, as confirmed by salt crystallization tests (see later).

The evaporation of the Na$_2$SO$_4$ solution with borax percolating through the stone was much slower than that of the pure solution (0.35 v. 0.88 g h^{-1}). The difference in evaporation rate seems to be mainly due to the massive blocking of pores by sodium sulphate within the stone pore network. The surface tension was 74.1 ± 0.4 mN m^{-1} in the control solution and 70.7 ± 0.6 mN m^{-1} in the solution with 0.1 M borax, while the water vapour pressure remained constant at 19.95 mbar in both the control and the borax-doped saline solutions. The lower the surface tension is, the lower the (negative) capillary pressure will be, thus enhancing the drying of the saturated porous support (Scherer 1990). However, the reduction in the evaporation rate due to pore blocking seems to mask this effect. MIP analyses showed how sodium sulphate mainly crystallized in the larger pores of the stone (Fig. 4b). Crystals both obstructed pore entries and

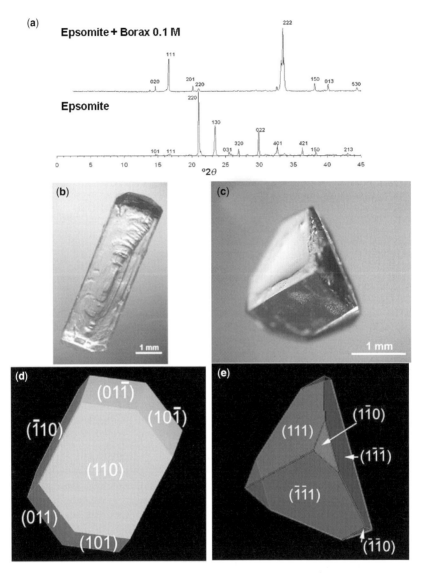

Fig. 3. Epsomite crystals precipitated in open-beaker crystallization experiments: (**a**) XRD patterns; optical microscopy images of (**b**) pure and (**c**) borax-doped epsomite crystals; (**d**) morphology of epsomite predicted using the BFDH algorithm; and (**e**) schematic of the overdeveloped {111} form.

diminished the pore volume, as can be deduced by the shift of the main peak in the pore-size distribution curve to a lower radius and by the reduction in the peak height (Angeli *et al.* 2008). This effect was enhanced in the presence of borax. In this latter case, crystallization also took place in the smaller pores of the stone. ESEM observations of calcarenite samples show sodium sulphate crystals completely filling cracks, fractures and pores when precipitation occurred in the presence of borax (Fig. 5). The observed crystallization promotion effect and the subsequent reduction in the crystallization pressure and damage to the stone (Fig. 4b) was further confirmed by the measured reduction in both the amount of efflorescence formed [from 31.71 (control) to 0.75 g (0.1 M borax)] and the amount of material lost during the crystallization experiment [from 7.69 (control) to 0.58 wt% (0.1 M borax)]. In summary, sodium sulphate crystallization is promoted in the presence of borax and takes place mainly within the stone, at a minimum supersaturation. Because crystallization pressure is

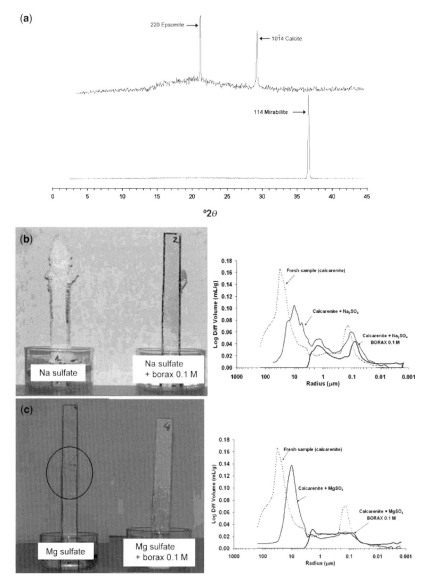

Fig. 4. Salt crystallization on and within calcitic supports: (**a**) *in situ* XRD patterns of mirabilite and epsomite crystals precipitated on Iceland spar crystals in the presence of 0.1 M borax (the main Bragg peak of calcite is due to a slight misalignment of the beam geometry). Photographs and mercury intrusion porosimetry analysis of porous calcareous stone slabs after crystallization tests (12 days): (**b**) sodium sulphate; and (**c**) magnesium sulphate.

proportional to supersaturation (Rodriguez-Navarro & Doehne 1999), the pressure exerted by the crystals growing against the pore walls is reduced and, as a consequence, so too is the damage to the stone support.

Epsomite crystallization within the calcarenite in the presence of borax occurred in a very similar way to mirabilite crystallization. This was unexpected because crystallization experiments using open beakers showed that borax acts as a crystallization inhibitor of epsomite. On the one hand, the formation of efflorescence on the stone surface was not detected as expected (Rodriguez-Navarro *et al.* 2002); but, on the other hand, massive blocking of both large and small pores was shown by the MIP results (Fig. 4c), which indicate that crystallization within the stone pore network was promoted by the additive. ESEM observations show that

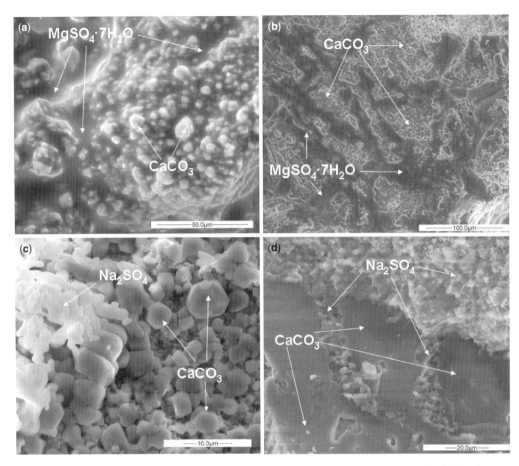

Fig. 5. ESEM photomicrographs of sulphates grown within biocalcarenite slabs. Epsomite formed in (**a**) the absence and (**b**) the presence of 0.1 M borax. Note the shapeless (wax-like) morphology of epsomite aggregates in the control and the prismatic habit developed in the presence of 0.1 M borax. Sodium sulphate (thenardite after mirabilite dehydration) formed in (**c**) the absence and (**d**) the presence of 0.1 M borax. Massive infilling of pores with salt is detected in the latter case.

epsomite formed in the absence of an additive and was characterized by the development of shapeless aggregates covering calcite grains (Fig. 5a). In contrast, large epsomite crystals with near-equilibrium morphologies precipitated within limestone pores in the presence of borax, sealing the void spaces between calcite grains (Fig. 5b). This latter crystallization pattern is in agreement with the hypothesis of crystallization at a low supersaturation promoted by the additive. *In situ* XRD crystallization experiments show epsomite crystallizing with $\{110\}$ faces parallel to $\{10\bar{1}4\}$ faces of Iceland spar crystals, as indicated by the intense 220 Bragg peak, which appears alone, without any other epsomite peak (Fig. 4a). No preferred orientation of epsomite crystals growing onto Iceland Spar crystals was detected in the absence of borax (i.e. the XRD pattern was similar to that corresponding to epsomite formed in the absence of borax, shown in Fig. 3a). As in the case of mirabilite, the formation of a 2D template of a borate ion on calcite may account for the oriented crystallization of epsomite; nucleation will thus take place at a lower supersaturation than in the absence of additive. The evaporation rate of $MgSO_4$ solutions flowing through the limestone porous network was not affected by the additive (0.19 g h^{-1}). This is striking, as borax enhanced the massive blocking of the pore system (Fig. 4c). Such a pore blocking should result in a reduction in the evaporation rate. This striking result can be explained by the reduction in the surface tension of the magnesium sulphate solution following the addition of borax (from 75.0 ± 0.7 to 62.6 ± 0.4 mN m^{-1}), which will enhance the evaporation of the solution, as has previously been stated.

The damage reduction was assessed by visually comparing the limestone slabs submitted to magnesium sulphate crystallization tests. In this case, it was not possible to quantify material loss as the damage pattern of magnesium sulphate is manifested by the formation and propagation of cracks, with no scale detachment or granular disintegration as in the case of sodium sulphate (Ruiz-Agudo *et al.* 2007). No cracks were detected on the slab surfaces when epsomite crystallization took place in the presence of borax (Fig. 4c). All in all, these results suggest that epsomite precipitation within limestone slabs is promoted by borax, reducing the damage to the porous material due to the crystallization of this otherwise damaging salt.

Conclusions

Borax has a critical effect on the crystallization of sodium and magnesium sulphate. While mirabilite crystallization is promoted, epsomite precipitation is clearly inhibited in open beakers. The interaction between borax and magnesium sulphate takes place through the adsorption or poisoning of borate ions on active sites of {111} epsomite surfaces. This changes epsomite morphology from prismatic to wedge-like. In the case of mirabilite, the effect of borax is mainly catalytic; mirabilite crystals grow on borax seeds, with a clear overdevelopment of {140} faces. This changes the mirabilite habit from bulky to plate-like. Epitaxial growth is considered because the structures of both compounds are similar enough to account for such a possibility. In the presence of calcite, the effect of borax as a promoter of mirabilite crystallization is enhanced. In addition to its role as a nucleating agent in the bulk solution, borate ions form a 2D template on calcite surfaces within the porous system of a limestone that contributes to the heterogeneous nucleation of mirabilite crystals at a low supersaturation. The formation of a 2D template may also explain the observed borax-induced promotion of epsomite crystallization within the tested porous limestone. All in all, crystallization promotion within the limestone pore network results in a reduction in the crystallization pressure exerted by the salt crystals when they grow within this confined geometry and, as consequence, in the damage to the stone support. This positive effect associated with the addition of borax may be used to minimize the deleterious effects of sodium and magnesium sulphates when they crystallize within porous limestones.

This research was financed by the European Commission VIth Framework Programme, under Contract No. SSP1-CT-2003-501571 (Saltcontrol project) and by the research group NRM-179 (Junta de Andalucía, Spain). The ESEM used is from the CEAMA (Junta de Andalucía-Universidad de Granada). We thank I. Sanchez-Almazo (CEAMA) for her assistance during the ESEM study and J. Martínez (University of Granada) for his help during surface tension measurements.

References

ADDADI, B. L. & WEINER, S. 1992. Control and design principles in biological mineralization. *Angewante Chemie, International Edition*, **31**, 153–169.

AL-JIBBOURI, C. & ULRICH, J. 2004. Impurity adsorption mechanism of borax for a suspension growth condition: A comparison of models and experimental data. *Crystal Research and Technology*, **39**, 540–547.

AL JIBBOURI, S., STREGE, C. & ULRICH, J. 2002. Crystallization kinetics of epsomite influenced by pH-value and impurities. *Journal of Crystal Growth*, **236**, 400–406.

ANGELI, M., BENAVENTE, D., BIGAS, J., MENÉNDEZ, B., HÉBERT, R. & DAVID, C. 2008. Modification of the porous network by salt crystallization in experimentally weathered sedimentary stones. *Materials and Structures*, **41**, 1091–1108; DOI: 10.1617/s11527-007-9308-z.

ASTM. 1997. ASTM C 88-90. Standard test method for soundness of aggregate by use of sodium sulphate or magnesium sulfate. *Annual Book of ASTM Standards*, **4**, 37–42.

BISWAS, D. R. 1977. Thermal energy storage using sodium sulphate decahydrate and water. *Solar Energy*, **19**, 99–100.

BRODALE, G. & GIAUQUE, W. F. 1958. The heat of hydration of sodium sulfate. Low temperature heat capacity and entropy of sodium sulphate decahydrate. *Journal of the American Chemical Society*, **80**, 2042–2044.

DONNAY, J. D. H. & HARKER, D. 1937. A new law of crystal morphology extending the law of Bravais. *American Mineralogist*, **22**, 446–467.

GOUDIE, A. S. & VILES, H. A. 1997. *Salt Weathering Hazards*. Wiley, Chichester.

GROTH, P. VON. 1908. Magnesiumsulfat–Heptahydrat = $SO_4Mg \cdot 7H_2O$. Rhombische Modification (nat. Epsomit, Bittersalt). *Chemische Kristallographie*, **2**, 429–432.

HOBB, M. Y. & REARDON, E. J. 1999. Effect of pH on boron coprecipitation by calcite: further evidence for nonequilibrium partitioning of trace elements. *Geochimica et Cosmochimica Acta*, **63**, 1013–1021.

ICHIKUNI, M. & KIKUCHI, K. 1972. Retention of boron by travertines. *Chemical Geology*, **9**, 13–21.

JIANG, W., PAN, H., TAO, J., XU, X. & TANG, R. 2007. Dual roles of borax in kinetics of calcium sulphate dihydrate formation. *Langmuir*, **23**, 5070–5076.

LEVY, H. A. & LISENSKY, G. C. 1978. Crystal structures of sodium sulphate decahydrate (Glaubert's salt) and sodium tetraborate decahydrate (Borax). Redetermination by neutron diffraction. *Acta Crystallographica B*, **34**, 3502–3510.

RODRIGUEZ-NAVARRO, C. & DOEHNE, E. 1999. Salt weathering: influence of evaporation rate, supersaturation and crystallization pattern. *Earth Surface Processes and Landforms*, **24**, 191–209.

RODRIGUEZ-NAVARRO, C., DOEHNE, E. & SEBASTIAN, E. 2000a. How does sodium sulphate crystallize? Implications for the decay and testing of building materials. *Cement and Concrete Research*, **30**, 1527–1534.

RODRIGUEZ-NAVARRO, C., DOEHNE, E. & SEBASTIAN, E. 2000b. Influencing crystallization damage in porous materials through the use of surfactants: experimental results using sodium dodecyl sulphate and cetyldimethylbenzylammonium chloride. *Langmuir*, **16**, 947–954.

RODRIGUEZ-NAVARRO, C., LINARES-FERNANDEZ, L., DOEHNE, E. & SEBASTIAN, E. 2002. Effects of ferrocyanide ions on NaCl crystallization in porous stone. *Journal of Crystal Growth*, **243**, 503–516.

RUBBO, M., AQUILANO, D. & FRANCHINI-ANGELA, M. 1985. Growth morphology of epsomite ($MgSO_4 \cdot 7H_2O$). *Journal of Crystal Growth*, **71**, 470–482.

RUIZ-AGUDO, E., MESS, F., JACOBS, P. & RODRIGUEZ-NAVARRO, C. 2007. The role of saline solution properties in porous limestone weathering by magnesium and sodium salts. *Environmental Geology*, **52**, 269–281.

RUIZ-AGUDO, E., RODRIGUEZ-NAVARRO, C. & SEBASTIAN-PARDO, E. 2006. Sodium sulfate crystallization in the presence of phosphonates: implications in ornamental stone conservation. *Crystal Growth and Design*, **6**, 1575–1583.

SCHERER, G. W. 1990. Theory of drying. *Journal of the American Ceramic Society*, **73**, 3–14.

SELWITZ, C. & DOEHNE, E. 2002. The evaluation of crystallization modifiers for controlling salt damage to limestone. *Journal of Cultural Heritage*, **3**, 205–216.

SHEIKHOLESLAMI, R. 2003. Nucleation and kinetics of mixed salts in scaling. *AIChe Journal*, **49**, 194–202.

TANTAYAKOM, V., FOGLER, H. S., CHAROENSIRITHAVORN, P. & CHAVADEJ, S. 2005. Kinetic study of scale inhibitor precipitation in squeeze treatment. *Crystal Growth and Design*, **5**, 329–335.

TELKES, M. 1952. Nucleation of supersaturated inorganic salt solutions. *Industrial and Engineering Chemistry*, **44**, 1308–1310.

Weathering effects in an urban environment: a case study of tuffeau, a French porous limestone

K. BECK & M. AL-MUKHTAR*

Centre de Recherche sur la Matière Divisée, Université d'Orléans – CNRS-CRMD, 1B rue de la Férrolerie 45067 Orléans Cedex 2, France

*Corresponding author (email: muzahim@cnrs-orleans.fr)

Abstract: A case study is carried out on a highly porous limestone called tuffeau located in an urban environment. Weathering effects are characterized by several complementary techniques: mechanical resistance (compressive test), imbibition (capillarity test), mercury intrusion porosimetry, chemical analysis [inductively coupled plasma-optical emission spectrometer (ICP-OES)], scanning electron microscope (SEM) image analysis and X-ray diffraction (XRD). The results show the different composition of the deteriorated zone of the stone and mainly the presence of high gypsum content in the black crust surface. The analysis also shows that the natural penetration of water into this stone under environmental conditions (atmospheric precipitation, relative humidity and temperature) is limited to 20 mm. Moreover, the presence of gypsum is only detected in this limited zone, which demonstrates the depth of alteration in this stone. Finally, this study clearly highlights the important role of water movements in the deterioration of stone.

The current interest in the preservation of our built heritage has generated numerous studies on the description and analysis of alteration developed on monuments (Samson-Gombert 1993; Moropoulou *et al.* 1995; Esaki & Jiang 1999; Maravelaki-Kalaitzaki & Biscontin 1999). The appearance, distribution and development of alteration on stonework depend not only on the lithology of the stone (mineral composition and texture), but also on environmental exposure, water transfer and the composition of the ambient air. The ageing process and deterioration of a stone is a natural evolution and is therefore inevitable. However, stonework presents discrepancies in its pattern of alteration, in kinetics and intensity, according to the nature of the stone. Fronteau *et al.* (1999) introduced the concepts of preferential weathering and developed alteration. In fact, each type of stone will behave according to its lithology, and will tend to generate specific so-called preferential alteration if exposed to the natural environment. However, the development of the alteration is triggered by the interaction between the stone and its environment (exposure, pollutant). The description of the various types of alteration encountered shows that the behaviour of the stones with respect to deterioration can be directly related to water and its movements within the rock. Indeed, the water can act directly (freezing, selective dissolution of minerals, etc.) and indirectly (transport of dissolved salts, crystallization of new phases, biological activity, etc.). The presence of water is also necessary to enhance the action of pollutants (Winkler 1966; Price 1995), where sulphur dioxide is one of the most important agents (Rodriguez-Navarro & Sebastian 1996; Böke *et al.* 1999).

This work is a case study of a highly porous limestone located in an urban environment. The studied building stone is the white tuffeau, which is a whitish sedimentary limestone and belongs to middle Turonien age (i.e. formed 90 Ma ago). This rock is plentiful in the Loire Valley and commonly used in the famous Loire castles in France. Depending on the quarries, there are different classes of tuffeau stones (Dessandier *et al.* 1997) that have differences in the mineral composition (e.g. calcite content can vary from 40 to 70%) and in the physical properties (e.g. porosity can vary from 30 to 50%). This porous limestone is very sensitive to environmental conditions (pollution, salts, water movement, etc.) and has numerous cases of weathering (Beck *et al.* 2003; Beck 2006). The studied stone sample comes from a building restoration site in downtown Orléans. The weathering effect was analysed according to the depth of the stone by several complementary techniques of characterization in a multi-scale approach: mechanical resistance (compressive strength test), capillary absorption (imbibition test), mercury intrusion porosimetry (MIP),

X-ray diffraction (XRD), chemical analysis and scanning electron microscopy (SEM) analysis. The aim of this study is to determine how urban environmental conditions cause stone alteration, and induce structural and mineralogical changes in the interior of the stone.

Macroscopic characterization of the studied stone

Exposure and visual description

The studied stone was taken from an *in situ* stone block located in the façade of a house in downtown Orléans. The façade of the house from which the samples were taken is oriented to the east in a narrow street. Thus, the exposure of this stone block is rather shaded and protected from driving rain. Photographs of the studied stone block are shown in Figure 1. This stone sample is in a rather advanced state of deterioration. The surface exposed to atmospheric conditions has a black crust of about 0.5 mm in thickness. This black crust has a rather powdery aspect, and exfoliates easily. In the first 2 mm behind this black crust, the stone is easily friable and has microcracks parallel to the exposed surface. One also notes the presence of an orange edging of about 2 mm in thickness and roughly parallel to the exposed surface. This edging has a rather consistent aspect and is evenly located at about 20 mm depth.

Mechanical properties

The mechanical properties of the stone were examined using the compressive strength test (Standard NF P94-420 2000) in order to determine the effect of the altered zone on the mechanical behaviour. The tests were carried out on a cylindrical sample (diameter 40 mm and height 80 mm). An increasing compressive force was applied with an Instron 4485 press at a loading rate of 0.05 mm min^{-1} along the axis of the cylindrical sample (i.e. loading direction) until rupture. All samples were tested in a dry state (for 24 h in an oven at 105 °C). These samples were then placed for 6 h in a desiccator with relative humidity close to 0%, controlled by a phosphorus pentoxide, P_2O_5,-saturated salt solution for a return to ambient temperature under dry conditions. Two samples were tested: one with the black crust that corresponds to the base of the sample, and one cut from the centre of the block, presumed to be unaltered. These samples were of the same size and were cut according to the same stratigraphic direction, which is parallel to the stone bed. The results are presented in Figure 2.

The compressive strength was 30% lower in the altered surface zone compared with the unaltered zone. This degradation of mechanical strength was more significant for the elastic modulus, with a 60% decrease. The stress–strain curves obtained are typical of a rock sample (Guégen & Palciauskas 1992). These curves can be divided into three stages:

- first, stresses increase slowly due to the applied strain up to 0.1–0.2%. This behaviour can be attributed to the crushing of not-perfectly-parallel sides in contact with the plates of the press;
- second, elastic behaviour represented by a linear curve (E is the static Young modulus given by the slope of the linear curve);
- and, finally, an elasto-plastic behaviour with opening and propagation of cracks until rupture (σ_c is the maximum stress at rupture).

There is an important difference in the compressive behaviour of both samples, mainly in the first part of

Fig. 1. Tuffeau sample extracted from a monument in an urban environment.

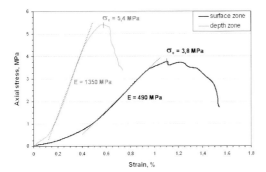

Fig. 2. Stress–strain curves for the compressive strength test.

the curve. The sample that includes an altered surface zone shows very high strain under a small value of applied stress. This is due to the black crust, which possesses a significant surface roughness, and especially to the presence of microcracks, which weaken the resistance of the surface zone. Indeed, the black crust area collapses considerably under applied axial stress. The elastic behaviour of the stone begins once all of the microcracks are closed, that is, at a strain of 0.5%. So, the surface zone can be considered as a deteriorated zone from a mechanical point of view.

Capillary properties

Capillary water absorption properties in a porous material such as stone are directly related to the size and shape of the pores, as well as to the connectivity of the pore network. The principle is that when a dry porous solid is in contact with liquid water, the non-wetting fluid (air) that filled the pores is substituted by the wetting fluid (water) without applying any external pressure.

The height of water rise in a vertical cylindrical tube of radius, r_c, is determined using the Washburn capillary model, which neglects the influence of gravity. With this model the height of the capillary front, h, and the mass uptake, m, per surface unit, S, during the elapsed time, t, can be determined (equations 1 & 2):

$$h = \sqrt{\frac{r_c \sigma}{2\eta}} \sqrt{t} = B\sqrt{t} \quad (1)$$

$$\frac{m}{S} = A\sqrt{t} \quad (2)$$

where σ and η are the superficial tension and the viscosity of water, respectively. This capillary tube model is more qualitative in nature but allows us to understand the nature of capillary imbibition curves obtained experimentally for porous material such as stone ($h = B\sqrt{t}$ and $m = A.S\sqrt{t}$), and indicates clearly the influence of the pore radius on the imbibition kinetics. The imbibition coefficients A and B correspond to the slopes of the curves of mass uptake and rise of the capillary front according to the square root of elapsed time.

In this test [Standard NF-EN 1925 (B10-613) 1999], previously dried cylindrical samples were placed in a hermetic tank, at the bottom of which the distilled water level was maintained as constant during the entire period of the test. The lower surface of the sample in contact with the water was the surface zone with the black crust. The mass of the wet samples was measured as the height of the capillary front was advancing at increasing intervals of times. The advancing capillary front height was measured using vernier calipers.

In order to determine whether the black crust and the orange edging modify the imbibition properties of the stone, imbibition tests were carried out on a sample cut from the block and again on the same sample after cleaning by leaching with water.

The results of the imbibition test are presented in Figure 3. The imbibition curve for the altered sample (before washing) was segmented into three parts. At first, the water rose by capillary action into the stone through the black crust without difficulty up to the orange edging (15–20 mm). It appears that the black crust had no observable effect on the imbibition properties. At the zone with orange edging, the rise of water slowed markedly and the imbibition coefficients almost halved (Table 1). Gradually, the matter forming the orange edging began to be dissolved and leached by the water, as shown in Figure 3. The imbibition kinetics are then similar to the first part prior to the orange edging, with a linear curve and imbibition coefficients close to the initial values. This test shows that the black crust had no influence on the imbibition kinetics, but the orange edging located 20 mm from the exposed surface caused a significant slowing of the imbibition kinetics. In this zone the orange matter reduces the natural water transfer in the stone because of its accumulation in this area.

The total leaching of water-soluble components from the sample was conducted using three successive lixiviations (washing). The first leachate was a clear yellow-orange colour, the second was very light coloured and the third colourless, reflecting the complete cleaning of the sample. Once cleaned and dried, the specimen was again subjected to the imbibition test. The imbibition curve is linear (Fig. 3) and the imbibition coefficients are nearly identical to those determined for the third part of the imbibition curves before leaching (Table 1). Therefore, the orange matter is responsible for the

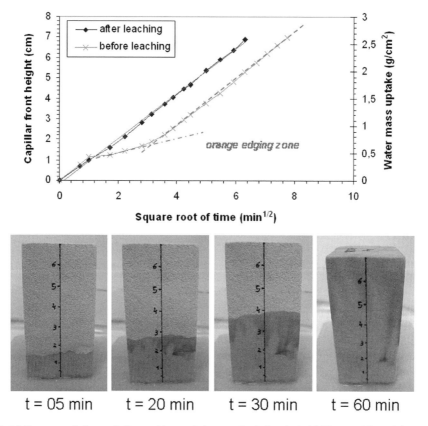

Fig. 3. Imbibition curves before and after washing, and photographs during the imbibition test (elapsed times at 5, 20, 30 and 60 min).

slowing of water penetration at 20 mm depth. Unfortunately, the matter giving this orange coloration has not been exactly identified within the stone because the quantity was too low to be detected. Nevertheless, the dissolved orange matter present in the leachate was recovered after removal of water. Thus, a very small quantity was extracted from the stone. According to XRD analysis and thermo-gravimetric analysis, this substance is not crystalline and is not made up of iron oxides, and it mainly breaks down close to 400 °C.

Table 1. *Results of the imbibition test*

	Coefficient A (g cm^{-2} min$^{-1/2}$)	Coefficient B (cm min$^{-1/2}$)
Before washing:		
First zone	0.36	1.15
Second zone (orange edging)	0.18	0.45
Third zone	0.40	1.14
After washing	0.39	1.10

It is highly probable that it is constituted organic matter. Moreover, the orange matter seems to be naturally present within the stone because the water even, drains this substance from the area behind the orange edge (cf. Fig. 3). This orange matter probably has a biological origin as a result of the natural infiltration into the rock of organic matter from the ground (Mertz pers. comm. 2006). The water flow would cause its accumulation in an area where its concentration becomes high enough to become visible. The existence of this edging is very interesting because the orange matter is coloured and transportable by water. This orange edging operates as a tracer that indicates the maximum depth of water penetration into the stone under the extant field conditions.

Microscopic changes in the studied stone

Pore space changes due to deterioration

The analysis of pore space was carried out by mercury intrusion porosimetry. In this test, bulk and skeletal densities, total porosity and pore size distribution of stone were determined. The principle was to inject liquid mercury under pressure (up to 210 MPa) into a degassed porous material. The volume of mercury injected corresponds to the cumulative volume of the pores accessible to mercury at a given pressure. Theoretically, the pores with a diameter of between 350 μm and 6 nm can be investigated with the equipment used: a Micrometics Porosizer 9320. Two samples were tested: one with the black crust (surface–5 mm depth), and one cut from the centre of the block (50 mm depth) and presumably unaltered. The results are presented in Figure 4 and in Table 2. This analysis was completed by scanning electron microscopy (SEM) observation of an area showing the first 2 mm in from the surface.

One can observe the presence of numerous microcracks about 100 μm wide and parallel to the black crust. The mercury intrusion curves were quite similar for both samples, with a total porosity of about 44% and a bulk density close to 1.40.

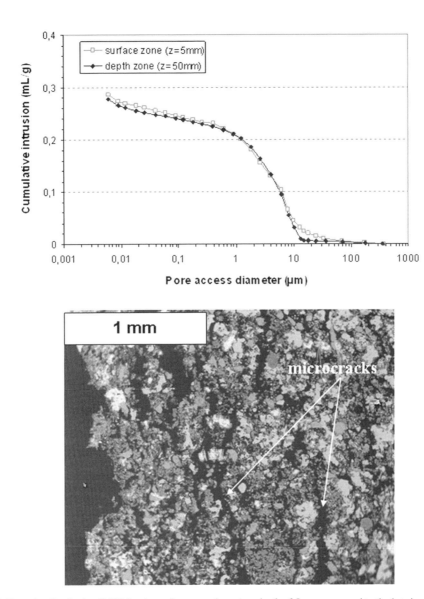

Fig. 4. Pore size distribution (MIP) for the surface zone down to a depth of 5 mm compared to the interior zone at a depth of 50 mm. A scanning electron microscopy (×50) photograph of the first 2 mm below the surface.

Table 2. *Densities and porosities for the surface zone down to 5 mm depth and for the interior zone at a depth of 50 mm*

	Depth 5 mm (surface zone)	Depth 50 mm (interior zone)
Dry bulk density, ρ_a (g cm^{-3})	1.37	1.42
Solid skeletal density, ρ_s (g cm^{-3})	2.41	2.54
Total porosity, N_{tot} (%)	43.3	44.0

Nevertheless, the skeletal density was lower in the surface zone compared with the interior zone. Therefore, the solid phase is different in the black crust compared with the mineralogical composition of Tuffeau. Moreover, the pore distribution is also different in the macroporous range in the surface zone and in the first 5 mm from the surface. There is a greater proportion of volume accessible to mercury for a range of pore access diameters of between 10 and 100 μm. This volume corresponds to the microcracks observed on the images obtained by SEM. The presence of large pores, such as microcracks present up to 2 mm in depth, shows that the block of tuffeau is deteriorated to a greater depth than the 0.5 mm depth corresponding to the thickness of the black crust.

Mineralogical changes due to deterioration

The stone has been subjected to a moderately aggressive urban environment. Indeed, Orléans is a moderately polluted city (data from Lig'Air, http://www.ligair.fr) with a SO_2 annual rate ranging from 2 to 5 μg m^{-3} (the annual limit value for standard health is 20 μg m^{-3}) and a rate of fine particulate matter smaller than 10 μm (PM_{10}) ranging from 15 to 30 μg m^{-3} (the annual limit value for standard health is 40 μg m^{-3}). As evidenced by the presence of a black crust, it is the area of the stone close to the surface and exposed to the outside atmosphere that has suffered the greatest damage. In order to identify the mineralogical modifications due to weathering, the stone was analysed according to the depth by several complementary techniques: XRD, chemical analysis and SEM. The analysis of mineralogical composition was carried out from the surface to a depth of 50 mm (the zone assumed to be unaltered). Powder sampling obtained every 5 mm starting from the external surface allowed the average composition of the stone according to depth to be measured.

Figure 5 presents the XRD patterns for the surface zone with the black crust and for a zone at a depth of 5 mm. A view of these zones is given with SEM images. The main crystalline phases are calcite ($CaCO_3$) and silica (SiO_2), which occur mainly in two crystalline forms: quartz and another form of silica named opal cristobalite-tridymite (opal-CT), which are particles with the morphology of spherule formed during the diagenesis of rock (Dessandier et al. 1997; Beck 2006). Moreover, in the tuffeau, the presence of minor minerals like clays (glauconite, smectite) and micas are also identified. The most interesting observation is the presence of XRD peaks characteristic of gypsum. The black crust is mainly composed of gypsum. The intensity of these peaks steadily decreases with depth and after a depth of 20 mm no trace of gypsum is detected. This specific depth corresponds to the orange edging. The SEM images show that the black crust is composed mainly of opal-CT spherules cemented by gypsum (Fig. 5). Then, up to a depth of 20 mm gypsum deposits are also present but more scarcely.

In order to observe the strictly quantitative mineralogical evolution from the exposed surface to the depth zone, chemical analyses were carried out on the same powder samples. Chemical analysis was carried out by optical emission spectrometry with inductively coupled plasma (ICP-OES). The ICP technique detects and quantifies chemical elements ranging from a few ppm to very high concentrations. The sulphur content was determined with a LECO analyser using the combustion of total sulphur at 1350 °C and the quantification of produced sulphur dioxide with an infrared detector. The results are presented in Tables 3 and 4. The high content of SiO_2 and CaO confirm that the main phases of this stone are calcite, quartz and opal, as shown by the XRD study. This tuffeau also contains small amounts of clayey minerals, which can be seen by the proportion of elements such as Al, Fe, Mg, Na and K. The results indicate that, whatever the analysed depth, the quantity of these elements does not vary significantly. The content of clay minerals is generally stable according to depth, as well as detrital minerals such as Ti-minerals (rutile TiO_2), which appear with a metal brightness in optical microscopy. The black crust (depth 0 mm) has a phosphorus content (Table 3). This shows biological activity (fungi, algae, etc.) on the surface exposed to the outside atmosphere; and microscopic organic elements can be observed by SEM at the surface of the stone (Beck 2006). Moreover, the black crust has a significant lead content (Table 3). This pollution by lead is detected to a depth of 20 mm. This chemical element confirms the assumption of atmospheric pollution and soot from car emissions. The measurements of sulphur are consistent with the XRD analysis. In the black crust the sulphur content is 9.5%, which corresponds to

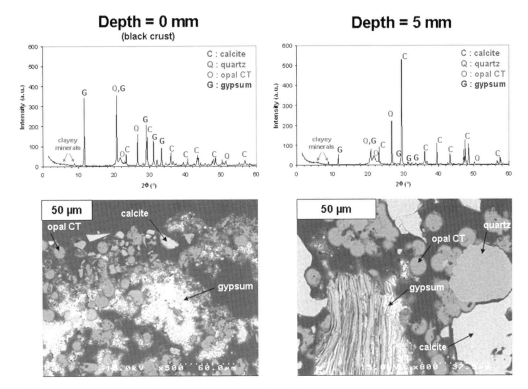

Fig. 5. XRD diffractograms and SEM photographs (BSE mode) of a surface zone near the black crust and a zone to 5 mm depth.

Table 3. *Results of chemical analysis determined by ICP-OES for major elements (%) and for lead (ppm) according to depth*

Depth (mm)	0	5	10	15	20	30	50
SiO_2 (%)	31.56	33.96	34.53	37.39	39.09	36.75	36.70
CaO (%)	25.37	34.90	35.27	34.15	32.43	33.53	34.07
Al_2O_3 (%)	1.76	1.52	1.52	1.52	1.56	1.60	1.57
Fe_2O_3 (%)	0.75	0.69	0.69	0.68	0.68	0.71	0.72
MgO (%)	0.12	0.15	0.21	0.16	0.19	0.15	0.21
Na_2O (%)	0.06	0.05	0.11	0.06	0.10	0.06	0.09
K_2O (%)	0.41	0.41	0.43	0.41	0.42	0.43	0.40
TiO_2 (%)	0.14	0.13	0.13	0.14	0.15	0.14	0.14
P_2O_5 (%)	0.05	0.00	0.00	0.00	0.00	0.00	0.00
Pb (ppm)	141	11	10	6	<2	<2	<2

Table 4. *Results of sulphur proportion determined by LECO analyser according to depth*

Depth (mm)	0	5	10	5	20	30	50
Sulphur (%)	9.51	0.85	0.32	0.19	0.12	0.04	0.03
Equivalent gypsum (%)	51.04	4.56	1.69	1.02	0.64	–	–

51% of gypsum $CaSO_4.2H_2O$ (Table 4). This gypsum is derived from the recombinant calcium from the calcite dissolved by water that penetrates into the stone, and the sulphur is brought from outside by air pollution and can be present for several millimetres (Brunet-Imbault et al. 2000). The sulphur content decreases strongly within the first 5 mm, and sulphur is practically absent beyond a depth of 20 mm. For the most part, the sulphur is attributed to gypsum and a very low content can be attributed to pyrite FeS_2, which is found only scarcely within the tuffeau in the form of small grains easily visible with optical microscopy and the naked eye. So the deterioration process is not limited to the black crust. Mineralogical characterization reveals that the gypsum content decreases from a high value at the surface (black crust) to much lower values in the interior of the stone. Gypsum is detected in a zone limited to 20 mm in depth, which can be considered as the weathered zone. In this stone the orange edging, resulting from the accumulation of movable matter, can be used as a natural tracer and shows the maximum depth for water movement. Thus, the results obtained allowed us to demonstrate that, in the urban environment in which this stone was used, water movement and the weathering effect are dependent and limited to 20 mm from the outside surface of this porous limestone.

Conclusion

This case study allowed us to determine the alteration type and the structural and mineralogical changes in the white tuffeau stone under urban environmental conditions. Several complementary techniques of characterization coupling macroscopic (imbibition, compressive strength) and microstructure analysis (MIP, SEM, XRD, ICP-OES) were used.

The results confirm first of all the major role of water in the development of stone alteration. Indeed, the alteration of this tuffeau taken from an urban environment exceeds the depth of the black crust (0.5 mm thick) on the surface and attains a thickness of several millimetres.

The black crust is essentially composed of gypsum (over 50%). This gypsum results from the combination of calcium from the previously dissolved calcite and sulphur transported by air pollution. Development of biological material (algae, fungi, etc.) occurs in the black crust, as shown by the presence of phosphorus in the analyses and SEM observations (Beck 2006). In addition, the effect of air pollution is clearly demonstrated by the presence of lead-rich particles. The content of gypsum mainly decreases from the black crust into the core until a depth of 20 mm, where only a very small amount of gypsum can be detected. Beyond 20 mm in depth no gypsum can be found: the stone is unaltered.

This limit of 20 mm in depth is highlighted by a very clear orange edging. This orange matter, probably made up of biomaterials, was originally distributed throughout the entire stone, but is now very scarcely disseminated. This orange edging is assumed to result from the water movements that transport the orange matter from the surface zone to concentrate them in the edging zone. This higher concentration renders the orange matter visible, in contrast to the unaltered zone where their natural concentration is too low to create the orange shade. In this case study the orange matter acts as a tracer to reveal the maximum boundary (extent) of water penetration resulting from rainwater exposure. The correlation between the occurrence of gypsum deposits up to a depth of 20 mm and the precise localization of the orange edging at 20 mm depth clearly demonstrates that the alteration of the tuffeau limestone is limited to the shallow zone affected by the flow of water.

References

BECK, K. 2006. *Etude des propriétés hydriques et des mécanismes d'altération de pierres calcaires à forte porosité*. PhD thesis, University of Orléans.

BECK, K., AL-MUKHTAR, M., ROZENBAUM, O. & RAUTUREAU, M. 2003. Characterization, water transfer properties and deterioration in tuffeau: building material in the Loire Valley. *Journal of Building and Environment*, **38**, 1151–1162.

BÖKE, H., GÖKTÜRK, E. H., CANER-SALTIK, E. N. & DEMIRCI, S. 1999. Effect of airborne particle on SO_2–calcite reaction. *Applied Surface Science*, **140**, 70–82.

BRUNET-IMBAULT, B., MULLER, F., RANNOU, I. & RAUTUREAU, M. 2000. Mineralogical phase analysis in monumental tuffeau limestone by X-ray powder diffraction. *Materials Science Forum*, **321–324**, 1010–1015.

DESSANDIER, D., ANTONELLI, F. & RASPLUS, L. 1997. Relationships between mineralogy and porous medium of the crai tuffeau (Paris Basin, France). *Bulletin de la Société Géologique de France*, **186**(6), 741–749.

ESAKI, T. & JIANG, K. 1999. Comprehensive study of the weathered condition of welded tuff from a historic stone bridge in Kagoshima, Japan. *Engineering Geology*, **55**, 121–130.

FRONTEAU, G., BARBIN, V. & PASCAL, A. 1999. Impact du faciès sédimento-diagénétique sur l'altération en œuvre d'un géomatériau calcaire. *Compte-Rendu de l'Académie des Sciences, Sciences de la terre et des planètes*, **328**, 671–677.

GUÉGEN, Y. & PALCIAUSKAS, V. 1992. *Introduction à la physique des roches*. Hermann Science, Paris.

MARAVELAKI-KALAITZAKI, P. & BISCONTIN, G. 1999. Origin, characteristics and morphology of weathered crusts on Istria stone in Venice. *Atmospheric Environment*, **33**, 1699–1709.

MOROPOULOU, A., THEOULAKIS, P. & CHRYSOPHAKIS, T. 1995. Correlation between stone weathering and environmental factors in marine atmosphere. *Atmospheric Environment*, **29**, 895–903.

PRICE, D. G. 1995. Weathering and weathering processes. *Quarterly Journal of Engineering Geology*, **28**, 243–252.

RODRIGUEZ-NAVARRO, C. & SEBASTIAN, E. 1996. Role of particulate matter from vehicle exhaust on porous building stones (limestone) sulfation. *Science of the Total Environment*, **187**, 79–91.

STANDARD NF P94-420. 2000. *Détermination de la résistance à la compression uniaxiale*. Association Française de Normalisation (AFNOR), Paris.

STANDARD NF-EN 1925 (B10-613). 1999. *Méthodes d'essai pour pierres naturelles – Détermination du coefficient d'absorption d'eau par capillarité*. Association Française de Normalisation (AFNOR), Paris.

SAMSON-GOMBERT, C. 1993. *Influences d'un environnement urbain et maritime sur les altérations d'un calcaire en oeuvre: la pierre de Caen*. PhD thesis, University of Caen.

WINKLER, E. M. 1966. Important agents of weathering for building and monumental stone. *Engineering Geology*, **1**(5), 381–400.

Approaches to the problem of limestone replacement in Greece

M. A. STEFANIDOU

Laboratory of Building Materials, Civil Engineering, Aristotle University of Thessaloniki, Greece (e-mail: stefan@civil.auth.gr)

Abstract: One of the most common stone types in building construction in Greece is limestone. Whole structures such as castles, palaces, fortresses and churches were built only with limestone blocks or limestone pieces combined with other types of stone. In this paper two types of limestone (a biogenic limestone and a travertine) used in the construction of monuments are tested and analysed in terms of their physical, mechanical and microstructural characteristics in order to record their properties. Their exposure to different environmental conditions and the pathology forms they present are also recorded. The possibility of their replacement is approached either by supplementing the missing parts, by finding a comparable new stone or by applying an artificial one. The results of the study performed in each case are also presented.

One of the most common stone types in building construction in Greece is limestone due to be its ability to be cut and shaped easily, its 'warm' colour and its abundance (Kouzeli 2004). Whole structures were built only with limestone or limestone blocks combined with other stone types. Under the term of 'limestone' a wide range of petrographically different stones are named. Some common characteristics of different limestones are the high porosity and the relatively low strength compared with metamorphic or igneous stones. The porous structure of limestone serves as a conveyor in order for humidity and other deteriorating agents to enter into their structure (Charola 2004). In order to approach the pathological problems that these stones present, two different trends are of value. The first concerns the replacement of the decayed part with a new compatible stone and, the second, the filling of the missing or damaged part with a type of artificial stone (Papayianni et al. 2004). In both cases in order to proceed to a successful intervention, the properties of the historic building material, as well as of the new repair material, should be well known. The methodology developed in the Laboratory of Building Materials at the Aristotle University of Thessaloniki (AUTH) for that purpose concerns the analysis of the mechanical properties (flexure and compression), the physical properties (porosity, capillary absorption, pore size distribution, water absorption), the microstructure properties and the chemical properties (composition, salt content) of the historic and the new repair material (Papayianni et al. 2005).

Methods

The approach in order to understand the nature of an old building material and to take measurements for its preservation consists of three discernible steps. The first is related to the characteristics of the old, historic material in terms of recording the mechanical, physical, chemical and microstructural properties in order to have a holistic opinion about its quality. The mechanical properties usually refer to the flexure and compressive strength, as well as the measurement of the modulus of elasticity according to relevant regulations (EN 1926 : 2006 for stones and EN 1015-11 : 1999 for mortars). The physical properties refer to water absorption measured according to RILEM CPC11.3 (RILEM TC 14-CPC : 1984) and capillary elevation according to EN 1015-18 : 2002. The capillary measurement is not always possible owing to the shape of the available sample taken from the historic structures. The chemical analysis is performed using X-ray diffraction (XRD) or X-ray Fluoresence (XRF) equipment in the case of stones and according to BS 4551 : 2005 for mortars. Microstructural analysis is performed according to RILEM TC 167-COM (Middendorf et al. 2005) for mortars and EN12407 : 2007 for stone samples.

The second step concerns the repair material, which should be designed according to the results derived from the analysis of the old, historic material. During this phase, laboratory tests are performed in order to test the properties of the new, repair material according to the above-mentioned properties. Compatibility and durability seem to be the required properties in order for the old and the new material to work harmonically into the structure. The third and crucial step is the application of the repair material during which trained masons and strict supervision by the person responsible for the building is required.

In this paper two examples relating to limestone are analysed. The first case is the limestone used as the main building material in the medieval city of Rhodes, which is exposed to a typical marine, mediterranean climate, and the other example concerns a travertine exposed to a rural environment.

Medieval city of Rhodes

Local limestone is used for the construction of the medieval monuments of Rhodes, which are included in the UNESCO World Heritage List. The stone is petrographically characterized as biogenic limestone from local deposits. Its genesis dates back to the Upper Pliocene–Lower Pleistocene and consists of rounded fossils joined together with microsparitic calcite cement (Fig. 1).

The monuments are exposed to marine environment and they are affected both by direct contact with the waves (especially those in the coast yard) and indirectly through wind action.

The alteration forms that are present are irregular material loss, efflorescence, exfoliation, biodeterioration, black crust and salts. In an effort to monitor the pathology forms and the extent of pathology, mapping of the section of the city walls was performed. Four different stages were separated ranging from I (healthy) to IV (heavily deteriorated), as seen in Figure 2.

The results of the *in situ* analysis showed that large parts of the city walls suffered from intense deterioration and that the need for preservation measurements was urgent. Under this status, more than 20 samples from the building stones were analysed in the laboratory in order to record their properties (Papayianni *et al.* 2004).

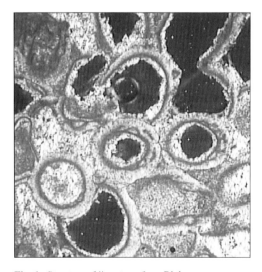

Fig. 1. Structure of limestone from Rhdoes (polarized ×65).

State of deterioration	In situ	Stereoscope (x8)
Stage I Compact structure		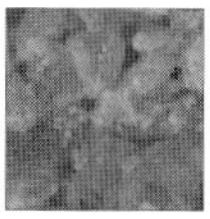
Stage II Material loss	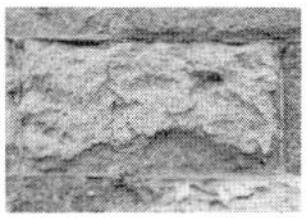	
Stage III Intense loss of material		
Stage IV Lost parts	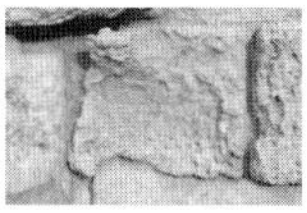	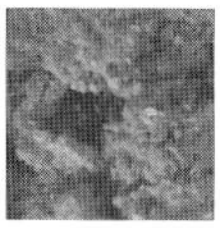

Fig. 2. Different stages of stone deterioration.

The results of the analysis are listed in Tables 1 and 2, while the pore size distribution is seen in Figure 3.

From the results it appears that the stone is soft and porous, and has a considerable salt content. The *in situ* observation and the results of the analysis indicate the need for protective measurements for the structure. The problem of sourcing compatible stone for restoring large pieces of the building stone was addressed by testing local stone samples from new quarries. Basic requirements are the aesthetic harmonization, equivalent porosity properties (open porosity, pore size distribution, behaviour under capillary elevation), similar mechanical properties and durability (freeze–thaw, salt cycles, wetting–drying cycles). The tests were similar to the ones mentioned for the historic stone samples and the following data were recorded (Table 3).

Microscopic examination of the new and old samples indicates small differences in the cement and the density of the samples, while the main pore volume (73%) in the quarried samples is in the area of 200–2000 μm (Fig. 4).

The results show that from the scientific point of view, the use of new local stone is feasible as repair material for the intervention works in the case of the medieval city of Rhodes.

Logos-Edessa

The history of the ancient city starts from the Hellenistic period and continues until the Byzantine era.

Table 1. *Physical and mechanical properties of biogenic limestone (variations between 20 samples)*

	Compressive strength (MPa)	Specific gravity	Water absorption (%)	Total porosity (%)	
				By O.M.* (5–2000 μm)	By B.E.T.† (10–2000 Å)
Biogenic limestone	0.5–1.5	2.01–1.74	10–25	12–20	3.19

*O.M., optical microscope; †B.E.T., nitrogen absorption.

Table 2. *Content in soluble salts measured (% by wt) (variations between 20 samples)*

Sample	Cl^-	NO_3^-	SO_4^{2-}
Biogenic limestone	0.85–2.94	0.13–0.25	0.15–1.20

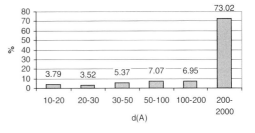

Fig. 4. Pore size distribution in travertine.

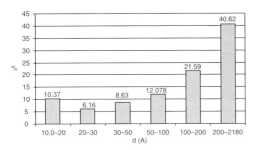

Fig. 3. Pore size distribution by nitrogen absorption on samples from the historic stone.

The life of the city was closely related to water, as the River Edesseos formed the landscape on which the city was built and also covered a part of the city with a thick layer of sediments after its retreat.

The site is located in a rural environment and is open to the air, with significant temperature variations. In the archeological site of Loggos local travertine from the Pleistocene was the main building material (Stefanidou & Papayianni 2007). The main problems of the stones presented at the site in Logos are cracks, probably due to frost action, biological growth, colour alterations due to black crust deposition and missing parts.

Analysis of mechanical and physical properties was carried out at the Laboratory of Building Materials, AUTH, as in the previous case. Fifteen samples were analysed in order to record their properties. The results are shown in Table 4.

Pore size distribution was measured using image analysis, as seen in Figure 5, and salt content was determined by liquid chromatography (Table 5).

The decision to replace the damaged stone pieces and reinforce the structures was combined with the lack of new quarries in the area, the strict legislation for re-opening old ones and the budget limitations. An alternative solution, of replacing the missing parts of the stones by the production of artificial stone, seemed to be easy, feasible and cost-effective. In this case the results of the analysis of the historic stones served as guidelines for the desired properties that were needed. The selection of the raw materials used for the production of the

Table 3. *Physical and mechanical properties of new quarried limestone (average from five samples)*

	Compressive strength (MPa)	Specific gravity	Water absorption (%)	Total porosity (%)	
				By O.M.* (10–1000 μm)	By B.E.T.† (10–2000 Å)
New limestone	1.0	2.207	11.5	10–15	1.49

*O.M., optical microscope; †B.E.T., nitrogen absorption.

Table 4. *Mechanical and physical properties of travertine from Loggos (variations between 15 samples)*

	Compressive strength (MPa)	Specific gravity	Water absorption %	Total porosity (%)	
				By O.M.* (5–2000 μm)	By B.E.T.† (10–2000 Å)
Travertine	1–10	1.55–1.05	18–22	19–22	3.54

*O.M., optical microscope; †B.E.T., nitrogen absorption.

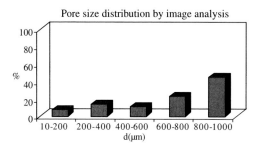

Fig. 5. Pore size distribution in travertine.

Table 5. *Content in soluble salts (% by wt) in stones from Logos (variations between 15 samples)*

Sample	Cl^-	NO_3^-	SO_4^{2-}
Travertine	0.0–0.06	0.0–0.7	0.05–0.10

artificial stone, and their percentage in the mixture, regulated the predefined physical and mechanical properties. The first step was to test the suitability of the raw materials in terms of their reactivity, clearness and gradation according to relevant regulations (EN 196-5:1994 and EN 196-21:1989). In this case white cement, low in alkalies, clay (for colour harmonization) and natural pozzolana were used as binders, and river sand in 0–4 mm gradation was added as aggregate. The properties of the used materials are seen in Table 6.

Trial mixtures were performed in the laboratory before applying *in situ*. The results are shown in Table 7.

From the results it seems that it is possible to produce artificial stone in the laboratory with compatible characteristics to the old, historic material.

Discussion

In the two cases studied the structural stones are a type of porous limestone that suffered under the

Table 6. *Properties of raw materials used for artificial stone*

	Specific gravity	Pozzolanic index (MPa)	Finesse	Salt content		
				Cl^-	SO_4^{2-}	NO_3^-
White cement	2.864	–				
Clay		0.5				
Pozzolana	2.220	10	10% residue in 45 μm	0.07	0.03	0.00
Sand	2.650	–	0–4 mm	0.003	0.00	0.001

Table 7. *Mechanical and physical properties of the artificial stones (variations between five samples)*

	Compressive strength (MPa)	Specific gravity	Water absorption (%)	Total porosity (%)	
				By O.M.* (5–2000 μm)	By B.E.T.† (10–2000 Å)
Artificial stone	4.5–15	1.95–2.35	11–15	7–11	1.22–1.56

*O.M., optical microscope; †B.E.T., nitrogen absorption.

local climatic conditions. The decision for replacing the damaged parts of the structure was approached by sampling and analysing the mechanical, physical, chemical and microstructural properties of the stone pieces in the laboratory. In the case of Rhodes, the laboratory tests indicated that the new, local quarries could provide stone compatible to that of the historic one.

In the case of Logos, this possibility was rejected from the beginning and an alternative solution was the production of artificial stone. The methodology used in this second case was the same as in the case of Rhodes. From the tests performed it was possible to use binders such as pozzolana, and cement and sand, in order to produce stone-like materials with properties close to the defined ones.

Conclusions

The soft and porous nature of limestone makes it prone to different decay patterns according to the exposed environment. The replacement of the old authentic building stones can be approached in different ways. All of them need special care as a significant part of our cultural heritage is constructed using limestones. In cases where intervention works are needed, the main demand is to put an old historic material to work with a new one in order to serve the continuity of the structure. Laboratory tests should be performed in order to prove the compatibility before applying any material to the structure. Using the results of the holistic analysis performed in the laboratory on the old historic material as guidelines for the production of the new material seems to be a safe way to approach stone replacement.

References

BS 4551:2005. *Mortar. Methods of Test for Mortar. Chemical Analysis and Physical Testing.* BSI British Standards, London.

CHAROLA, A. E. 2004. Stone deterioration in historic buildings and monuments. *In*: KWAITKOWSKI, D. & LOFVENDAHL, R. (eds) *10th International Congress on Deterioration and Conservation of Stone, Stockholm.* ICOMOS, Sweden, 3–14.

EN 196-5:1994. *Methods of Testing Cement. Pozzolanicity Test for Pozzolanic Cements.* European Committee for Standardization, Brussels.

EN 196-21:1989. *Methods of Testing Cement: Determination of the Chloride, Carbon Dioxide and Alkali Content of Cement.* European Committee for Standardization, Brussels.

EN 1015-11:1999. *Methods of Test for Mortar for Masonry. Determination of Flexural and Compressive Strength of Hardened Mortar.* European Committee for Standardization, Brussels.

EN 1015-18:2002. *Draft Document – Methods of Test for Mortar for Masonry – Part 18: Determination of Water Absorption Coefficient due to Capillary Action of Hardened Rendering Mortar.* European Committee for Standardization, Brussels.

EN 1926:2006. *Natural Stone Test Methods. Determination of Uniaxial Compressive Strength.* European Committee for Standardization, Brussels.

EN 12407:2007. *Natural Stone Test Methods – Petrographic Examination.* European Committee for Standardization, Brussels.

KOUZELI, K. 2004. Fossiliferous limestones used in Ancient Greek monuments: the influence of their specific features on their durability. *In*: KWAITKOWSKI, D. & LOFVENDAHL, R. (eds) *10th International Congress on Deterioration and Conservation of Stone, Stockholm.* ICOMOS, Sweden, 123–130.

MIDDENDORF, B., HUGMES, J. J., CALLEBAUT, K., BARONIO, G. & PAPYIANNI, I. 2005. RILEM TC 167-COM: investigative methods for the characterisation of historic mortars – Part I: mineralogical characterisation. *Materials and Structures*, **38**, 761–769.

RILEM TC 14-CPC. 1984. CPC 11.3: Absorption d'eau par immersion sous vide. [Absorption of water by immersion under vacuum.] *Materials and Structures*, **17**, 391–394.

STEFANIDOU, M. & PAPAYIANNI, I. 2007. Analysis of the old mortars and proposals for the conservation of the archeological site of Logos-Edessa. *In: 10th International Conference on Studies, Repairs and Maintenance of Heritage Architecture, Czech Republic, STREMAH.* WIT Press, Southampton, 261–266.

PAPAYIANNI, I., STEFANIDOU, M. & PACHTA, V. 2005. Proposals for the restoration of stones in the castle of Hagios Nikolaos based on the analysis of the authentic stones. *In: 1st National Congress, ETEPAM, Thessaloniki, Greece.* Graphica, 112–119.

PAPAYIANNI, I., THEOULAKIS, P. & STEFANIDOU, M. 2004. The mechanical and physicochemical characteristics of stone masonries of the medieval monuments of Rhodes. *In*: KWAITKOWSKI, D. & LOFVENDAHL, R. (eds) *10th International Congress on Deterioration and Conservation of Stone, Stockholm.* ICOMOS, Sweden, 643–650.

Historic lime mortars: potential effects of local climate on the evolution of binder morphology and composition

KARA R. DOTTER

School of Geography, Archaeology and Palaeoecology, Queen's University Belfast, Belfast, Northern Ireland BT7 1NN, UK (e-mail: kdotter01@qub.ac.uk)

Abstract: This research explores preliminary observations regarding the potential effects of local climate conditions on the evolution of binder morphology and the composition or mineralogy of historic lime mortars. Samples were collected from historic buildings, one group dating to *c.* late nineteenth century and an older group from *c.* Sixteenth–Seventeenth century, with both groups representing distinct climates. The samples were then prepared and subjected to polarized light microscopy (PLM), point count, scanning electron microscopy with energy-dispersive spectroscopy (SEM-EDS) and backscatter election imaging (BSE) analysis to investigate differences in porosity, environmental contaminants and morphological characteristics of the lime binder. By examining the effects of climate on lime mortars, conservators can improve identification and condition assessment of historic lime mortars, better understand causes of damage and decay attributed to previous mortar repairs, and formulate more appropriate conservation mortars to improve protection and conservation of our architectural heritage.

Lime-based mortars cure through carbonation and/or chemical processes, depending upon the clay content of the limestone from which the quicklime was derived or any added clays or pozzolans. Either way, the mortar interacts with its environment during curing, as evinced by its usefulness in the radiocarbon dating of historic structures (Folk & Valastro 1976; Nawrocka *et al.* 2005; Lindroos *et al.* 2007). The environment also plays a role in the evolution and decay of lime mortars as they age. The RILEM Technical Committee 167-COM emphasized the role of the following environmental factors in the decay of historic mortars: moisture, salt, air pollution, temperature extremes and variations, exposure to fire, dynamic load, and soil settlement (van Hees *et al.* 2004).

In addition to the environment, several factors may influence the evolution of lime mortar, including craftsmanship, type of lime (putty, quicklime, natural hydraulic lime), and aggregate shape, size and grading. When examining historic mortar, the influence of craftsmanship can be difficult to quantify. The type of lime can be identified through chemical and petrographic analysis, with the latter being able to discern the method of production: lime blebs indicate a 'hot mix', where quicklime is dry-slaked with aggregate and then applied with little to no ageing (Callebaut *et al.* 1999), while 'stringers' in the binder indicate the use of lime putty (Price 2005). The aggregate itself influences the physical and mechanical strength of mortars, and can potentially react with the lime binder (Bloem 1962; Baronio *et al.* 1999; Lanas & Alvarez 2003). Yet, as with carbonate rocks, lime-based binders dissolve and reprecipitate over time, which in turn influences the gain or loss of binder material as well as its morphology (Leslie & Gibbons 1999; Leslie & Hughes 2002). Thus, despite the influence of craftsmanship, binder type and aggregate characteristics on the mortar, the changes ensuing from decades or centuries of weathering processes can overprint the initial impact of these variables. As such, understanding, or at least recognizing, the impact of climate on mortar evolution becomes an integral aspect of comprehending extant historic mortars. Given the importance of diagnosing the causes of degradation when assessing historic structures and the reliance of successful repair mortar design upon understanding the influence of environmental factors (van Hees 1999), it is important to understand the impact of the climate region in which a structure exists in order to better address conservation of that structure.

The factors of moisture, salt and temperature are directly influenced by the climate regime of the structure's environment. In order to explicate the influence of the natural environment upon lime-based mortars, samples of original material were collected from spatially and temporally distinct sites in the United States and Europe. The samples were then subjected to a range of analytical techniques to explore the impact of environmental conditions upon the porosity and pore structure, the presence of environmental contaminants, and the morphological characteristics of the lime binder.

Methodology

Two groups of mortar samples were collected from nine structures in the United States, Northern Ireland and Latvia. Group A represents buildings constructed *c*. mid-nineteenth–early twentieth century and located in three distinct climate zones. Group B represents buildings constructed during the Sixteenth and Seventeenth centuries, but located in climate zones similar to, albeit slightly cooler than, the climate zones for Group A (Fig. 1).

The local climate regions and site-specific region identification are based on the updated world map of the Köppen–Geiger climate classification system. The original Köppen classification system, inspired by Grisebach's 1866 global vegetation map and Köppen's own expertise in plant sciences, logically focused upon changes in vegetation as the chosen method for delineating different climate regions (Wilcock 1968). Thus, one of the complaints frequently levied against using the Köppen classification is its exclusion of temperature and precipitation data. Peel *et al.* (2007) therefore redrew the Köppen–Geiger using global long-term monthly precipitation and temperature station data for the entire period of record. The mean annual precipitation and mean annual temperature provided below represent climate data reported at or near the buildings' locations.

Group A: climate

The focus of the first group of mortar samples was to investigate possible changes in the lime binder due to climatic conditions. The samples were collected from a selection of buildings of similar age, but located in three distinct climate regions across the United States. The climate regions and their associated buildings were as follows.

- Arid Desert Cold (Bwk): mean annual precipitation 22.4 cm, mean annual temperature 17.3 °C. Three buildings (*c*. 1900) constructed of local sandstone; Fort Bliss US Army Air Defense Artillery Center in El Paso, Texas, USA.
- Temperate Wet Hot Summer (Cfa): mean annual precipitation 166.4 cm, mean annual temperature 18.8 °C. Mortar collected from brick foundation pier and a brick chimney hidden in the attic space; Phillips House (*c*. 1840) in Bay St Louis, Mississippi, USA.
- Cold Wet Hot Summer (Dfa): mean annual precipitation 98.7 cm, mean annual temperature 13 °C. Two brick houses (*c*. 1840 and *c*. 1900) in Staunton, Virginia, USA, and the Grandma Moses House (brick, *c*. 1840) in Verona, Virginia, USA.

Group B: temporal

The second set of mortar samples were intended to explore climate-related temporal changes in lime mortars. The samples were collected in climates similar to the initial sample set, but from buildings of significantly older ages. Note that the following climate classifications are based upon the current regime, and do not reference past climatic changes, such as the Little Ice Age. These samples include the following.

- Temperate Wet Warm Summer (Cfb), Sixteenth–Seventeenth century: modern mean annual precipitation 110 cm, modern mean annual temperature 9.5 °C. Bonamargy Friary (built *c*. 1500, second phase *c*. 1621; documented

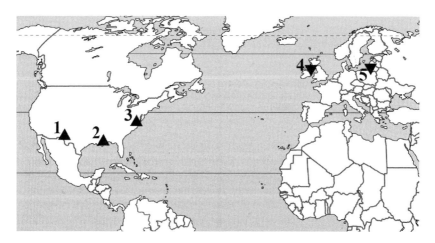

Fig. 1. Map of sample locations. Group A: (1) Ft Bliss, Texas, USA; (2) Bay St Louis, Mississippi, USA; (3) central Virginia, USA. Group B: (4) Ballycastle, Northern Ireland, UK; (5) Cesis, Latvia.

fire in 1584) constructed of local Fairhead sandstone near Ballycastle, Northern Ireland, UK.
- Cold Wet Warm Summer (Dfb), Sixteenth century: modern mean annual precipitation 72.7 cm, modern mean annual temperature 4.5 °C. Cesis Castle (constructed of local stone, c. 1540) in Cesis, Latvia.

The samples were then prepared for the selected analytical techniques. Small pea-sized chunks of samples were mounted onto metal stubs and then gold sputter-coated for scanning electron microscopy with energy-dispersive spectra analysis (SEM-EDS). In addition, blocks of the samples were vacuum-impregnated with blue-dyed epoxy, and then thin sectioned for polarized light microscopy (PLM) and point count analysis. Point count analysis enabled calculation of the percentage of aggregate, binder and macroporosity contained in each sample. For the sake of analysis, aggregate included the carbonate grains, whether derived from the aggregate source or from incompletely burnt binder material; lime blebs were included in the binder fraction; and the macroporosity comprised oversized pores (those larger than typical aggregate size in the sample), cracks and curvilinear vesicles. It should be noted that the Bonamargy sample represents a mortar designed for use as render, whereas the other samples were bedding mortars. After PLM and point count analysis, the thin sections were carbon-coated for backscatter electron imaging (BSE).

Results

Group A: climate

Regarding the climate region samples, the analyses offered intriguing results (Table 1, Fig. 2). Point counts reveal the percentage of total macroporosity comprised by desiccation cracks increased with increasing aridity (Fig. 3). The Arid Desert Cold samples and Cold Wet Hot Summer samples average 63.7 and 62%, respectively, of total macroporosity derived from desiccation cracks, whereas the Temperate Wet Hot Summer samples average only 41.6% total macroporosity from desiccation cracks. It should be noted that the two Temperate Wet Hot Summer samples exhibit markedly different pore structures. The sample collected from the exterior foundation pier contained much fewer desiccation cracks, accounting for 21.6% of total macroporosity; whereas the sample collected from the attic chimney had 68.2% of total macroporosity ascribed to desiccation cracks. This difference is attributed to the location of the samples; one being exposed to natural environmental conditions, while the other was from an interior location

Table 1. *Results of point count analysis (in per cent)*

Climate	Sample	Aggregate						Binder					Pores			
		Quartz	Feldspar	VRF	Chert	MRF	Limestone	Total	Binder	Bleb	Calcite	Acicular cement	Total	Oversize	Desiccation	Vesicle	Total
Bwk	FB 273	24.6	6.3	6.6	3.7	0.0	0.0	41.2	39.1	8.0	0.0	0.0	47.1	4.6	7.1	0.0	11.7
	FB 2019	21.1	2.6	8.9	1.4	0.0	2.3	36.3	48.5	8.6	0.0	0.0	57.1	1.7	4.9	0.0	6.6
	FB 2022	34.0	5.7	8.3	3.1	0.9	0.3	52.3	37.7	2.3	0.0	0.0	40.0	3.1	4.6	0.0	7.7
Cfa	PH-pier	37.7	0.0	0.0	0.0	0.0	0.0	37.7	36.0	1.1	0.0	0.0	37.1	19.7	5.5	0.0	25.2
	PH-chimney	50.0	0.0	0.0	0.0	0.0	0.0	50.0	31.1	0.0	0.0	0.0	31.1	6.0	12.9	0.0	18.9
Dfa	ROB-1	46.0	0.9	1.1	0.3	0.9	0.0	49.2	34.6	1.1	0.0	0.0	35.7	5.4	9.7	0.0	15.1
	ROB-2	36.9	0.0	0.0	5.1	0.9	0.0	42.9	47.4	5.4	0.0	0.0	52.8	1.4	2.9	0.0	4.3
	ROB-3	45.4	1.7	0.0	0.0	0.0	0.0	47.1	49.7	0.0	0.0	0.0	49.7	1.7	1.5	0.0	3.2
Cfb	BF-lime	3.8	0.0	0.0	0.0	0.0	0.0	3.8	64.7	3.3	0.0	0.0	68.0	11.6	7.5	9.1	28.2
Dfb	Cesis-1	25.1	4.0	1.4	0.3	0.3	3.1	34.2	24.0	0.6	9.7	14.0	48.3	17.2	0.3	0.0	17.5
	Cesis-2	31.4	1.7	3.7	0.0	0.0	17.4	54.2	33.1	0.9	4.0	0.0	38.0	5.2	2.6	0.0	7.8

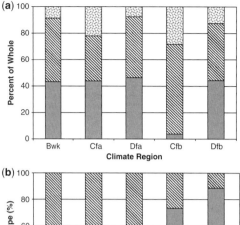

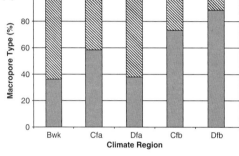

Fig. 2. Results from point count analysis, by climate region: (**a**) average percentage of aggregate (solid sections), binder (hashed sections) and macroporosity (spotted sections) in the mortar samples; and (**b**) percentage of macropores contributed by desiccation cracks (hashed sections) and oversized pores (solid sections). Note: for the Cfb sample, oversized pores include the curvilinear vesicles.

exposed to heat and pollutants derived from use of the fireplace.

Furthermore, petrographic analysis revealed the majority of aggregates consisted of fine–medium

Fig. 3. Black linear voids of desiccation cracks as seen in an Arid Desert climate sample (arrows; BSE).

sands of subrounded–subangular grains. The major constituent was quartz, with relatively minor amounts of feldspar, chert, volcanic rock fragments (VRF), metamorphic rock fragments (MRF) and limestone. These minor constituents were more common in the Arid Desert Cold samples. The Temperate Wet Hot Summer samples were the exception, with aggregate composed of fine-grained, well-rounded quartz sand. The results indicate that the aggregates found in the majority of the samples were largely sourced as sand from local stream deposits, whereas the Temperate Wet Hot Summer samples were locally sourced from beach sands (Table 1).

SEM images revealed differing binder crystal morphologies. The Arid Desert Cold samples' predominant morphology was small anhedral crystals, between 1 and 5 μm in size. The Temperate Wet Hot Summer and Cold Wet Hot Summer samples contained platy, interconnected crystals between 5 and 20 μm, which were typically observed along pore pathways. In one of the Staunton, Virginia samples, SEM analysis showed evidence of new calcite crystal growth in binder microcracks.

EDS results indicated the presence of environmental contaminants, in this case chlorides, can be related to the climate regime as well. Four of the nine samples contained chlorides, and two of those samples only contained trace amounts. The two samples from the Temperate Wet Hot Summer climate, near the Gulf of Mexico, contain a percentage of chloride in their binder that is nearly two orders of magnitude greater than in the other samples. EDS also highlighted further differences between the two Temperate Wet Hot Summer samples, in that the chimney sample contained significant quantities of sulphur, which comprised approximately 20% of the binder's elemental constituents.

Group B: temporal

Examination by SEM and PLM revealed changes in crystal morphology over time. Larger, more platy (10–20 μm) binder crystals, increasing in size with increasing proximity to macropores, were much more prevalent in the Sixteenth and Seventeenth century samples than in their younger climatic counterparts (Fig. 4). In addition, the Cesis Castle samples contained isopachous prismatic binder crystals measuring approximately 10 μm in length.

Point count analysis showed that the older samples contained an average of 12.6% macropore space by volume, on par with the younger samples that ranged from 3.1 to 25.1% macropore space by volume. However, the pore structures were markedly different. The younger samples from the Arid

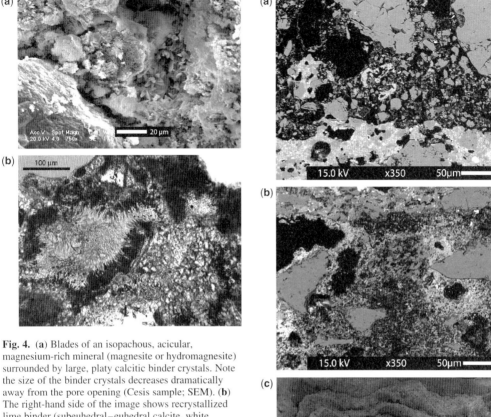

Fig. 4. (**a**) Blades of an isopachous, acicular, magnesium-rich mineral (magnesite or hydromagnesite) surrounded by large, platy calcitic binder crystals. Note the size of the binder crystals decreases dramatically away from the pore opening (Cesis sample; SEM). (**b**) The right-hand side of the image shows recrystallized lime binder (subeuhedral–euhedral calcite, white arrow), while the left-hand side of the image contains an oversized pore filled with generations of secondary cements (black arrow): an isopachous, acicular, magnesium-rich mineral (magnesite or hydromagnesite) followed by anhedral calcite (Cesis sample; PLM, plane light).

Desert, Temperate Wet and Cold Wet climates contained a greater percentage of desiccation cracks (63.7, 41.6 and 62.0% of total macropore space, respectively), whereas the macroporosity of the older Temperate Wet and Cold Wet climate samples was dominated by oversized pores (98.4 and 88.6% of total macropore space, respectively).

PLM and EDS analysis also revealed the growth of new mineral species within the mortars as age progresses, leading to a series of secondary pore-filling cements such as euhedral calcite or isopachous acicular magnesite or hydromagnesite. In addition, BSE analysis showed light-grey aureoles around black voids in the binder, indicating calcium-rich mineral enrichment along pore networks (Fig. 5). Finally, PLM analysis results revealed the presence of distinctively shaped curvilinear, semi-cylindrical vesicles within the binder of the Bonamargy Friary sample.

Fig. 5. (**a**) Pores (white arrows) exhibiting little to no increased cementation at their periphery (Virginia sample, *c.* 1900; BSE). (**b**) Entrenched pores (white arrows) with increased cementation (black arrows), believed to cause liesegang band-like weathering patterns (Cesis sample, *c.* 1540; BSE). (**c**) Liesegang band-like weathering of historic lime mortar (medieval wall in Mdina, Malta).

Discussion

Climate-related changes

Moisture and temperature play pivotal roles in the environmental effect on lime binders. Lime binders cure through carbonation, which is facilitated by exposure to moisture. Lack of moisture, due to climatic or enclosed conditions, encourages drying-out of the binder and formation of desiccation, or shrinkage, cracks in the lime binder (Leslie & Hughes 2002). Shrinkage cracks can form as a result of inadequate water supply in mixing the mortar, drying out during curing or inappropriate aggregate grading. However, the consistent presence and shear abundance of desiccation cracks in lime mortar from structures in dryer Arid Desert and Temperate Warm Wet Summer climates indicates that there exists a strong correlation between prevailing climate conditions and the presence and extent of desiccation crack porosity.

Changes to the pore network, such as formation of desiccation cracks, impacts the movement of salts through mortar and substrates. Research has shown that when pore size is greater in plasters than in the adjoining substrate, salts preferentially move into the plaster from the substrate. As drying action is enhanced by an increase in external temperature or decreasing relative humidity, such conditions will encourage movement of salts into the plaster and, possibly, to its exposed surface. Conversely, if pore size is greater in the substrate, salts and moisture become trapped at the plaster–substrate interface or within the substrate itself (Lubelli et al. 2006; Petković et al. 2007). Using this as an analogue for salt and moisture transport in mortars and their substrates, this appears to imply that the larger pore spaces formed by desiccation cracks would facilitate moisture migrating out of the system and that any salts present in the substrate would move into the mortar. The key environmental parameter in the Arid Desert region is the dearth of atmospheric moisture, although the high annual temperature experienced would increase the speed of any moisture and salt migration, ironically feeding back into the lack of moisture problem causing the desiccation cracks.

With regard to environmental pollutants, moisture percolating through the binder may deposit various contaminants. This can impact on the mineralogy of the binder by encouraging growth of destructive minerals, such as calcium or sodium chloride. The samples in the Temperate Hot Wet Summer region, a climate classification resulting in part from the effects of proximity to the warm Gulf of Mexico, exhibit marked increases in the amount of chloride present. Yet, the mortars collected from buildings in climate areas not dominated by the influence of large nearby water bodies contained only trace to no levels of chlorides. While it is relatively common knowledge that sea spray and aerosols supply soluble salts to nearby materials, it is important to consider the impact of the soluble salts upon materials in historic structures in relation to the prevailing climatic conditions.

The presence of moisture also encourages a unique characteristic of lime mortars: autogenous healing. As water percolates through the binder dissolving calcite, the water becomes saturated and eventually reprecipitates calcite and other minerals, which in turn can encourage healing through crystal formation in microcracks. If enough atmospheric moisture is present, this self-healing mechanism allows buildings to absorb and adjust to stress caused by settlement, or even experience traumatic events such as mild earthquakes, without sacrificing long-term stability (Anderegg 1942). Evidence of autogenous healing was not observed in the Arid Desert samples, implying that the low levels of atmospheric moisture do not facilitate such self-healing processes. It also appears this mechanism is not sufficient to counteract the formation of desiccation cracks in regions of moderate rainfall, as evinced by the Cold Wet Hot Summer samples that contain both a large percentage of desiccation cracks and indications of autogenous healing as seen in SEM images.

Temporal-related changes

Moisture and fire both play key roles in the temporal changes observed in the mortar samples. Over time, percolating pore waters deposit minerals in the interstices of the medium through which it passes. This leads to localized enrichment along pore paths, shown in the BSE image as light-grey aureoles around black voids in the binder. Such localized enrichment can lead to concentric zoning, or Liesegang banding, of more highly cemented areas alternating with less cemented areas within lime-based mortar and render over time, which in turn results in a distinctive weathering pattern (Fig. 5). Rodriguez-Navarro et al. (2002) suggest that the formation of Liesegang bands in lime mortars occur when the binder is an aged lime putty, as it reportedly carbonates faster than other types of lime binders. Thus, a Liesegang band weathering pattern may indicate that aged lime putty was the binder medium of choice for the mortar mix in question.

Age and available moisture also impact on the observed shift in binder crystal morphologies. With time, and under conducive climatic conditions, the binder crystals grow and recrystallize to form a distinct evolution of crystal morphologies from anhedral to platy to prismatic shapes observed in the late nineteenth–early twentieth century, mid-nineteenth century and sixteenth–seventeenth century mortars, respectively. The smaller, anhedral crystals dominate the late nineteenth–early twentieth century Arid Desert Cold samples, while larger platy crystals are observed in six of the seven mortars from the Temperate Wet and Cold Wet regions (the exception being the chimney sample). A preponderance of large platy crystals, combined with prismatic calcite, was noted in the sixteenth and seventeenth century samples. This evolution of crystal morphologies, spanning younger to older mortars in different climate regions, indicates that the natural progression in crystal growth of ageing lime mortars is impacted by climatic influences as well.

An impact of fire upon historic lime binders is the increase of porosity due to removal of fillers. Organic fillers, such as animal hair or straw, were common additives in historic lime mortars. The filler imparted resistance against shrinkage cracking while reducing permeability of the mortar or render. When subjected to fire, an increasingly likely event as a building ages, these organic fillers exceeded their ashing point temperature and were therefore reduced to ash. The ash was then washed out of the binder, either through natural weathering processes or during sample preparation, leaving behind the distinctively curvilinear, semi-cylindrical vesicles observed in the Bonamargy sample. This process effectively increases the porosity and permeability, and alters the pore size distribution, resulting in marked changes to the dynamic equilibrium of moisture ingress and egress. As mentioned previously, the pore sizes of mortar and substrate influence the salt and moisture transport regimes. With the increase in pore size attributable to the destruction of organic fillers by fire, moisture and salts most probably move preferentially into or through the mortar from the substrate. This may trigger the blossoming of salts upon mortars previously free from visible efflorescence, or increase efflorescence where already an issue. Furthermore, the fire-related porosity changes may also impact on the flow regime of pore waters and therefore the reprecipitation of minerals, perhaps leading to an abundance of oversized pores and larger, more distinct crystal growth habits of pore-lining cements, such as observed in the Cesis Castle samples.

Conclusions

From this research several preliminary conclusions can be drawn. Climate possibly influences the evolution of lime mortars by increasing the quantity and extent of desiccation cracks in more arid environments; altering the porosity distribution; and sourcing contaminants that alter mineralogy. Climate also appears to impact temporal variations by affecting binder crystal morphology; entrenching pore networks; changing porosity distribution and permeability; and facilitating or impeding growth of secondary cements. Understanding the potential impact of climate upon the evolution of lime mortars, and their resulting weathering patterns, will further facilitate development of compatible repair mortars for conservation projects, thus encouraging improved preservation of our built heritage. Ongoing research seeks to better explain these research results, and to further explore the roll of local climate on lime-based mortar morphology and mineralogy.

Analyses were carried out at the University of Texas at Austin in the Jackson School of Geosciences under the guidance of Dr K. Milliken, with the exception of the Bonamargy Friary sample, which was analysed at Queen's University Belfast. Funding was provided in part by the Jackson School of Geosciences and a Queen's University Scholarship. Thanks to G. Thompson (UT-Austin) and Pat McBride (QUB) for preparing thin sections. The author also wishes to thank A. Lapin, R. O. Byrne, Dr S. McCabe and H. Gardea for providing mortar samples from faraway locations.

References

ANDEREGG, F. O. 1942. Autogeneous healing in mortars containing lime. *ASTM Bulletin*, **116**, 22.

BARONIO, G., BINDA, L. & SAISI, A. 1999. Mechanical and physical behaviour of lime mortars reproduced after the characterisation of historic mortar. *In*: BARTOS, P., GROOT, C. & HUGHES, J. J. (eds) *International RILEM Workshop on Historic Mortars: Characteristics and Tests, University of Paisley, Scotland, 12–14 May*. RILEM Publications S.A.R.L., 307–326.

BLOEM, D. L. 1962. *Effects of Aggregate Grading on Properties of Masonry Mortar*. NSGA Circular, **89**. American Society for Testing and Materials.

CALLEBAUT, K., VIAENE, W., VAN BALEN, K. & OTTENBURGS, R. 1999. Petrographical, mineralogical, and chemical characterisation of lime mortars in the Saint-Michel's Church (Leuven, Belgium). *In*: BARTOS, P., GROOT, C. & HUGHES, J. J. (eds) *International RILEM Workshop on Historic Mortars: Characteristics and Tests, University of Paisley, Scotland, 12–14 May*. RILEM Publications S.A.R.L., 113–124.

FOLK, R. L. & VALASTRO, S. 1976. Successful technique for dating of lime mortar by Carbon-14. *Journal of Field Archaeology*, **3**, 203–208.

LANAS, J. & ALVAREZ, J. I. 2003. Masonry repair lime-based mortars: factors affecting the mechanical behavior. *Cement and Concrete Research*, **32**, 1867–1876.

LESLIE, A. B. & GIBBONS, P. 1999. Mortar analysis and repair specification in the conservation of Scottish historic buildings. *In*: BARTOS, P., GROOT, C. & HUGHES, J. J. (eds) *International RILEM Workshop on Historic Mortars: Characteristics and Tests, University of Paisley, Scotland, 12–14 May*. RILEM Publishing S.A.R.L., 273–280.

LESLIE, A. B. & HUGHES, J. J. 2002. Binder microstructure in lime mortars: implications for the interpretation of analysis results. *Quarterly Journal of Engineering Geology and Hydrogeology*, **35**, 257–263.

LINDROOS, A., HEINEMEIER, J., RINGBOM, Å., BRASKÉN, M. & SVEINBJÖRNSDÓTTIR, Á. 2007. Mortar dating using AMS 14C and sequential dissolution: Examples from medieval, non-hydraulic lime mortars from the Åland Islands, SW Finland. *Radiocarbon*, **49**, 47–67.

LUBELLI, B., HEES, R. P. J. v. & GROOT, C. J. W. P. 2006. Sodium chloride crystallization in a 'salt transporting' restoration plaster. *Cement and Concrete Research*, **36**, 1467–1474.

NAWROCKA, D., MICHNIEWICZ, J., PAWLYTA, J. & PAZDUR, A. 2005. Application of radiocarbon method for dating of lime mortars. *Geochronometria*, **24**, 109–115.

PEEL, M. C., FINLAYSON, B. L. & MCMAHON, T. A. 2007. Updated world map of the Köppen–Geiger climate classification. *Hydrology and Earth Systems Sciences*, **11**, 1633–1644.

PETKOVIĆ, J., HUININK, H. P., PEL, L., KOPINGA, K. & HEES, R. P. J. v. 2007. Salt transport in plaster/substrate layers. *Materials and Structures*, **40**, 475–490.

PRICE, J. 2005. *Discussion of mortar analysis techniques, PLM observations, lime mortar technology, and conservation mortar mix designs*. September 2005. Virginia Lime Works, Madison, Heights, VA.

RODRIGUEZ-NAVARRO, C., CAZALLA, O., ELERT, K. & SEBASTIAN, E. 2002. Leisegang pattern development in carbonating traditional lime mortars. *Journal of the Royal Society of London*, **458**, 2261–2273.

VAN HEES, R. P. J. 1999. Damage diagnosis and compatible repair mortars. *In*: BARTOS, P., GROOT, C. & HUGHES, J. J. (eds) *International RILEM Workshop on Historic Mortars: Characteristics and Tests, University of Paisley, Scotland, 12–14 May*. RILEM Publications S.A.R.L., 27–36.

VAN HEES, R. P. J., BINDA, L., PAPAYIANNI, I. & TOUMBAKARI, E. 2004. Characterisation and damage analysis of old mortars. *Materials and Structures*, **37**, 644–648.

WILCOCK, A. A. 1968. Köppen after fifty years. *Annals of the Association of American Geographers*, **58**, 12–28.

Crushed limestone as an aggregate in concrete production: the Cyprus case

I. IOANNOU[1]*, M. F. PETROU[1], R. FOURNARI[1], A. ANDREOU[1,3], C. HADJIGEORGIOU[2], B. TSIKOURAS[3] & K. HATZIPANAGIOTOU[3]

[1]*University of Cyprus, Department of Civil and Environmental Engineering, 75 Kallipoleos Avenue, P.O. Box 20537,1678 Nicosia, Cyprus*

[2]*Ministry of Agriculture, Natural Resources and Environment, Geological Survey Department, 1415 Nicosia, Cyprus*

[3]*University of Patras, Department of Geology, 26500 Patras, Greece*

*Corresponding author (e-mail: ioannis@ucy.ac.cy)

Abstract: Limestones in Cyprus are mainly quarried for the production of coarse and fine aggregates to be used in concrete. The objective of this paper is to examine the properties of crushed limestone aggregates. The petrographical and physico-mechanical properties of these aggregates are described and their suitability for concrete production is examined. The coarse crushed limestone aggregates from Cyprus have water absorption values exceeding 3.3%, which is considered high for concrete applications. Their abrasion resistance (Los Angeles) values are consistently above 23%, while their weathering coefficients generally range between 10 and 30%. The fine crushed limestone aggregates show significantly lower water absorption values (less than 2.2%) and higher weathering coefficients (above 35%) than the coarse aggregates. The weathering coefficient of crushed limestone aggregates increases with a decrease in the fraction size up to 5 mm, after which it remains fairly constant. The physico-mechanical properties of crushed limestone aggregates are distinctly variable irrespective of the fact that they belong to the same geological formation and show relatively similar petrography.

Limestones are sedimentary rocks that may be formed either inorganically or by biochemical processes. Their primary component is calcium carbonate ($CaCO_3$) in the form of calcite or aragonite. Limestones may also contain variable amounts of magnesium carbonate, silica, clay, silt, iron oxides and hydroxides, organic materials, and sand. These impurities determine the colour of the limestone.

In Cyprus, limestones are mainly quarried for the production of coarse and fine aggregates to be used in concrete. Crushed limestone aggregates correspond to approximately 27–29% of the total production of aggregates on the island. The physico-mechanical and index properties of these aggregates influence considerably the quality of concrete, since they are one of its essential components. For this reason it is important to investigate the properties of aggregates before adding them to concrete mixtures.

According to the European Standards (EN 12620) (European Committee for Standardization 2002), in order to assess the suitability of aggregates for concrete production it is necessary to perform a series of laboratory tests, including the determination of their water absorption, their geometrical properties and their resistance to fragmentation. It is also important to assess experimentally their weathering properties.

The objective of this paper is to examine the properties of crushed limestone aggregates quarried in Cyprus. The petrographical and physico-mechanical properties of these aggregates are described and their suitability for concrete production is examined.

Geological setting of Cyprus

Cyprus is one of the larger Mediterranean islands, with a surface area of 9251 km². It lies in the NE corner of the Mediterranean Sea in a complex tectonic active zone between the African lithospheric plate to the south and the Eurasian lithospheric plate to the north (Ben-Avraham *et al.* 1988).

Cyprus comprises four main geological units: (1) the Late Triassic–Late Cretaceous Mamonia metamorphic terrane to the south and SW; (2) the central, uplifted, Cretaceous Troodos ophiolite; (3) the Permian–Eocene, strongly deformed Kyrenia sedimentary terrane that includes the Pentadaktylos

Range to the north; and (4) the Circum-Troodos sedimentary terrane at the peripheral areas of the Troodos ophiolite (Ben-Avraham et al. 1988; Robertson et al. 1991; Robertson & Xenophontos 1993; Konstantinou et al. 1997; Van der Meer et al. 1997; de Coster et al. 2004).

The last terrane (Fig. 1) consists, from bottom to top, of the Kannaviou Formation that includes volcanistic sandstones with bentonitic clays, followed by the Moni and Kathikas formations, which contain a mélange of Triassic–Cretaceous fragments of sandstones, siltstones, serpentinites and debris flows derived from the Mamonia Terrane and the Troodos ophiolite. Carbonate sedimentation started with Palaeocene marl and chalk of the Lefkara Formation, succeeded by the Miocene Pakhna reefal formation and Upper Miocene evaporates of the Kalavasos Formation. The Pakhna Formation is subdivided in a lower Terra Member and an upper Koronia Member (Robertson et al. 1991). The present study focuses on samples coming from outcrops in the last member, which reflects the second phase of growth of reefal, creamy–off-whitish limestones (recrystallized bioclastic, bioherms and biostromes) in the Pakhna Formation; these rocks are comparable with similar fringing reefs of monospecific, poritid corals commonly developed around the Mediterranean Basin in the Tortonian–Early Messinian (Follows & Robertson 1990). The stratigraphic section of the Circum-Troodos sedimentary terrane is completed by detrital, calcareous and fanglomerate deposits of Pliocene (Nicosia Formation) to Quaternary.

Materials and methodology

Nine different crushed limestone aggregate samples from the regions of Xylofagou, Mitsero and Androlikou were used in this study. All three regions belong to the Pakhna Formation of the Circum Troodos Sedimentary Terrane (Fig. 1). Six of the samples were coarse-grained aggregates (8–20 mm), while the other three were fine grained (0.3–5 mm).

The petrography of the bedrock was studied in each case using optical microscopy. Polished thin sections were prepared first by thoroughly drying the samples and then by vacuum impregnating them with epoxy resin clear. The samples were then cut with a slow-speed diamond saw to minimize damage, mounted on glass slides, ground to 30 μm thickness and polished. The polished thin sections were examined using an optical microscope (Leica DM 2500P), with reflected and transmitted light, and a camera phototube.

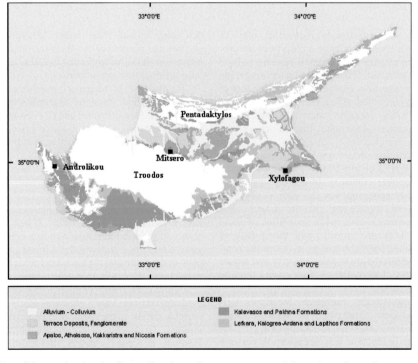

Fig. 1. Map of Cyprus showing the Circum Troodos sedimentary terrane and the regions of sampling.

The physico-mechanical (water absorption, resistance to fragmentation) and geometrical (particle shape) properties of the samples were examined according to the procedures described in the European Standards (EN 933-3:1997, 1097-2:1998, 1097-6:2000). The weathering properties of the aggregate samples were also examined using the European Standard for the soundness of aggregates (EN 1367-2:1998).

For the determination of the water absorption, the pyknometer method was used (EN 1097-6:2000). The samples of coarse aggregates were initially washed through 31.5 and 4 mm test sieves in order to remove very-coarse and very-fine particles, respectively. They were then placed in a pyknometer with water at 22 ± 3 °C and immersed in a water bath at the same temperature for 24 ± 0.5 h. At the end of the soaking period the pyknometer was overfilled by adding water and its weight was recorded (M_2). The aggregates were then removed from the pyknometer and allowed to drain on a dry cloth until all visible films of water were removed. The weight of saturated and surface dried aggregates was recorded (M_1), and the aggregates were allowed to dry in an oven at a temperature of 110 ± 5 °C until the samples had reached a constant mass (M_4). The water absorption, WA (%), of the samples was calculated in accordance to equation (1):

$$\text{WA} = \frac{(M_1 - M_4)}{M_4} \times 100. \quad (1)$$

For the fine aggregates, the same procedure was followed. However, in order to drain the aggregates following the soaking period, the samples were spread in a uniform layer and were exposed to a gentle current of air until the surface moisture evaporated. A metal cone mould was used to determine whether the surface dried stage had been achieved. This was overfilled with the test portion and a tamper was used to tap the surface through the hole at the top of the mould 25 times. The mould was then lifted and the extent of collapse of the aggregate cone was noted. If the aggregate cone collapsed totally, then the surface dried stage had been achieved.

The geometrical properties (particle shape) of coarse aggregates were determined in accordance with EN 933-3:1997, which describes a method for measuring the flakiness index of aggregates. Samples of mass 2.5 kg were gathered from the bulk and dried in an oven at 110 ± 5 °C for 24 ± 0.5 h. The samples were then allowed to cool down to room temperature before being sieved through a series of specific sieves with square openings. The masses (M_1) of each fraction were recorded and the parts that were generated were sieved separately using a proper striped mesh. The material that passed through the sieves was measured each time and the masses (M_2) of the particles were recorded. The flakiness index, FI (%), of the aggregates was calculated using equation (2):

$$\text{FI} = \frac{M_2}{M_1} \times 100. \quad (2)$$

For the test for the determination of resistance to fragmentation of the aggregates (EN 1097-2:1998), a sample of about 15 kg was sieved through 10 and 14 mm sieves. From this sample a percentage of 60–70% passed through the 12.5 mm sieve, while a percentage of 30–40% passed through the 11.2 mm sieve. The test portion was then washed and dried in an oven at a temperature of 110 ± 5 °C. After cooling down to room temperature, the aggregates were mixed and a sample of mass 5 kg (± 5 g) was collected. The sample was introduced to the drum of a Los Angeles abrasion machine, together with 11 steel balls of approximate diameter 50 mm, and the machine was rotated for 500 revolutions at a steady speed. At the end of this procedure the sample was washed through a 1.6 mm sieve and the remaining portion was dried to constant mass in an oven at 110 ± 5 °C. The Los Angeles abrasion value, LA (%), of the aggregates was calculated from equation (3):

$$\text{LA} = \frac{(5000 - M)}{5000} \times 100 \quad (3)$$

where M is the mass (in grams) retained on the 1.6 mm sieve at the end of the procedure.

The weathering properties of the aggregates were determined in accordance with EN 1367-2:1998 using magnesium sulphate ($MgSO_4$) salt. For this test two samples with precise original mass (M_1) were sieved through specific sieve sizes (depending on the tested fraction) before being immersed in a $MgSO_4$ solution of density 1.284–1.300 g ml^{-1} for 17 ± 0.5 h. Following the immersion in the salt solution, the samples were dried in an oven at 110 ± 5 °C for 24 ± 1 h. They were then allowed to cool down and the procedure was repeated four more times. At the end of the fifth cycle the samples were washed to remove excess salt and then they were dried in an oven for 24 ± 1 h. After drying they were sieved through the original sieves and the mass (M_2) retained on these was recorded. The weathering coefficient, MS (%), was calculated using equation (4):

$$\text{MS} = \frac{M_1 - M_2}{M_1} \times 100 \quad (4)$$

Results and discussion

Petrography

The limestone that is used for the production of fine and coarse aggregates in the regions of Mitsero and Xylophagou comes from the reef limestone of the Koronia Member. Microscopic investigation suggests that the limestone from Mitsero contains mainly micritic calcite, which in restricted areas shows transitions to microspar. Fossils are abundant, with their skeletal fragments mostly composed of recrystallized, sparry calcite; infrequent siliceous fossils are also observed. Frequently, there are micritic endoclasts, while sparse non-carbonate grains include quartz, cherts and spinel dispersed in the matrix. Traces of anhydrite were also observed filling micropores. Significant primary intraparticle porosity occurs occassionally within partially unfilled fossils or between endoclasts. Elongated pores rimmed with restricted, recrystallized spar crystals have been formed in places after enlargement of boring porosity and local fenestral fabrics due to dissolution of calcite (Fig. 2). According to the classification scheme of Dunham (1962) it comprises a wackestone, while according to the classification of Folk (1962) it is a sparse biomicrite.

Locally, this rock type contains reddish areas with abundant carbonate and non-carbonate allochems. It contains micrite, which in several places is recrystallized to microsparry calcite. Sparsely, recrystallization resulted in the formation of columnar calcite crystals. The allochems include both intraclasts and extraclasts, fossils and fragments of chert, ophicalcite and grains of quartz, K-feldspar, chlorite, opaque minerals, and rare clay minerals. Locally, opaque minerals show oxidation with the formation of Fe oxides and hydroxides, which impregnate the micritic matrix. The fossil skeletal fragments are mainly of sparry calcite. However, micritic shells also occur. Siliceous fossils (radiolaria) are observed as well. Rare anhydrite fills micropores. The existing porosity is similar to that described earlier, with primary intraparticle pores owing to partially unfilled fossils or voids between the allochems. Solution-enhanced boring porosity is frequent. Typical microsparry calcite grains grow along the rims of pores of both the intraparticle, between the allochems (excluding the unfilled fossils), and the boring types, hence slightly reducing the primary porosity via cementation. According to the classification scheme of Dunham (1962) it comprises a packstone, while according to the classification of Folk (1962) it is a packed biomicrite.

The limestone from the region of Xylofagou is composed mostly of microspar and, locally, pseudospar. Micrite remnant areas are rare. It also contains numerous allochem grains; fossils (Fig. 3) are dominant, and usually angular biosparitic endoclasts occur throughout the sample. The fossil fragments are commonly micritic, but some of them have been recrystallized to sparry calcite. Rare glauconite, chlorite and clay minerals occur in the matrix. The investigated samples present fabric-selective boring porosity, which locally has been enhanced by dissolution. Micropores, a few microns in size, among the microspar crystals also create an intercrystal type of porosity in this sample. On the classification scheme of Dunham (1962) this limestone comprises a grainstone, while in the Folk's (1962) classification it is classed as an unsorted biosparite.

The limestone from Androlikou comes from the reef limestone of the Terra Member (Lower Miocene) of the Pakhna Formation. The Terra Member represents the first phase of reef growth on Cyprus, which occurred in the lower part of the Pakhna Formation. These reefs were formed in the late Aquitanian–early Burdigalian time on isolated stable basement highs (Banner et al. 1999). They were cemented by Mg-calcite and botryoidal aragonite shortly after their growth. Uplift and extensional faulting caused fracturing of the brittle reefs, and fissures were locally enlarged by karst-forming solutions. The fissures were eventually filled with fine-grained carbonate sediments, which were prone to dolomitization. The limestone from the Terra Member is generally hard, massive, relatively porous and has a creamy–off-white colour.

The study of the thin sections suggests that the limestone from the Androlikou area contains mainly carbonate and limited non-carbonate allochems. The allochems include intraclasts, microfossils (Fig. 4), grains of chert, quartz, feldspar, opaque minerals and rare clay minerals. Locally,

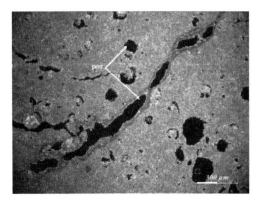

Fig. 2. Photomicrograph of a limestone test specimen showing the solution-enlarged boring porosity, as well as local fenestral fabrics with elongated partially spar-filled pores (por: porosity). Crossed nicols.

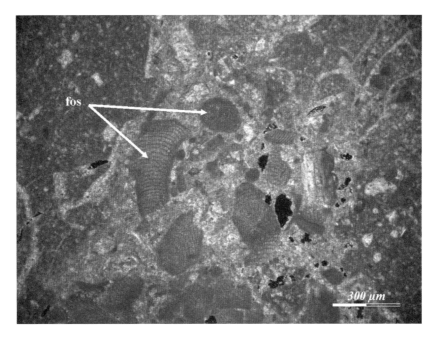

Fig. 3. Photomicrograph of a limestone test specimen showing the fossils and angular biosparitic endoclasts (fos: fossil). Parallel nicols.

Fig. 4. Photomicrograph of a limestone test specimen showing fossils (fos: fossil) and intraparticle porosity (por). Crossed nicols.

opaque minerals show oxidation, with the formation of Fe oxides and hydroxides that impregnate the micritic matrix. The fossil skeletal fragments, such as mollusks, gastropods and corals, are mainly recrystallized to sparry calcite. Micritic shells, however, also occur.

Limited primary intraparticle porosity occurs in this limestone owing to partially unfilled fossils or voids (Fig. 4) between the allochems. In places solution-enlarged boring porosity, caused by karstic processes, as well as local fenestral fabrics with elongated partially spar-filled pores are present. Typical microsparry calcite grains grow along the rims of pores of both intraparticle and boring type porosity, thus slightly reducing via cementation the primary porosity. On the classification scheme of Dunham (1962) this limestone comprises a wacketstone, while in the Folk's (1962) classification it is classed as an intrabiomicrite.

Physico-mechanical and weathering properties

The results for the coarse aggregates are summarized in Table 1.

The water absorption values are markedly high and are consistently above 3.3% (except for Lst 3). Such values are rarely reported in the literature (Brandes & Robinson 2006) and are usually considered unsuitable for producing good-quality concrete. However, the lack of better-quality limestone aggregates in Cyprus makes their use unavoidable. Since these aggregates absorb a lot of water, it is important to monitor their stage of saturation in order to calculate correctly the exact amount of water required in concrete mix design. Aggregates with such high water absorption values may also be utilized in concrete production as a means of slow water release. This may lead to internal

Table 1. *Physico-mechanical and weathering properties of coarse aggregates*

Sample	Location	MS (%)		WA (%)	LA (%)	FI (%)
		10–14 mm	14–20 mm			
Lst 1	Androlikou	17	12	3.3	28	7
Lst 2	Androlikou	22	17	3.5	26	4
Lst 3	Androlikou	5	3	1.6	23	7
Lst 4	Mitsero	29	26	5.6	–	2
Lst 5	Mitsero	14	11	4.4	28	1
Lst 6	Mitsero	14	13	3.5	28	4

curing of concrete, which helps to develop a more durable and stronger material. Research investigation into this topic is currently being undertaken at the University of Cyprus by the authors.

The abrasion resistance (Los Angeles) values are consistently above 23%. This is also considered high for concrete production. The weathering coefficients for the coarse aggregates range between 11 and 29%, with the exception of Lst 3, which has a value of approximately 4%. The flakiness index is the only property with low values, varying between 1 and 7%. The low values of flakiness index are due to the softness of the aggregate material, which depends on the degree of cementing and the porosity of the raw material, as well as on the mechanical means used in the production of the aggregates. A significant portion of softer limestone, coming from the deeper levels in the quarry, is usually used in the production of aggregates in Cyprus.

From the results it is clearly evident that there is a correlation between the different measured properties. The water absorption of the aggregates is linearly related to the weathering coefficient (Fig. 5). This has been confirmed for two different fractions of the coarse limestone aggregates (10–14 and 14–20 mm). A similar correlation is reported in the literature by various researchers (Kazi & Al-Mansour 1980; Koukis *et al.* 2007) for different types and fractions of aggregates with considerably lower water absorption values. However, for higher water absorption values, a correlation with the weathering coefficient has only been reported once (Brandes & Robinson 2006). It is worth noting that the correlation is independent of the salt used in the weathering test. According to European Standard specifications (EN 1367-2:1998), magnesium sulphate is the required reagent used in the test, while sodium sulphate is also an option in the ASTM Standard specifications (ASTM C-88-05).

A linear relationship is also valid for the correlation of the abrasion resistance (Los Angeles) with the water absorption values (Fig. 6). Such a relationship has also been reported by Kazi & Al-Mansour (1980) for coarse plutonic rock aggregates with considerably lower water absorption values (between 0.9 and 1.5%).

The value of the weathering coefficient of the aggregates appears to fall with increasing flakiness index (Fig. 7). However, the correlation is not very strong. This is in line with the comment made previously regarding the generally low values of flakiness indices measured. Aggregates that are prone to weathering, and hence show high MS values, have low flakiness indices due to their modified edges, as opposed to durable aggregates that show low MS values and high flakiness

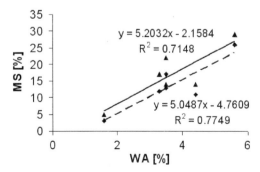

Fig. 5. Correlation between the weathering coefficient (MS) and water absorption (WA). (▲, 10–14 mm; ♦, 14–20 mm).

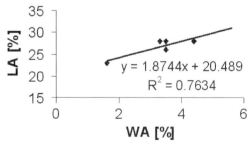

Fig. 6. Correlation between the Los Angeles abrasion resistance (LA) and water absorption (WA).

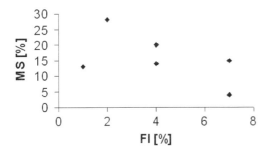

Fig. 7. Correlation between the weathering coefficient (MS) and flakiness index (FI).

indices. The abrasion resistance coefficient, on the other hand, seems to be independent of the flakiness index (Fig. 8), as depicted by the low value of the correlation coefficient, $R^2 = 0.24$.

The fine aggregates (Table 2) show significantly lower water absorption values and higher weathering coefficients than the coarse aggregates. Other researchers (Kazi & Al-Mansour 1980; Goswami 1984) have also reported lower water absorption values for the fine-grained aggregates. This result can be explained due to the lack of large pores in the smaller fractions. Such pores disappear during the fragmentation of the rock.

The increased weathering coefficient values of the fine aggregates, however, contradict the trend reported in the literature (Goswami 1984) and noted earlier for coarse aggregates. This trend shows that the water absorption and weathering coefficient values are linearly correlated. For the fine aggregates tested in this study, it is also evident (Fig. 9) that the weathering coefficient is almost constant and independent of the fraction size. A possible explanation lies in the fact that some of the fine particles consist of the cementing material, which is easily weathered during the testing procedure.

It is also interesting to note the trend of the weathering coefficient for a sample from the Mitsero area that consists of both fine and coarse aggregates. Figure 10 shows that the weathering coefficient increases with a decrease in the fraction size up to 5 mm, after which it remains fairly constant. This is possibly due to the repetitive fracture of the bedrock to produce the different aggregate fractions. Smaller fractions are the result of a larger number of fracture cycles, which results in a higher probability of cracked surfaces that are prone to weathering. Such an effect has implications on the testing procedure. The ASTM Standard method (ASTM C88-05) for testing the soundness of aggregates requires the use of a weighted average in the calculation of the percentage weight loss, whereas the European Standard (EN 1367-2:1998) requires individual fraction testing. It is also worth noting that the European Standard suggests the testing of the 10–14 mm fraction only. It is therefore obvious that results from the two tests cannot be compared easily. From the results (Table 3) it is also evident that the weathering coefficient, water absorption and Los Angeles abrasion resistance values of the different fractions of the tested sample are lower than the equivalent coefficients of other samples from the same quarry zone (Table 1). This is not unreasonable as the samples come from different quarry zones and depths. Samples from deeper levels in the quarry may be significantly softer, and, hence, they may show higher weathering coefficient, water absorption and abrasion resistance values.

Conclusions

Based on the experimental results, the following conclusions can be drawn.

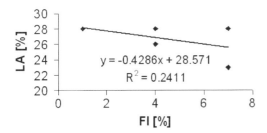

Fig. 8. Correlation between the Los Angeles abrasion resistance (LA) and flakiness index (FI).

Table 2. *Water absorption and weathering coefficients of fine aggregates*

Sample	Location	MS (%)					WA (%)
		5.00–3.35 (mm)	3.35–2.36 (mm)	2.36–1.18 (mm)	1.18–0.60 (mm)	0.60–0.30 (mm)	
Lst 7	Xylofagou	37	41	43	43	43	1.6
Lst 8	Xylofagou	80	81	72	76	72	1.0
Lst 9	Mitsero	35	40	37	40	40	2.2

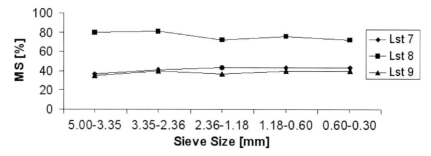

Fig. 9. Magnesium sulphate weight loss for fine limestone aggregate.

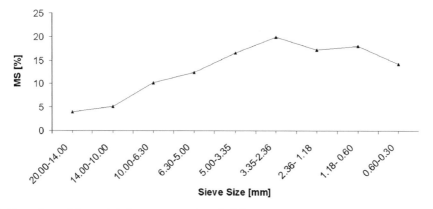

Fig. 10. Magnesium sulphate weight loss aggregates from the Mitsero area.

- The coarse crushed limestone aggregates from Cyprus have water absorption values exceeding 3.3%, which is considered high for concrete applications. Their abrasion resistance (Los Angeles) values are consistently above 23%, while their weathering coefficients generally range between 10 and 30%. The flakiness index is the only property with low values, which vary between 1 and 7%.
- The fine crushed limestone aggregates show significantly lower water absorption values (less than 2.2%) and higher weathering coefficients (above 35%) than the coarse aggregates. The low water absorption values can be explained by the lack of large pores in the smaller fractions. The high weathering coefficients, on the other hand, may be due to the fact that some fine particles consist of the cementing material, which is easily weathered during the testing procedure.
- The water absorption of the coarse aggregates is linearly related to the weathering coefficient and abrasion resistance values. While this relationship is not unusual, it has only rarely been reported for such high water absorption values. The weathering coefficient of the coarse aggregates appears to fall with increasing flakiness index. However, the correlation is not very strong. The abrasion resistance coefficient, on the other hand, seems to be independent of the flakiness index.

Table 3. *Weathering coefficients, water absorption and Los Angeles abrasion values for a crushed limestone aggregate sample from the Mitsero area*

Sieve size (mm)	MS (%)	WA (%)	LA (%)
20–14	4	2.8	22
14–10	5	2.8	20
10–6.3	10	2.6	–
6.3–5.0	12	2.6	–
5.0–3.35	17	1.0	–
3.35–2.36	20	1.0	–
2.36–1.18	17	1.0	–
1.18–0.60	18	1.0	–
0.60–0.30	14	1.0	–

- The weathering coefficient of crushed limestone aggregates increases with a decrease in the fraction size up to 5 mm, after which it remains fairly constant.
- The physico-mechanical properties of crushed limestone aggregates are distinctly variable, irrespective of the fact that they belong to the same geological formation and show relatively similar petrography.

The authors would like to acknowledge financial support from the Cyprus Research Promotion Foundation.

References

ASTM C88-05. 2005. *Standard Test Method for Soundness of Aggregates by Use of Sodium Sulfate or Magnesium Sulfate*. American Society for Testing and Minerals (ASTM) Committee on Standards (ASTM C88-05), Philadelphia, PA.

BANNER, F. T., LORD, A. R. & BOUDAGHER-FADEL, M. K. 1999. The Terra Limestone Member (Miocene) of western Cyprus. *Greifswalder Geowissenschaftliche Beitrage*, **6**, 503–515.

BEN-AVRAHAM, Z., KEMPLER, D. & GINZBURG, A. 1988. Plate convergence in the Cyprean arc. *Tectonophysics*, **146**, 231–240.

BRANDES, H. G. & ROBINSON, C. E. 2006. Correlation of aggregate test parameters to hot mix asphalt pavement performance in Hawaii. *Journal of Transportation Engineering*, **132**, 86–95.

DE COSTER, M., ZOMENI, Z., PANAYIDES, I., PETRIDES, G. & BERSKOY, O. 2004. *Seismic Hazard and Risk Assessment of the Greater Nicosia Area*. Nicosia, UNOPS (CD-ROM).

DUNHAM, R. J. 1962. Classification of carbonate rocks according to depositional structure. *In*: HAM, W. E. (ed.) *Classification of Carbonate Rocks*. AAPG Memoirs, **1**, 108–121.

EN 933-3:1997. *Test for Geometrical Properties of Aggregates. Determination of Particle Shape. Flakiness Index*. European Committee for Standardization, Brussels.

EN 1097-2:1988. *Tests for Mechanical and Physical Properties of Aggregates – Part 2: Methods for the Determination of Resistance to Fragmentation*. European Committee for Standardization, Brussels.

EN 1097-6:2000. *Tests for Mechanical and Physical Properties of Aggregates – Part 6: Determination of Particle Density and Water Absorption*. European Committee for Standardization, Brussels.

EN 1367-2:1998. *Tests for Thermal and Weathering Properties of Aggregates – Part 2: Magnesium Sulphate Test*. European Committee for Standardization, Brussels.

EN 12620:2002. *Aggregates for Concrete*. European Committee for Standardization, Brussels.

FOLK, R. L. 1962. Spectral subdivision of limestone types. *In*: HAM, W. E. (ed.) *Classification of Carbonate Rocks*. AAPG Memoirs, **1**, 62–84.

FOLLOWS, E. J. & ROBERTSON, A. H. F. 1990. Sedimentology and structural setting of Miocene reefal limestones in Cyprus. *In*: MALPAS, J., MOORES, E. M., PANAYIOTOU, A. & XENOPHONTOS, C. (eds) *Ophiolites: Oceanic Crustal Analogues*. Department of Geological Survey, Ministry of Agriculture, Nicosia, Cyprus, 207–216.

GOSWAMI, S. C. 1984. Influence of geological factors on soundness and abrasion resistance of road surface aggregates: a case study. *Bulletin of Engineering Geology and the Environment*, **30**, 59–61.

KAZI, A. & AL-MANSOUR, Z. R. 1980. Influence of geological factors on abrasion and soundness characteristics of aggregate. *Engineering Geology*, **15**, 195–203.

KONSTANTINOU, G., PANAYIDES, I., XENOPHONTOS, C., APHRODESIS, S., MICHAELIDES, P. & KRAMBIS, S. 1997. *Information Document No. 10: Geology of Cyprus*. Department of Geological Survey, Ministry of Agriculture, Natural Resources and Environment, Nicosia, Cyprus.

KOUKIS, G., SABATAKAKIS, N. & SPYROPOULOS, A. 2007. Resistance variation of low-quality aggregates. *Bulletin of Engineering Geology and the Environment*, **66**, 457–466.

ROBERTSON, A. H. F. & XENOPHONTOS, C. 1993. Development of concepts concerning the Troodos ophiolite and adjacent units in Cyprus. *In*: PRICHARD, H. M., ALABASTER, T. & HARRIS, T. (eds) *Magmatic Processes and Plate Tectonics*. Geological Society, London, Special Publications, **70**, 85–120.

ROBERTSON, A. H. F., EATON, S., FOLLOWS, E. J. & MCCALLUM, J. E. 1991. The role of local tectonics versus global sea-level change in the Neogene evolution of the Cyprus active margin. *In*: MACDONALD, D. I. M. (ed.) *Sedimentation, Tectonics and Eustasy – Sea Level Changes at Active Margins*. International Association of Sedimentologists, Special Publications, **12**, 331–369.

VAN DER MEER, F., VAZQUEZ-TORRES, M. & VAN DIJK, P. M. 1997. Spectral characterization of ophiolite lithologies in the Troodos Ophiolite complex of Cyprus and its potential in prospecting for massive sulphide deposits. *International Journal of Remote Sensing*, **18**, 1245–1257.

On the use of eggshell lime and tuffeau powder to formulate an appropriate mortar for restoration purposes

KÉVIN BECK[1], XAVIER BRUNETAUD[1], JEAN-DIDIER MERTZ[2] & MUZAHIM AL-MUKHTAR[1]*

[1]*CRMD, UMR 6619 CNRS, 1 bis rue de la Férollerie, 45071 Orléans, France*

[2]*LRMH, 29 rue de Paris, 77420 Champs sur Marne, France*

*Corresponding author (e-mail: muzahim@cnrs-orleans.fr)

Abstract: Preservation of cultural heritage, especially historic monuments, is a vital task. In addition, waste recovery is an essential goal of sustainable development. In this study the properties of a food waste, eggshells, and a quarry waste, powder obtained from stone-extraction operations, are evaluated in terms of their possible use in restoration mortar.

An excellent lime has been developed based on eggshell calcination. A lime-based mortar exclusively composed of this eggshell lime and tuffeau powder was developed and characterized in order to demonstrate its compatibility with restoration works. To pursue this objective, microstructural, physical and mechanical properties were investigated and compared to those of the tuffeau limestone.

The tested formulation using tuffeau powder was found to meet compatibility requirements up until the hygrometric environment lowered the water content of the mortar. The desiccation of the mortar can then generate shrinkage cracks leading to mortar crumbling. As a consequence, hygrometric susceptibility of the mortar is an essential parameter to be investigated in order to study the durability of lime-based mortars made out of limestone powder.

The mortars in stone constructions are essential constituents of ancient masonry. In several cases extensive damage to the ancient masonry as a result of the mortar used being incompatible with the old materials has been clearly established (Rodriguez-Navarro *et al.* 1998; Moropoulou *et al.* 2000). For this reason, it is important to ensure physico-chemical compatibility between the stone and the mortar. The compatibility of mortars with stones remains a prime factor in the process of restoration and preservation of monuments (Degryse *et al.* 2002; Mosquera *et al.* 2002).

A variety of mortars have been used in the construction of historical monuments and other structures. Lime-based mortars are more commonly used than other mortars (Belin *et al.* 1995). Numerous civilizations used aerial lime (Furlan & Bissegger 1975; Biscontin *et al.* 2002; Fassina *et al.* 2002) or hydraulic lime (Alvarez *et al.* 1995); and it is not unusual, even today, to find vestiges of Roman mortars still in good condition (Khanoussi & Maurin 2000). For economic reasons and ease of use in construction, the use of Portland cement as mortar rapidly exceeded the use of lime. During the second half of the twentieth century, as the restoration of a lot of historical monuments became necessary, the use of cement-based mortars increased, and the effect was devastating – the degradation of architectural monuments constructed with limestones in France is often related back to the use of mortars that were not compatible with the stones (Rautureau 2001; Beck 2006).

These numerous cases of unsuccessful restorations marked a major change in the methods of conservation. The need to use a mortar whose chemical composition is compatible with stone, especially for soft limestone, is now recognized (Beck 2006). Moreover, the manner in which restoration materials age must be appropriate with respect to the ancient materials, time and environmental exposure.

The objective of this study was to design an appropriate lime-based mortar for restoration purposes using a special lime and a special aggregate that both come from waste recovery. The special lime is from eggshells; the use of this common food waste would provide a high value-added opportunity. The special aggregate comes from the waste powder produced from sawing limestone in a quarry. The studied stone is tuffeau, which is a typical French stone used in construction of castles and houses in the Loire Valley region (Beck *et al.* 2003). First, the aerial lime obtained from calcination of eggshell must show similar properties compared with commercial aerial lime. Second, the resulting mortar must meet compatibility requirements with respect to tuffeau stone in terms of physical, chemical and mechanical properties.

Eggshell lime

Characterization of eggshells

Eggshells, which are known to be very rich in calcium carbonate, could provide a new source of aerial lime production. For example, the annual egg production in France reaches 16 billion (where 1 billion $= 1 \times 10^9$) for food purposes; about 1 billion are wasted as not meeting quality standards with respect to shell defects. The mass of a medium-size hen egg is about 60 g, of which the proportion of the shell is about 10 wt%. Hence, from annual production of hen eggs, more than 100 000 tonnes (t) of calcium carbonate in the form of eggshells is potentially usable for the production of lime. The transformation of this large amount of waste into value-added products like aerial lime has not yet been studied.

Eggshells were crushed and then analysed using X-ray diffractometry (XRD) and thermogravimetric analysis (TGA). X-ray diffractometry of powders ($\lambda_{Cu} = 1.5406$ Å) enables identification of the present crystalline structures and, subsequently, the qualitative mineralogical composition of the powder. The principle of TGA is to record the mass loss of a given sample under controlled temperature ramp under an argon atmosphere. The temperature range is 20–1000 °C and the heating rate 100 °C h^{-1}. The calcite proportion can be determined quantitatively by the mass loss around 800 °C resulting from calcium carbonate, $CaCO_3$, decomposition into calcium oxide, CaO. Hen eggshells produce very pure crystals of calcite arranged in an organic matrix resulting from the biological calcification. Indeed, the proportion of calcite reaches about 97% and the organic matter that vanishes at 300 °C does not exceed 3%.

Making of eggshell lime (ESL)

The making of lime produced from eggshells includes a stage of calcination and a stage of slaking (Fig. 1). Eggshells are coarsely crushed and quicklime is then obtained from the decarbonization of eggshells at 1000 °C.

When adding water to quicklime for slaking, a large amount of water is wasted in vapour form because of the exothermic reaction between water and calcium oxide ($\Delta H° = -65.5$ kJ mol^{-1}) which produces a lot of heat. As a consequence, it is necessary to add an excess of water to ensure that all of the quicklime is converted into slaked lime. Water and decarbonized eggshell are then mechanically mixed to ensure good homogeneity.

Fig. 1. Different steps in the making of eggshell lime.

This exothermic reaction pulverizes the resulting calcium hydroxide, so-called portlandite or aerial lime, into powder. Because of the added water, the resulting mix is a paste that must be dried at 105 °C to eliminate excess water. This aerial lime powder is then stored in a hermetic box to avoid any natural carbonation or humidification.

Characterization of eggshell lime (ESL) and comparison with commercial calcic lime (CL)

Tables 1 and 2 present the main physical characteristics of CL and ESL limes, and their chemical characterization by inductively coupled plasma-optical emission spectrometry (ICP-OES), respectively.

The density and the fineness of ESL and commercial calcic lime (class CL-90, according to European Standard EN 459-1: 2001) are very similar (Table 1).

Commercial lime is obtained by the calcination of calcareous stones, which always contain some impurities; however, the only mineral phase of eggshell is calcium carbonate. As a consequence, ESL is purer than CL with respect to portlandite proportion and the resulting colour is even whiter (Fig. 2). Indeed, lime is obtained from the decarbonization of eggshells at 1000 °C, and contains relatively pure calcium oxide with some minor traces of magnesium, sodium and phosphoric oxides that come from the organic matter (Table 2).

According to laser granulometry, the grain size ranges from 0.3 to 100 μm for both limes, and the majority of the particles are between 8 and 30 μm. The fineness and the specific surface of the CL are slightly higher. An interesting point is that even if

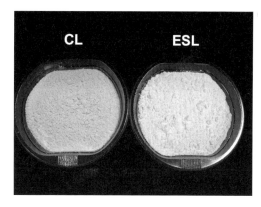

Fig. 2. Comparison of the visual aspects of CL and ESL.

the ESL was not mechanically ground, the resulting fineness was very similar to that of the CL, which was mechanically ground. This difference represents a significant saving in energy and, consequently, a significant economic advantage for the ESL. However, no energy balance of ESL can be estimated as it mostly depends on gathering and transport processes that have not yet been defined.

Given the average weight of a medium-size hen egg, the proportion of shell in a egg, and the yield of the calcination process that produces lime from calcium carbonate, which is 65%, the 17 billion wasted eggshells could theoretically be used to produce up to 78 000 tons of lime per year. This represents 3% of the annual production of lime in France, which was 2 480 000 t in 2001 (Ecole d'Avignon 2003).

Table 1. *Main physical characteristics of CL and ESL limes, and tuffeau powder*

	Commercial aerial lime (CL)	Eggshell lime (ESL)	Tuffeau powder
Portlandite content Ca(OH)$_2$ (TGA measurements)	92.2% (± 0.7%)	97.1% (± 0.5%)	–
Solid density (g cm^{-3})	2.22 (± 0.02)	2.26 (± 0.02)	2.57 (± 0.02)
Bulk dry density (g cm^{-3})	0.47 (± 0.05)	0.41 (± 0.05)	0.76 (± 0.05)
Blaine specific surface (cm^2 g^{-1})	11900 (± 500)	9550 (± 500)	3070 (± 500)

Table 2. *Chemical analysis of CL and ESL limes, and tuffeau powder*

	SiO$_2$ (%)	CaO (%)	Al$_2$O$_3$ (%)	Fe$_2$O$_3$ (%)	MgO (%)	Na$_2$O (%)	K$_2$O (%)	P$_2$O$_5$ (%)	TiO$_2$ (%)	LOI[*] (%)
ESL lime	0.00	72.33	0.00	0.00	0.84	0.16	0.07	0.23	0.00	24.22
CL lime	0.24	73.70	0.11	0.21	0.35	0.00	0.01	0.00	0.00	24.81
Tuffeau powder	41.41	29.75	2.14	0.87	0.50	0.05	0.64	0.00	0.16	24.00

[*]LOI, lost on ignition.

Tuffeau powder

The selected tuffeau limestone comes from the quarry at Usseau (Vienne, France). Aggregates used for mortar formulation are quarry waste because they come from powder obtained from stone-extraction operations. Tables 1 and 2 give the main physical characteristics of tuffeau powder and their chemical characterization by ICP-OES, respectively.

Grain-size distribution tests performed on this tuffeau powder show that the particle size is between 10 and 300 µm. The grains present in tuffeau powder are calcite ($CaCO_3$) and silica (SiO_2) in the form of opal cristobalite–tridymite and quartz. Other minerals are present in the stone as micas, clays and detritic minerals such as TiO_2.

Limestone mortar made out of tuffeau powder

Formulation and making of samples

As the physical and chemical characteristics of both CL and ESL limes are similar, only the results of mortars designed with ESL lime are presented.

In order to quantify the effect of the lime content, the binder/aggregate ratios prepared were 10, 15, 20, 25 and 30 wt%. Because the workability corresponding to the liquid limit (Beck & Al-Mukhtar 2008) was satisfactory, this limit was used to set the water content of the mortar. The resulting water/solid ratio varies according to the binder content, ranging from 44% for 10 wt% lime upto 47% for higher proportions. Both ESL and CL lime were used as binder. Because ESL and CL mortars behave identically, only ESL mortar results are presented here.

The setting time of these lime-based mortars is 4 days (Beck & Al-Mukhtar 2008). As a consequence, samples were kept in their mould for 4 days before turning out. $40 \times 40 \times 160$ mm prisms were prepared for every test: flexural and compressive strength, shrinkage and hydrous dilation. Each sample was stored at 100% RH for 28 days before testing. Mechanical tests were carried out at a saturated state after immersion in water for 24 h.

Microstructure

The microstructure of pore space was analysed by scanning electron microscopy (SEM) (Fig. 3) and mercury intrusion porosimetry (MIP) (Fig. 4). These data are essential for a comparative study with the stone. Samples were dried at 105 °C for both the SEM and MIP investigations. In the case of SEM, samples were impregnated with a resin before polishing. The SEM micrographs were obtained using backscattered electrons (BSE) on the polished surface for two different magnifications: ×100 and ×500.

The resulting densities and porosities of the mortars, measured by MIP, do not significantly change according to lime content. The apparent dry density of the mortars varies from 1.25 to 1.30 g cm^{-3}, and their total porosity from 46 to 48%; these values are comparable to those for the tuffeau (i.e. density 1.38 g cm^{-3} and porosity 46%).

Because of the complementary fineness of the lime and tuffeau powder, an increment in the lime content from 10 to 20% tends to fill the smallest pores partially, which can be observed by SEM at

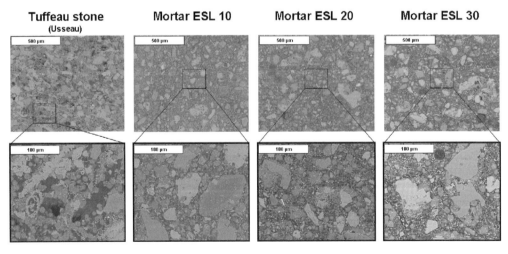

Fig. 3. SEM images of tuffeau and ESL mortars.

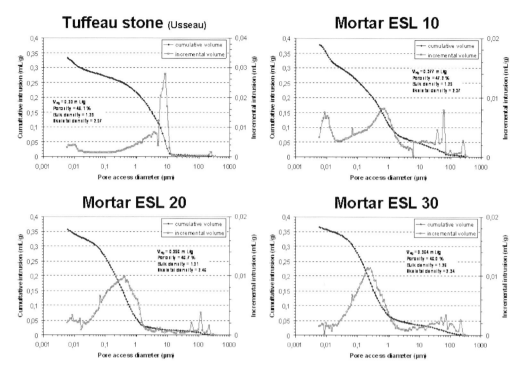

Fig. 4. Pore-size distribution of tuffeau and ESL mortars.

×500 magnification, i.e. between 1 and 5 μm (Fig. 3), whereas no significant difference can be observed at ×100 magnification (Fig. 3). Moreover, an increment in the lime content from 10 to 20% changes the microstructure from a two-peak pore-size distribution to a one-peak distribution (Fig. 4).

The peak corresponding to the pore access diameters of 0.01 μm is believed to come from the tuffeau powder, because an identical peak appears in the pore-size distribution of the tuffeau limestone (Fig. 4). These smallest pores correspond to clay mineral particles and to the roughness of opal spheres. The disappearance of these pores can be attributed to a pozzolanic reaction between portlandite and the soluble silica from clay minerals and opals. Lime reactivity measurements (Chapelle test) were carried out on tuffeau powder and confirmed that 20 wt% lime reacts with soluble silicate particles from tuffeau powder (Beck & Al-Mukhtar 2008).

Hydrous dilation tests (first and second cycle)

Hydrous expansion coefficient measurements were performed on samples using a system composed of a dilatometric displacement sensor LVDT (linear variable differential transformer) with an accuracy of 0.4 μm over a range of ±1 mm. During this test, the reference state is saturated after immersion for 24 h and the comparative state is dried at 105 °C.

This drying shrinkage measurement is conducted several times consecutively to study the role of the first cycle compared with the others, which were revealed to be similar to the second cycle. Results are presented in Table 3.

The hydrous expansion coefficient of tuffeau is of the order of 0.6×10^{-3} m m^{-1}. The first drying cycle led to a coefficient of mortars close to 3.0×10^{-3} m m^{-1}, whereas from the second cycle this coefficient falls to about 1.2×10^{-3} m m^{-1}. This hysteresis in hydrous behaviour corresponds to an irreversible shrinkage during the first drying.

First, the significant discrepancy between mortar and tuffeau coefficients could lead to an accelerated ageing process of the stonework owing to the wetting and drying exposure. As a consequence, the effect of such a discrepancy in the hydrous expansion coefficient must be studied at the structural scale, by digital simulation for example, in order to establish a reliable degree of compatibility between tuffeau and these lime mortars. To overcome this potential problem, one solution is to limit the water susceptibility of these lime mortars by optimizing the design.

Table 3. *Hydrous dilation of ESL mortars during the first drying cycle, and during the second and subsequent drying cycles*

	Tuffeau	ESL 10%	ESL 15%	ESL 20%	ESL 25%	ESL 30%
Hydrous dilation during the first drying cycle (m m^{-1})	$0.6 \times 10^{-3} \pm 0.1 \times 10^{-3}$	$3.0 \times 10^{-3} \pm 0.2 \times 10^{-3}$	$3.1 \times 10^{-3} \pm 0.1 \times 10^{-3}$	$2.9 \times 10^{-3} \pm 0.2 \times 10^{-3}$	$3.0 \times 10^{-3} \pm 0.2 \times 10^{-3}$	$3.1 \times 10^{-3} \pm 0.2 \times 10^{-3}$
Hydrous dilation during the second and next subsequent drying cycles (m m^{-1})	$0.6 \times 10^{-3} \pm 0.1 \times 10^{-3}$	$1.4 \times 10^{-3} \pm 0.3 \times 10^{-3}$	$1.3 \times 10^{-3} \pm 0.1 \times 10^{-3}$	$1.2 \times 10^{-3} \pm 0.1 \times 10^{-3}$	$1.3 \times 10^{-3} \pm 0.1 \times 10^{-3}$	$1.2 \times 10^{-3} \pm 0.1 \times 10^{-3}$

Second, the hysteresis in hydrous behaviour was revealed to be concurrent with the openinig of cracks (Fig. 5). As a result the mechanical consequences of these shrinkage cracks were investigated to check on the damage level achieved by the mortar. This effect cannot be attributed to thermal shock at 105 °C as samples dried at room temperature showed similar cracks.

Mechanical resistance (before and after the first drying cycle)

Results of compressive and flexural strength tests (performed according to European normative EN 1015-11:2000) are presented in Tables 4 and 5, respectively. Results are presented for undamaged samples before the first drying cycle and for samples submitted to the first drying cycle. For each mortar, three $40 \times 40 \times 460$ mm prismatic samples were tested to provide consistent results.

The flexural strength of undamaged ESL mortars, which does not significantly vary according to the lime content, was found to be slightly higher than for that of the tuffeau. The compressive strength of undamaged ESL mortars is slightly lower compared with tuffeau limestone. It should be noted that pozzolanic reactions take much longer than 28 days to achieve significant strength (Lanas & Alvarez 2003). As a consequence, performances will increase with time. The behaviour of undamaged ESL mortars lead us to expect a good mechanical compatibility between these lime mortars and tuffeau, at least as long as they remain undamaged.

For damaged specimens (i.e. after the first drying cycle) the compressive strength decreases, whereas the flexural strength becomes almost insignificant. As a consequence, such a mortar could not provide any bond at the interface between the mortar and the stone in the case of restoration works. This means that the design of this mortar has to be improved in order to significantly decrease damage generated during drying.

Discussion and conclusion

The aim of this work was to evaluate the use of limestone powder and eggshell lime as exclusive components of a mortar suitable for restoration. Both materials come from waste recovery; this innovative combination represents a relevant opportunity for sustainable development.

To reach optimal performances with respect to multiple environmental stresses, it is necessary to use a mortar whose properties are compatible with those of the stone used in the construction. Thus, it becomes critical to assess such compatibility

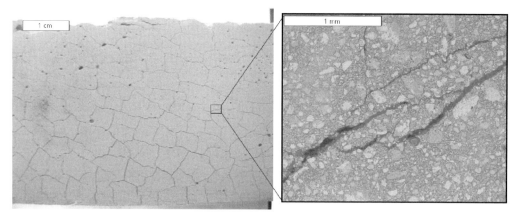

Fig. 5. Photographs of shrinkage cracks observed just after drying (ESL30 prismatic sample).

Table 4. *Compressive strength of tuffeau and ESL mortars before and after the first drying cycle*

	Tuffeau	ESL 10%	ESL 15%	ESL 20%	ESL 25%	ESL 30%
Compressive strength **before** the first drying cycle (MPa)	5.53 ± 0.33	3.59 ± 0.11	3.92 ± 0.06	3.58 ± 0.09	3.83 ± 0.19	3.73 ± 0.10
Compressive strength **after** the first drying cycle (MPa)	5.53 ± 0.33	2.43 ± 0.19	2.12 ± 0.12	1.71 ± 0.23	2.55 ± 0.06	2.80 ± 0.08

Table 5. *Flexural strength of tuffeau and ESL mortars before and after the first drying cycle*

	Tuffeau	ESL 10%	ESL 15%	ESL 20%	ESL 25%	ESL 30%
Flexural strength **before** the first drying cycle (MPa)	1.27 ± 0.11[3]	1.71 ± 0.09	2.42 ± 0.18	2.03 ± 0.07	2.38 ± 0.18	2.42 ± 0.18
Flexural strength **after** the first drying cycle (MPa)	1.27 ± 0.11[3]	0.06 ± 0.02	0.12 ± 0.01	0.07 ± 0.02	0.13 ± 0.02	0.16 ± 0.03

between both materials, where the term compatibility refers to different characteristics: physico-chemical, mechanical and hydraulic properties. For this reason a specific methodology was developed.

- The properties of the eggshell lime (i.e. hydrated lime purity, whiteness) must be at least as high as standard commercial hydrated lime.
- The behaviour of fresh mortar needs to be evaluated to provide a mortar whose consistency is compatible with its subsequent use in restoration works.
- Compression and flexion tests are extremely important to prove that both materials have similar mechanical resistance.
- In order to assess the susceptibility of mortar with respect to water transfer, tests of drying cycles were applied to mortars to quantify the consequences in terms of deformation and cracking.

An aerial lime was obtained from eggshells using a simple and efficient process. The purity of the resulting lime achieved 97%, which is higher than that of classic commercial aerial lime. Hence, this common food waste can be effectively valued as aerial lime. The sawing powder from the tuffeau stone quarry was mixed into the mortar paste as aggregate. A good consistency of the mortar was obtained by adding approximately 50 wt% water; this result suggests that this combination of eggshell lime and tuffeau powder can be effectively used for restoration work.

Mortars made out of eggshell lime were prepared and analysed after rigorous study, and show properties that are similar to that of commercial aerial lime mortar.

The characterization of the prepared mortars reveals the following.

- The porosity of the prepared mortar is similar to that of tuffeau stone.
- Mechanical resistance of undamaged mortar (flexural and compressive strength) approaches that of tuffeau. This behaviour ensures a good continuity in the structure and avoids generating a weak point in joint mortar.
- Hydrous expansion of undamaged mortar is two – three times higher than that of tuffeau. This could represent a source of deterioration when the structure is subjected to multiple wetting and drying cycles. This potential incompatibility will be investigated during further accelerated ageing tests.

Moreover, because this mortar is exclusively composed of tuffeau powder, aerial lime and water (i.e. no resin, no cement, no adjuvant), a perfect chemical compatibility with tuffeau limestone is expected. Hence, as the main physical, mechanical and chemical properties of the prepared mortar are similar to those of the tuffeau limestone, a good compatibility between both materials is expected.

This partial conclusion would be relevant if the mortars did not present a deleterious susceptibility to water transfer. Indeed, the mortar irreversibly shrank during its first drying. when the drying cycle was applied to the mortars multiple cracks were generated that could be observed from the surface to the core, as revealed by SEM observations. These cracks drastically reduced the mechanical properties, especially the flexural strength. As a consequence, such a mortar would not provide any bond at the interface with the stone. All of this means that the design of the mortar has to be improved to avoid such water susceptibility. The irreversible shrinkage could be lowered by reducing flocculation during the mixing process.

Moreover, these statements prove that drying tests are essential in providing relevant information about lime mortar durability. Without these drying cycle tests a mortar could be incorrectly asserted to be as durable and compatible with tuffeau.

This paper describes results from the research project 'National Research Programme on the Knowledge and the Conservation of Cultural Heritage Material'. The authors would like to thank the French Ministry of Culture, which financed this study, the Maquignon and Frères quarry at Usseau for the tuffeau, and Calcia for the commercial lime.

References

ALVAREZ, J. I., MARTÍN, A. & GARCÍA CASADO, P. J. 1995. Historia de los morteros. *Boletín infomativo del Instituto Andaluz de Patrimonio Histórico*, **13**, 52–59.

BECK, K. 2006. *Etude du comportement et des mécanismes d'altération de pierres calcaires á forte porosité*. PhD thesis, University of Orléans.

BECK, K. & AL-MUKHTAR, M. 2008 Formulation and characterization of an appropriate lime-based mortar for use with a porous limestone. *Environmental Geology*, **56**, 715–727; doi:10.1007/s00254-008-1299-8.

BECK, K., AL-MUKHTAR, M., ROZENBAUM, O. & RAUTUREAU, M. 2003. Characterization, water transfer properties and deterioration in tuffeau: building material in the Loire Valley – France. *International Journal of Building and Environment*, **38**, 1151–1162.

BELIN, P., BORKOWSKI, M., LARPIN, D. & MERTZ, J.-D. 1995. La chaux dans les mortiers anciens. In: *La chaux et les mortiers: natures, propriétés, traitements / (organisé par) Section française du Conseil international des monuments et des sites (ICOMOS) et le Laboratoire de recherche des monuments historiques, Direction du patrimoine, Ministère de la culture et de la francophonie, Paris, 24 janvier 1995*. Icomos, France, 105–107.

BISCONTIN, G., BIRELLI, M. P. & ZENDRI, E. 2002. Characterization of binders employed in the manufacture of Venetian historical mortars. *Journal of Cultural Heritage*, **3**, 31–37.

DEGRYSE, P., ELSEN, J. & WAELKENS, M. 2002. Study of ancient mortars from Salassos (Turkey) in view of their conservation. *Cement and Concrete Research*, **32**, 1457–1563.

ECOLE D'AVIGNON. 2003. *Techniques et pratique de la chaux*. Eyrolles, Paris.

EN 459-1:2001. Chaux de construction – Partie 1: Définition, spécification et critère de conformité. [*Building Lime – Part 1: Definition, Specifications and Conformity Criteria.*] European Committee for Standardization, Brussels.

EN 1015-11:2000. *Méthodes d'essai des mortiers pour maçonnerie – Partie 11: détermination de la résistance en flexion et en compression du mortier durci.* [*Methods of Test for Mortar for Masonry – Part 11: Determination of Flexural and Compressive Strength of Mardened Mortar.*] European Committee for Standardization, Brussels.

FASSINA, V., FAVARO, M., NACCARI, A. & PIGO, M. 2002. Evaluation of compatibility and durability of a hydraulic lime-based plasters applied on brick wall masonry of historical buildings affected by rising damp phenomena. *Journal of Cultural Heritage*, **3**, 45–51.

FURLAN, V. & BISSEGER, P. 1975. Les mortiers anciens. Historique et essais d'analyse scientifique. *Revue suisse d'Art et D'Archéologie*, **32**, 1–14.

KHANOUSSI, M. & MAURIN, M. 2000. *Dougga, fragments d'histoire*. Ausonius Publications, Bordeaux-Tunis.

LANAS, J. & ALVAREZ, J. I. 2003. Masonry repair lime-based mortars: Factors affecting the mechanical behaviour. *Cement and Concrete Research*, **33**, 1867–1876.

MOROPOULOU, A., BAKOLAS, A. & BISBIKOU, K. 2000. Investigation of the technology of historic mortars. *Journal of Cultural Heritage*, **1**, 45–58.

MOSQUERA, M. J., BENITEZ, D. & PERRY, S. H. 2002. Pore structure in mortars applied on restoration. Effect on properties relevant to decay of granite buildings. *Cement and Concrete Research*, **32**, 1883–1888.

RAUTUREAU, M. 2001. *Tendre comme la pierre, ouvrage collectif sous la direction de Michel Rautureau.* Conseil régional Centre & Université d'Orléans, (available online on http://www.culture.fr/culture/conservation/fr/biblioth/biblioth.htm).

RODRIGUEZ-NAVARRO, C., HANSEN, E. & GINELL, W. S. 1998. Calcium hydroxide crystal evolution upon aging of lime putty. *Journal of American Ceramic Society*, **81**, 3032–3034.

Physical changes of porous Hungarian limestones related to silicic acid ester consolidant treatments

Z. PÁPAY & Á. TÖRÖK*

Department of Construction Materials & Engineering Geology, Budapest University of Technology and Economics, Budapest, Hungary

Corresponding author (e-mail: torokakos@mail.bme.hu)

Abstract: Porous limestones are widely used in the monuments of Hungary, and are often treated using stone consolidants on site during restoration works. Two types of porous Miocene limestones from Sóskút were treated with dilute and concentrated silicic acid ester under laboratory conditions. The aim of the experiments was to assess the performance of the consolidants of different concentrations on fine- and medium-grained limestones, and to detect physical changes caused by consolidation. The pore-size distribution and fabric of the limestones were also different with initial porosities of 37 and 23%, respectively. In the open pores only a few per cent (3.2–5.2%) of silica gel was precipitated even under vacuum saturation. The loss of porosity was higher when concentrated consolidant was used. The decrease in porosity was found not to be proportional to the changes in strength. Indeed, the dilute consolidant caused a greater increase in strength than the concentrated one. The increase of tensile strength was higher for the less porous medium-grained limestone than for the fine-grained one. The differences in strength between non-consolidated and treated specimens were also detectable using ultrasonic pulse velocities. The tests have shown that the efficacy of silica acid ester treatments in terms of strength is less influenced by the concentration of the consolidant, but rather its ability to reach the micropores of porous limestones.

Porous limestones are widely used in monuments throughout Europe, with known examples from the UK (Viles 1994; Smith & Viles 2006), France (Beck & Al-Mukhtar 2004), Italy (Adriani & Walsh 2002), Greece (Stefanidou & Papayianni 2008), Malta (Cassar 2002), Czech Republic (Přykril et al. 2002), Slovakia (Holzer et al. 2004) and Austria (Kieslinger 1949). Many of these limestone structures now show severe signs of decay that significantly affects the aesthetic appearance of, and cause structural damage to, the stones. Climatic factors and human-activity-induced air pollution are the triggers for these changes. To maintain the monuments and to slow down these processes various methods are known. One of these methods is the *in situ* application of stone consolidants. The treatment with stone consolidants is aimed at slowing down these harmful processes and/or strengthening the already weathered stones during repair works. The consolidation of porous limestone is not a common practice in many countries, but several examples are known where consolidants were successfully used for monuments in Vienna (Austria) and Budapest (Hungary). There are various methods for applying consolidants *in situ*, such as spraying, brushing or applying it in a poultice, but more recently a *in situ* vacuum consolidation method seems to work quite effectively (Pummer 2008; Török 2008). When using *in situ* vacuum consolidation injection tubes are inserted into drilled holes and the area for consolidation is sealed using plastic foil. The consolidant is injected via these holes under a vacuum, and the vacuum is kept for several hours or for a whole day to ensure a full saturation. Besides the selection of the proper consolidation methodology, it is essential to understand the behaviour of the consolidant and the interaction of the stone with the applied material. In this paper the various methods of applying consolidants are not discussed in detail, but, instead, the physical changes caused by various types of consolidants on porous limestones are investigated. To understand the mechanical changes caused by consolidants it was necessary to apply a method where the possible maximum saturation by consolidant could be achieved. This method is vacuum impregnation. It has been applied previously for testing the performance of various consolidants under laboratory conditions (Steinhäuser & Wendler 2004; Stück et al. 2008).

In this study two types of porous Miocene limestones from Sóskút (Hungary) were treated with silicic acid ester consolidants under laboratory conditions. The tested stone types were selected because these are the most common lithologies in Budapest and many monuments that were constructed from these stones now show signs of rapid deterioration (Török 2002, 2003, 2007;

Smith *et al.* 2003). The consolidants were applied in two different concentrations to help in the selection of consolidants for future *in situ* applications, and to understand the strengthening mechanism and effect of dilute or concentrated compounds. Silicic acid ester has previously been used for consolidating various stone types, especially sandstones (Malaga *et al.* 2004; Meinhardt-Degen & Snethlage 2004), volcanic tuffs (Wendler *et al.* 1996; Forgó *et al.* 2006; Stück *et al.* 2008) and limestones (Lukaszewicz 2004). Testing and comparative studies of various consolidants and limestones are also known from earlier studies (Alvarez de Buergo & Fort 2002; Steinhäuser & Wendler 2004; Ahmed *et al.* 2006). The present paper is focused on assessing the performance of silicic acid ester on Hungarian porous limestone using rock mechanical tests. The limestone specimens were treated with the consolidants, and physical parameters such as density, tensile strength, porosity and ultrasonic pulse velocity were measured on non-consolidated and consolidated test specimens prior to and after the consolidation. The results are compared with previous test findings and are evaluated with reference to the previous consolidation trials of limestones (Kertész 1988; Ahmed *et al.* 2006; Cnudde *et al.* 2007; Pápay & Török 2007).

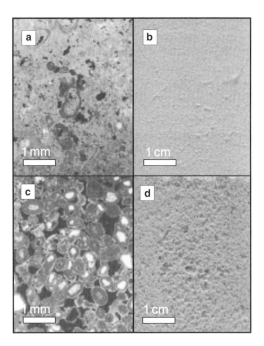

Fig. 1. Micro- and macrofabric of (**a**, **b**) fine-grained oolitic limestone and (**c**, **d**) medium-gained limestone.

Materials and methods

Stones

Two types of porous Miocene limestones (a medium-grained oolitic type and a fine-grained more micritic variety) were used for tests (Fig. 1). These stone types are probably the most important raw materials of the Hungarian cultural heritage, as emblematic buildings such as the House of Parliament, St Stephan's Basilica and Mathias Church in Budapest are constructed from these limestones (Kertész 1982). Very similar Miocene limestones are also known from Austria. The Leitha Mountains provided the construction material for the ashlars and ornaments of numerous monuments in Vienna (Opera House, St Stephans's Dome, Schönbrun Palace), castles of southern Moravia (Přykril *et al.* 2004; Török *et al.* 2004*a*) and even of Bratislava (Holzer *et al.* 2004).

The fine-grained limestone contains micritized peloids and small ooids with peloidal–ooid grainstone–packstone fabric. A few larger bioclasts can also be viewed under the microscope. The fabric of the medium-grained oolitic limestone according to Dunham's (1962) classification is an ooid grainstone. The limestone contains well-rounded ooids and micro-oncoids, 0.1–0.3 mm in diameter. Pores are mostly intergranular ones with mean pore sizes of less than 0.01 mm.

According to mercury porosimetry, the pore-size distribution of the two studied limestone types is different as in the medium-grained rock the pores below 10 μm are dominant, while in the fine-grained limestone pores larger than 10 μm are the most common (Figs 2 & 3). The number of micropores is very similar in both types.

Methods

For consolidation trials, cylindrical test specimens were drilled and cut from stone blocks obtained from the Sóskút quarry. The specimens, 40 mm in

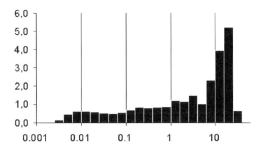

Fig. 2. Pore-size distribution of fine-grained limestone, porosity is 37.1% (vertical axis is % and horizontal axis is pore radius in μm).

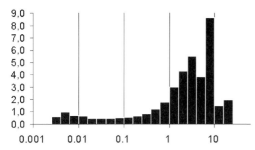

Fig. 3. Pore-size distribution of medium-grained limestone, porosity is 23.1% (vertical axis is % and horizontal axis is pore radius in μm).

diameter and 20 mm in height, were divided into testing groups on the basis of non-destructive tests (density, ultrasonic pulse velocity) prior to consolidation. All together 36 samples were obtained. Samples were divided into test groups containing six samples each. For reference, one set of samples from both rock types was left untreated, while the rest of the specimens were treated. The vacuum impregnation technique was applied to obtain full saturation of the samples. This method was selected because it provides information on the maximum physical changes caused by consolidants, and was also used in recent *in situ* conservation trials (Török 2008). The applied silicic acid ester (SAE) consolidants are well known and widely used (Snethlage 1997). The tests were performed using a diluted and a concentrated form of the same consolidant. Type A consisted of 100 mass% of silicic acid ester, while Type B contained only 20 mass% of silicic acid ester. The solvent for the latter was an aliphatic carbohydrate. The density of the first consolidant was 0.99 g cm^{-3}, while that of the dilute one was 0.79 g cm^{-3}.

The binding mechanisms of silicic acid esters are well known and have been studied in detail (Hilbert 1999). In the presence of water silicic acid ester transforms into silica gel, which fills the pores and forms silicate bonds between the particles. The amount of consolidant was tested by measuring the weight of the specimens throughout the consolidation trials. Therefore it was possible to assess the amount of absorbed liquid-phase consolidants, and after a period of 2 months the amount of precipitated and consolidated solid-phase silica gel too. These values provide information on the efficacy of consolidants in terms of pore-filling ability and increase in dry weight of consolidated specimens relative to non-consolidated ones. The micro-fabric of the limestone and the precipitated consolidants were viewed using scanning electron microscopy (SEM). Other physical tests were also performed according to European Norms. Ultrasonic sound (US) velocity (EN 14579:2005), and density and open porosity (EN 1936:2000), were measured both on non-consolidated and consolidated specimens in order to detect the physical changes. This method was applied as previous experiments had shown that US pulse velocities could be used to evaluate the efficacy of conservation (Bellopede & Manfredotti 2006; Myrin & Malaga 2008). It has been also proved that there is a decrease in US pulse velocity with increasing rate of weathering (Weiss *et al.* 2002). Efficiency of Type A and Type B silicic acid ester consolidant was evaluated by comparing the indirect tensile strength (Brazilian test) of untreated and silicic acid ester treated specimens. The standard deviation, mean and maximum values of each set of results were also calculated.

Results

The test results are summarized in Table 1. There is a significant difference between the apparent densities of the two limestones: the fine-grained one had a density of 1.646 g cm^{-3}, while the mean density of the medium-grained limestone was 19% greater (1.957 g cm^{-3}). These density differences reflect the differences in porosities rather than a difference in mineralogical composition, as both limestones are primarily composed of calcite while the contribution of other minerals is less than 1% (Török 2002). The measured values of ultrasonic pulse velocities and tensile strength also show the same tend; the medium-grained oolitic limestone has a greater strength and the ultrasonic wave propagation is faster in these samples. It was possible to detect the amount of silica gel that had been formed within the pores. Within the medium-grained limestone 3.8 and 3.2 vol.%, silica gel, while within the fine-grained limestone 5.2 and 4.0 vol.%, was precipitated from the concentrated and dilute consolidants, respectively. As a consequence the densities of the limestones became greater than before the consolidation. The augmentation of densities of both of the treated limestones was higher when the concentrated (Type A) consolidant was applied (Table 1). The porosities of consolidated samples v. non-consolidated samples clearly show an opposite trend to the density (Fig. 4) and reflect the amount of silica gels precipitated in pores. It was possible to visualize the precipitated silica gel within the pores using SEM. The silica gel was precipitated in the pores but did not occlude entirely the pore space; rather, it formed smaller bridges between carbonate particles. The microcracks within the silica gel provide pathways and allow a partial evaporation of the water vapour (Fig. 5).

The tensile strength of non-treated samples varied between 2.44 (medium grained) and

Table 1. *Selected physical properties in the treated and untreated condition: mean values (standard deviations are given in brackets)*

Analytical groups		Apparent density (g cm^{-3})	Ultrasonic pulse velocity (km s^{-1})	Open porosity (vol.%)	Amount of silica gel (vol.%)	Tensile strength (MPa)
Medium-grained limestone	Untreated	1.957 (0.026)	2.773 (0.094)	23.12		2.44 (0.44)
	Treated with Type A	2.031 (0.015)	2.958 (0.081)	19.37	3.8	2.99 (0.48)
	Treated with Type B	2.008 (0.037)	3.088 (0.077)	19.89	3.2	3.34 (0.56)
Fine-grained limestone	Untreated	1.646 (0.055)	2.359 (0.127)	37.14		1.32 (0.35)
	Treated with Type A	1.789 (0.041)	2.534 (0.088)	31.95	5.2	1.58 (0.36)
	Treated with Type B	1.749 (0.045)	2.704 (0.174)	33.11	4.0	1.66 (0.26)

1.32 MPa (fine-grained limestone), with higher standard deviations than the US pulse velocities (Table 1). In all of the consolidations the tensile strength increases, whereas the degree of the change depends on the type of consolidant and the rock fabric. Both lithologies treated with dilute SAE (Type B) showed higher values than the ones treated with concentrated SAE (Type A). The highest increase in tensile strength was recorded for medium-grained limestone treated with Type B consolidant (36.9%), meanwhile Type A consolidant only caused a 22.5% of increase in strength. Thus, there is 12% difference in the affectivity of dilute (Type B) and concentrated SAE (Type A) in medium-grained limestone, while the difference is only 5% for fine-grained limestone (Fig. 6). A similar trend is observed when ultrasonic pulse velocity values of non-treated and treated samples are compared, namely ultrasonic waves move faster in dilute consolidant-treated samples (Fig. 7). With reference to the non-treated samples, the greatest change in wave propagation was found in fine-grained limestone. In contrast, the tensile strength increase is higher in the medium-grained limestone (cf. Figs 6 & 7). The measured velocity values were found to have very low standard deviations (Table 1). A fairly good correlation was found between ultrasonic pulse velocity and tensile

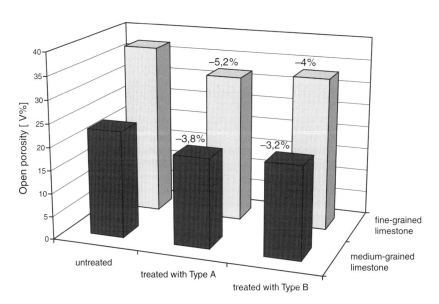

Fig. 4. Open porosity of untreated and treated limestones marked by bars (the pore space occluded by precipitated silica gel with respect to the volume of the entire specimen is given on the top of each bar).

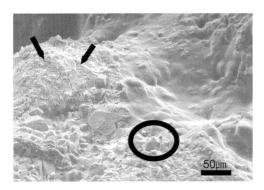

Fig. 5. SEM image of the consolidant-treated porous limestone; it clearly shows the envelope-like behaviour of silicic acid ester and the secondary microcracks that are formed within the silica gel (arrows); the uncovered rhombohedral calcite crystals (circled) are also visible.

strength measured on treated and non-treated specimens. A smaller scatter and better correlation was found for non-consolidated specimens (Fig. 8).

Discussion

Consolidated samples with reference to untreated samples are markedly different from the point of view of their physical properties. Despite the difference in effective porosities of medium-grained (23.1%) and fine-grained limestone (37.1%) (Figs 2 & 3) the amount of precipitated silica gel from both consolidants is very similar in both limestones, and is of the order of 3.2–5.2%. The higher values have been found in the more porous limestone (fine grained) when concentrated consolidant was used. Nevertheless, the major difference in the silicic acid ester content of the two tested consolidants – 100% (Type A) and 20% (Type B) – is not entirely reflected in the amount of precipitated solid silica gel. In medium-grained limestone the difference in silica gel content of dilute and concentrated SAE is 19%, while in fine-grained one only an additional 30% of silica gel was formed from the concentrated consolidant (Type A) (Table 1). This suggests that silica gel formation in pores is not proportional to the amount of consolidant, but it is, instead, governed by pore-size distribution, the available reaction surface of the stone, as well as by the penetration ability of the consolidants into the pores. It is in accordance with previous observations that the pore-filling ability of consolidants is controlled by the viscosity of the consolidant (Hilbert 1999; Snethlage 1997; Ahmed et al. 2006) and, similar to water uptake (Fitzner & Basten 1994), by pore-size distribution (Stück et al. 2008). In the present case the fine-grained limestone has a greater number of pores larger than 10 μm than that of the medium-grained limestone (Figs 2 & 3). Thus, both higher porosity and greater number of macropores allow a better saturation of fine-grained limestone by consolidant compared to medium-grained limestone. The changes in density values (Table 1) and in ultrasonic pulse velocities (Fig. 7) also mark these trends. It has been shown that SAE is able to penetrate into micropores

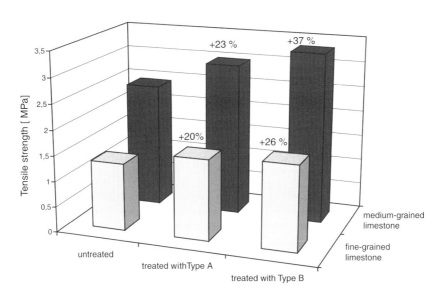

Fig. 6. Tensile strength of untreated and treated fine- and medium-grained limestones (percentages of differences with respect to the untreated specimens are given on the top of each bar).

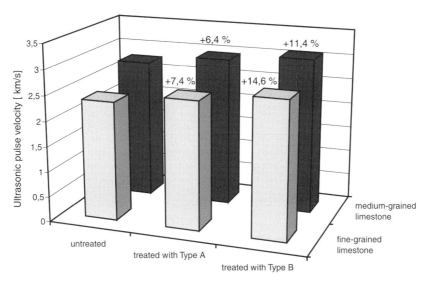

Fig. 7. Ultrasonic pulse velocities of untreated and treated limestones (percentages of differences with respect to the untreated specimens are given on the top of each bar).

and cause a reduction in microporosity that was documented for volcanic tuffs (Steindlberger 2004; Stück et al. 2008) and limestones (Cnudde et al. 2007). This is related to the fact that the molecule size of the chemical agent (SEA) is of the order of 3 nm, which enables the consolidant to migrate into micropores (Hilbert 1999). By comparing porosities prior to conservation and after treatment (Fig. 4) it is evident that only a very small part of the open porosity was occluded by the precipitated silica gel. Previous studies (Stück et al. 2008) and SEM observations (Fig. 5) have demonstrated that besides the reduction of porosity new pores are also formed within the consolidant. This secondary porosity in the form of microfissures is generated due to prolonged hydrolysis of precipitated SAE (Hilbert 1999). Thus, pore-size distribution is very probably shifted in two contrasting ways: micropores were reduced by silica gel precipitation and new micropores were also formed within the silica gel (Fig. 5). As a consequence, it is difficult to judge how these processes modified the porosity. Nevertheless, the small amount of precipitated consolidant suggests that, similar to

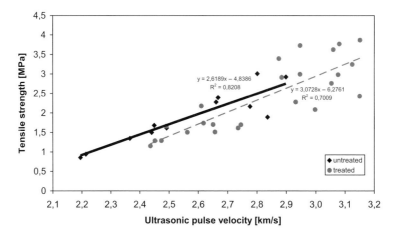

Fig. 8. Correlation of ultrasonic pulse velocity and the tensile strength of untreated (squares, continuous line) and treated specimens (circles, dashed line).

previous experiments (Wendler 2001; Forgó et al. 2006; Stück et al. 2008), only thin films of silica gel is found between the carbonate grains. Despite the small percentage of precipitated silica gel (3.2–5.2%), a fairly good increase in strength (19.7–36.8%) was measured. The consolidation decreased the porosity of the medium-grained limestone by 16.2% (concentrated SAE) and 12.2% (diluted SAE), which caused a strength increase of 22.5 and 36.8%, respectively. Similarly, a smaller loss in porosity (14.0 and 10.9%) and a relatively higher augmentation of tensile strength (19.7 and 25.8%) was documented for fine-grained samples. Hence, the porosity loss was not proportional to strength increase. This feature has also been reported when sandstones (Malaga et al. 2004) and tuffs (Wendler et al. 1996; Stück et al. 2006) were treated with SAE. The dilute consolidant (Type B) caused a more significant increase in tensile strength of both limestone types (Fig. 6). This means that it can penetrate more easily into smaller pores or can form silica gel bridges between carbonate grains. As a consequence, it is not the concentration of the consolidant, but rather its ability to reach the micropores that is the key factor in increasing strength. For the difference between laboratory test results and results of *in situ* applications of consolidant tests had previously been performed. These *in situ* vacuum consolidation trials, when the same type of consolidant were applied, have shown that, depending on porous limestone fabric and thus on pore-size distribution, a 20–23% increase in tensile strength, a 15–54% increase in uniaxial compressive strength and a 37–70% augmentation in modulus of elasticity were measured (Török 2008).

To judge the efficiency of consolidation, especially at historic sites, it is very important to use non-destructive test methods. In this study ultrasonic pulse velocity was tested with reference to tensile strength. The obtained data set shows that it is possible – to a certain extent – to evaluate the strength values by applying ultrasonic pulse testing methods (Fig. 8). It is also necessary to note that prior to application of consolidants more detailed laboratory testing is required, especially in terms of the long-term behaviour of consolidated stones and changes in pore-size distribution and durability. In addition, in many circumstances vacuum impregnation is not available, thus traditional methods such as brushing or spraying are applied for *in situ* consolidation. By applying these methods the saturation of pores and the penetration of consolidants are limited nearer to the surface of the stone and thus the changes of physical properties are restricted to the near-surface zones. As a consequence, the test results obtained by full immersion, brushing or spraying cannot be directly converted to data obtained by vacuum impregnation. Although there are previous studies that have tried to compare the various methods of consolidation or treatment (Wendler 2001; Alvarez de Buergo & Fort 2002; Török et al. 2004b), further studies are needed to compare these test results and to draw conclusions with reference to various methods of stone consolidations.

The experiments have also shown that it is necessary to adjust the concentration of SEA to obtain the best consolidation. It is also important to note that further tests are needed to understand the effect of various consolidants with different silicic acid ester content on the behaviour of porous limestones.

Conclusions

Consolidation by silica acid ester increased the density and the ultrasonic pulse velocity, and decreased the porosity, of the medium- and fine-grained limestone. The augmentation of these parameters is not proportional to the amount of chemicals present in the consolidant. These trends can be also expressed in numerical values. The concentrated silica acid ester was able to fill a greater percentage of pores in both limestones, but neither the concentrated one nor the dilute one was able to fill entirely the open pores, even under vacuum saturation. The concentrated consolidant occluded around 14.0% of the pores of the fine-grained limestone and 16.2% of the pores of the medium-grained limestone. Concurrently, the porosity loss was less for the dilute consolidant for both limestones (12.2 and 10.9%). Although a lesser amount of dilute consolidant was precipitated in the pores, the present laboratory tests have shown that the diluted silica acid ester (Type B) was able to strengthen these porous limestones more effectively. This can be explained by the more viscous nature of the dilute consolidant and its ability to reach smaller pores. It has been also documented from fabric properties that pore-size distribution is the main controlling factor for the efficacy of consolidation. SEM analyses have shown that silica acid ester forms silica gel envelopes in the pores and that these binding gels are responsible for the increase in strength. The envelopes are discontinuous and secondary microcracks provide pathways for water vapour. The increase in tensile strength can be documented by destructive tests, but can also been calculated using ultrasonic pulse velocities, as a fairly good correlation was found between these values. The standard deviations of ultrasonic pulse velocity test results were of the order of a few per cent, which indicates the reliability of these test methods.

This work was financed by the Hungarian Scientific Research Fund (OTKA, grant no. K63399). The help of E. L. Árpás and G. Emszt with the laboratory tests was also appreciated. The reviews of Dr J. Hughes and an anonymous reviewer helped in improving the quality of the paper.

References

ADRIANI, G. F. & WALSH, N. 2002. Physical properties and textural parameters of calcarenitic rocks: qualitative and quantitative equations. *Engineering Geology*, **67**, 5–15.

AHMED, H., TÖRÖK, Á. & LŐCSEI, J. 2006. Performance of some commercial stone consolidating agents on porous limestones from Egypt. *In*: FORT, R., ALVAREZ DE BUEGO, M., GOMEZ-HERAS, M. & VAZQUEZ-CALVO, C. (eds) *Heritage Weathering and Conservation*. Volume **2**. Taylor & Francis/Balkema, London, 735–740.

ALVAREZ DE BUERGO, M. & FORT, R. 2002. Characterizing the construction materials of a historic building and evaluating possible preservation treatments for restoration purposes. *In*: SIEGESMUND, S., WEISS, T. & VOLLBRECHT, A. (eds) *Natural Stones, Weathering Phenomena, Conservation Strategies and Case Studies*. Geological Society, London, Special Publications, **205**, 241–254.

BECK, K. & AL-MUKHTAR, M. 2004. The mechanical resistance properties of two limestones from France, Tuffeau and Sébastopol. *In*: PŘIKRYL, R. (ed.) *Dimension Stone*. Balkema, Rotterdam, 97–101.

BELLOPEDE, P. & MANFREDOTTI, L. 2006. Ultrasonic sound test on stone: comparison of indirect and direct methods under various test conditions. *In*: FORT, R., ALVAREZ DE BUERGO, M., GOMEZ-HERAS, M. & VAZQUEZ-CALVO, C. (eds) *Heritage Weathering and Conservation*. Volume **2**. Taylor & Francis/Balkema, London, 539–546.

CASSAR, J. 2002. Deterioration of the Globigerina Limestone of the Maltese Islands. *In*: SIEGESMUND, S., WEISS, T. & VOLLBRECHT, A. (eds) *Natural Stones, Weathering Phenomena, Conservation Strategies and Case Studies*. Geological Society, London, Special Publications, **205**, 33–49.

CNUDDE, V., DIERICK, M., VLASSENBROECK, J., MASSCHAELE, B., LEHMANN, E., JACOBS, P. & VAN HOOREBEKE, L. 2007. Determination of the impregnation depth of siloxanes and ethylsilicates in porous material by neutron radiography. *Journal of Cultural Heritage*, **8**, 331–338.

DUNHAM, R. J. 1962. Classification of carbonate rocks according to depositional structure. *In*: HAM, W. E. (ed.) *Classification of Carbonate Rocks*. AAPG Memoirs, **1**, 108–121.

FITZNER, B. & BASTEN, D. 1994. Gesteinsporosität–Klassifizierung, messtechnische Erfassung und Bewertung ihrer Verwitterungsrelevanz. *In*: SNETHLAGE, R. (ed.) *Jahresberichte Steinzerfall–Steinkonservierung 1992*. Ernst & Sohn, Berlin, 19–32.

FORGÓ, L. Z., STÜCK, H., MARÓTHY, E., SIEGESMUND, S., TÖRÖK, Á. & RÜDRICH, J. 2006. Materialverhalten von natürlichen und modellhaft konsolidierten Tuffen. *In*: AURAS, M. & SNETHLAGE, R. (eds) *Denkmalgestein Tuff*. Institut für Steinkonservierung, Mainz, Bericht, **22**, 65–75.

HILBERT, G. 1999. Natursteinkonservierung: Mittel und Durchführung. *Naturstein*, **2**, 44–49.

HOLZER, R., LAHO, M. & DURMEKOVÁ, T. 2004. Ancient building stone sources of Bratislava's monuments. *In*: PŘIKRYL, R. (ed.) *Dimension Stone*. Balkema, Rotterdam, 51–56.

KERTÉSZ, P. 1982. A Műemléki kőanyagok bányahelyeinek kutatása. *Építés-Építészettudomány*, **1–2**, 193–228 (in Hungarian).

KERTÉSZ, P. 1988. Decay and conservation of Hungarian building stones. *In*: MARINOS, P. G. & KOUKIS, G. C. (eds) *The Engineering Geology of Ancient Works, Monuments and Historical Sites*. IEAG Conference Proceedings, Athens, Volume **2**. Balkema, Rotterdam, 755–761,

KIESLINGER, A. 1949. *Die Steine von Sankt Stephan*. Herold, Wien.

LUKASZEWICZ, J. W. 2004. The efficiency of the application of tetraethoxysilane in the conservation of stone monuments. *In*: KWIATKOWSKI, D. & LÖFVENDAL, R. (eds) *Proceedings of the 10th International Congress on Deterioration and Conservation of Stone*. Volume **1**. ICOMOS Sweden, Stockholm, 479–486.

MALAGA, K., MYRIN, M. & LINDQVIST, J. E. 2004. Consolidation of Gotland sandstone. *In*: KWIATKOWSKI, D. & LÖFVENDAL, R. (eds) *Proceedings of the 10th International Congress on Deterioration and Conservation of Stone*. Volume **I**. ICOMOS Sweden, Stockholm, 447–454.

MEINHARDT-DEGEN, J. & SNETHLAGE, R. 2004. Durability of hydrophobic treatment of sandstone facades – investigation of the necessity and effects of re-treatment. *In*: KWIATKOWSKI, D. & LÖFVENDAL, R. (eds) *Proceedings of the 10th International Congress on Deterioration and Conservation of Stone*. Volume **I**. ICOMOS Sweden, Stockholm, 347–354.

MYRIN, M. & MALAGA, K. 2008. Evaluation of consolidation treatments of sandstone by use of ultrasound pulse velocity. *In*: LUKASZEWICZ, J. & NIEMCEWICZ, P. (eds) *Proceedings of the 11th International Congress on Deterioration and Conservation of Stone*. Volume **I**. Nicolaus Copernicus University Press, Torun, 441–448.

PÁPAY, Z. & TÖRÖK, Á. 2007. The effect of stone consolidation on the physical properties of porous limestone. A rock mechanical approach. *In*: SOUSA, L. R. E., OLALLA, C. & GROSSMANN, N. F. (eds) *11th Congress of the International Society for Rock Mechanics*. Volume **1**. Taylor & Francis, London, 465–467.

PŘYKRIL, R., SVOBODOVÁ, J. *ET AL.* 2002. Weathering of limestone cladding above the waterproofing layer: salt action due to previous restoration of the Colonnade (Lednice–Valtice area, Czech Republic). *In*: PŘYKRIL, R. & VILES, H. (eds) *Understanding and Managing Stone Decay*. Carolinum Press, Prague, 209–221.

PUMMER, E. 2008. Vacuum–circulation–process innovative stone conservation. *In*: LUKASZEWICZ, J. & NIEMCEWICZ, P. (eds) *Proceedings of the 11th*

International Congress on Deterioration and Conservation of Stone. Volume **I**. Nicolaus Copernicus University Press, Torun, 481–488.

SMITH, B. J., TÖRÖK, Á., MCALISTER, J. J. & MEGARRY, J. 2003. Observations on the factors influencing stability of building stones following contour scaling: a case study of the oolitic limestones from Budapest, Hungary. *Building and Environment*, **38**, 1173–1183.

SMITH, B. J. & VILES, H. A. 2006. Rapid, catastrophic decay of building limestones: thoughts on causes, effects and consequences. *In*: FORT, R., ALVAREZ DE BUEGO, M., GOMEZ-HERAS, M. & VAZQUEZ-CALVO, C. (eds) *Heritage Weathering and Conservation*. Volume **1**. Taylor & Francis/Balkema, London, 191–197.

SNETHLAGE, R. 1997. *Leitfaden Steinkonservierung*. Fraunhofer IRB, Stuttgart.

STEFANIDOU, M. & PAPAYIANNI, I. 2008. The porosity of limestone and its behaviour to salt cycles. *In*: LUKASZEWICZ, J. & NIEMCEWICZ, P. (eds) *Proceedings of the 11th International Congress on Deterioration and Conservation of Stone*. Volume **I**. Nicolaus Copernicus University Press, Torun, 283–290.

STEINDLBERGER, E. 2004. Volcanic tuff from hesse (Germany) and their weathering behaviour. *Environmental Geology*, **46**, 378–390.

STEINHÄUSER, U. & WENDLER, E. 2004. Conservation of limestone by surfactants and modified ethylsilicates. *In*: KWIATKOWSKI, D. & LÖFVENDAHL, R. (eds) *10th International Congress on Deterioration and Conservation of Stone – Stockholm 2004*. ICOMOS Sweden, Stockholm, 439–446.

STÜCK, H., FORGÓ, L. Z., SIEGESMUND, S., RÜDRICH, J. & TÖRÖK, Á. 2008. The behaviour of consolidated volcanic tuffs: weathering mechanisms under simulated laboratory conditions. *Environmental Geology*, **56**, 699–713.

TÖRÖK, Á. 2002. Oolitic limestone in polluted atmospheric environment in Budapest: weathering phenomena and alterations in physical properties. *In*: SIEGESMUND, S., WEISS, T. & VOLLBRECHT, A. (eds) *Natural Stones, Weathering Phenomena, Conservation Strategies and Case Studies*. Geological Society, London, Special Publications, **205**, 363–379.

TÖRÖK, Á. 2003. Surface strength and mineralogy of weathering crusts on limestone buildings in Budapest. *Building and Environment*, **38**, 1185–1192.

TÖRÖK, Á. 2007. Morphology and detachment mechanism of weathering crusts of porous limestone in the urban environment of Budapest. *Central European Geology*, **50**, 225–240.

TÖRÖK, Á. 2008. The application of silica acid ester under vacuum conditions for *in situ* consolidation of porous limestone monument, a case study from Hungary. *In*: LUKASZEWICZ, J. & NIEMCEWICZ, P. (eds) *Proceedings of the 11th International Congress on Deterioration and Conservation of Stone*. Volume **II**. Nicolaus Copernicus University Press, Torun, 1085–1091.

TÖRÖK, Á., ROZGONYI, N., PŘIKRYL, R. & PŘIKRYLOVÁ, J. 2004a. Leithakalk: the ornamental and building stone of Central Europe, an overview. *In*: PŘIKRYL, R. (ed.) *Dimension stone*. Balkema, Rotterdam, 89–93.

TÖRÖK, Á., GÁLOS, M. & KOCSÁNYI-KOPECSKÓ, K. 2004b. Experimental weathering of rhyolite tuff building stones and effect of an organic polymer conserving agent. *In*: SMITH, B. J. & TURKINGTON, A. V. (eds) *Stone Decay, Its Causes and Controls*. Donhead, Shaftesbury, Dorset, 109–127.

VILES, H. A. 1994. Observations and explanations of stone decay in Oxford, UK. *In*: THIEL, M. J. (ed.) *Conservation of Stone and Other Materials*, Volume **I**. *Causes of Disorders and Diagnosis*. E & FN Spon–RILEM, London, 115–120.

WEISS, T., RASOLOFOSAON, P. N. J. & SIEGESMUND, S. 2002. Ultrasonic wave velocities as diagnostic tool for the quality assessment of marble. *In*: SIEGESMUND, S., WEISS, T. & VOLLBRECHT, A. (eds) *Natural Stones, Weathering Phenomena, Conservation Strategies and Case Studies*. Geological Society, London, Special Publications, **205**, 149–164.

WENDLER, E. 2001. Elastifizierte Kieselsäreester als mineralische Bindemittel für unterschiedliche Konservierungsziele. Praktische Erfahrungen mit dem KSE-Modulsystem. *In*: *Natursteinkonservierung: Grundlagen, Entwicklungen und Anwendungen*, Volume **23**. WTA-Schriftenreihe, Aedificatio, Freiburg, 55–78.

WENDLER, E., CHAROLA, A. E. & FITZNER, B. 1996. Easter Island Tuff: laboratory studies for its consolidation. *In*: RIEDERER, J. (ed.) *Proceedings of the 8th International Congress on Deterioration and Conservation of Stone, Berlin*. Volume **2**, Möller Druck und Verlag, Berlin, 1159–1170.

Pore structure and durability of Portuguese limestones: a case study

C. FIGUEIREDO[1]*, R. FOLHA[1], A. MAURÍCIO[1], C. ALVES[2] & L. AIRES-BARROS[1]

[1]*Centre for Petrology and Geochemistry, CEPGIST, IST, Av. Rovisco Pais, 1049-001, Lisbon, Portugal*

[2]*University of Minho, Centro de Investigação Geológica e Valorização de Recursos, DCT, Campus de Gualtar, 4710-057 Braga, Portugal*

*Corresponding author (e-mail: carlos.m.figueiredo@ist.utl.pt)

Abstract: Exposed stone surfaces containing complex systems of pores, fractures and grain boundaries provide the surfaces where chemical, physical and biological deterioration processes take place. The pore space represents the preferred area for physical, chemical and biological weathering processes. It plays a significant role in the behaviour of porous materials. A full understanding of pore-channel network morphology, size and connectivity is important in stone decay assessment and conservation works. A contribution to the understanding of the role played by the pore system in controlling fluid-related properties and resistance to salt crystallization of limestones is presented. Optical microscopy, scanning electron microscopy (SEM) and mercury injection porosimetry (MIP) were used to characterize the pore structure of two Portuguese dimension stones ('Semi-rijo' and 'Moca-Creme') widely used for pavements and the cladding of buildings. Fluid migration physical tests (open and free porosity, capillary imbibition, and Hirschwald coefficient) were also performed, according to European (EN 1925:1999; EN 1936:1999) and French (N FB 10-504:1973) Standards. The resistance to salt crystallization was determined using European Standard EN 12370:1999. An integrated analysis facilitated comparison between durability results with stone pore network characteristics, fluid transport properties and petrographical features, suggesting the influence of available porosity and bedding.

Materials in nature continually deteriorate as result of physical, chemical and biological processes (Amoroso & Fassina 1983). Building stones in monuments are porous materials, and are usually characterized by a wide range of porosity, pore size and pore-size distribution (Fitzner 1993; Winkler 1997). Exposed stone surfaces containing complex systems of pores, fracture surfaces and grain boundaries provide the surfaces where chemical, physical and biological deterioration processes can take place (Siegesmund *et al.* 2002). The presence of moisture on and in stones is one of the most significant environmental factors contributing to stone decay. Usually related to the presence of water, salt decay is a major concern widely recognized in construction, cultural heritage, and monument stone decay assessment and conservation works. Salt crystallization inside the pore network is a major cause of damage to stone and other porous materials (Amoroso & Fassina 1983; Arnold & Kueng 1985; Arnold & Zehnder 1985; Arnold & Zehnder 1989; Trujillano *et al.* 1996; Smith & Kennedy 1998; Doehne 2002; Andriani & Walsh 2007; Angeli *et al.* 2007; Flatt *et al.* 2007; Rothert *et al.* 2007; Warke & Smith 2007).

Besides rock-forming minerals and their textural relations, pores and fissures are the most significant petrographical component governing the durability of natural stones. The pore space represents the preferred area for physical, chemical and biological weathering processes (Fitzner 1993; Smith & Kennedy 1998; Pérez-Bernal & Bello 2002; Benavente *et al.* 2007). A full understanding of pore-channel network geometrical characteristics (morphology, size and connectivity) is, therefore, important, and a classification of such a network system should be included in any technological characterization of stones (Winkler 1997). This understanding is also useful for conservation purposes (Iñigo *et al.* 1996; Carò & Giulio 2007).

This paper aims at providing a new contribution to the study of the role played by the pore system in controlling/determining the fluid-related properties and the resistance to salt crystallization of limestones. The studied stones are two Portuguese dimension stones ('Semi-rijo' and 'Moca Creme') widely used mainly in pavement and cladding inside and outside of buildings. New data regarding their petrography, pore structure, fluid-transport-related properties and durability (salt crystallization tests) are presented. An integrated analysis of all data has allowed comparison of the results of durability tests with pore network characteristics, fluid transport properties and petrographical features of the stones.

Materials

General features

The 'Semi-rijo' (SR) and 'Moca Creme' (MC) stones are calciclastic limestones representing different facies sourced from the same Bathonian (Middle Jurassic) 'Valverde' Formation that belongs to the Calcareous Massif of Estremadura, which is located in the Meso-Cenozoic Occidental Border of Portugal (*Catálogo de Rochas Ornamentais Portuguesas* (C.R.O.P.) 1983, 1984, 1985; Manuppella *et al.* 1985; Costa *et al.* 1988; Portuguese Natural Stones (P.N.S.) 1995). The Calcareous Massif of Estremadura is the main Portuguese centre of ornamental limestones, with a production of 630 000 tonnes (t) year^{-1} in 2000 (Carvalho *et al.* 2000). It covers an area of 900 km^2 in the centre of Portugal, about 100 km north of Lisbon. It is a thick sequence of carbonate Mesozoic rocks, structurally elevated, that were deposited in a sedimentary basin along the continental Portuguese margin (Manuppella *et al.* 1985; Carvalho *et al.* 1998, 2000; Moura 2001).

Regarding both stone types, some active quarries are located in the central region of Portugal (Fig. 1): District of Leiria, Municipality of Porto de Mós, for the 'Semi-rijo' (SR); and District of Santarém, Alcanede Parish, for the 'Moca Creme' (MC). For the 'Semi-rijo' (SR) variety, blocks of medium size are available and no preferential plane is taken into account to achieve a good ornamental pattern. Regarding the 'Moca Creme' (MC), its best ornamental aspect is achieved by cutting the blocks normal to bedding planes (C.R.O.P. 1983, 1984, 1985; Costa *et al.* 1988; Carvalho 1995).

Chemically, both stones are pure limestones, with values of CaO generally higher than 55% and silica content lower than 1% (Pettijohn 1975; Mason & Moore 1982; C.R.O.P. 1983, 1984, 1985; Figueiredo 1999; Carvalho *et al.* 2000; Moura 2001). Having compressive strength values of between 48 and 54 MPa ('Semi-rijo', SR) and from 75 to 92 MPa ('Moca Creme', MC), these stones have low (SR) to medium (MC) resistance to mechanical stress (Carmichael 1989).

Methodology

From two samples, of size 15 × 15 × 100 cm, of both lithotypes several test specimens were cut parallel (∥, H) and perpendicularly (⊥, V) to the bedding planes in order to evaluate the role of pore structure in controlling the durability of both limestones, and to get a visual and a quantitative characterization of their geometric features. Ordinary stereo and polarizing microscopes were used for macro- and micropetrographic analyses. Thin sections were prepared for detailed petrographic characterization by optical microscopy. To obtain a visual picture of the pore space and its structural relations with the rock-forming minerals and allochemical components by scanning electron microscopy (SEM), horizontal sections previously cut through the cylindrical specimens for petrophysical characterization were also analysed.

For each stone type and according to European (EN 1925:999; EN 1936:1999) and French (N FB 10-504:1973) Standards, 12 cylindrical specimens (50 mm diameter × 50 mm long) were used for the determination of open and free porosity, capillary imbibition and Hirschwald coefficient. Six of these were cut perpendicular to the bedding planes. In addition, 14 cylindrical-shape specimens (25 mm diameter × 25 mm long) were studied using mercury injection porosimetry (MIP). To perform the analysis a Micromeritics 33 000 psia (228 MPa) mercury porosimeter AutoPore III 9400 was used. This equipment enabled analysis of pore diameters ranging approximately from 360 to 0.005 μm (Micromeritics 1997). To evaluate the resistance of the limestones to salt crystallization, six 40 mm cubes of each stone were tested according to the European Standard EN 12370:1999. Samples were soaked in a 14% solution of sodium sulphate decahydrate (Na$_2$SO$_4$ · 10H$_2$O) for 2 h, and were then dried for 16 h in an ARALAB climatic chamber FITOCLIMA 300EDTU. As outlined in EN 12370:1999, the temperature was then raised to 105 ± 5 °C over a period of 12 hours.

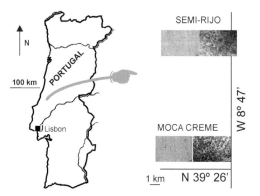

Fig. 1. Map of Portugal showing the locations of some active quarries the 'Semi-rijo' (SR) and 'Moca Creme' (MC) stones.

The temperature remained at this level for 4 h, after which heating was stopped and the samples allowed to cool to ambient temperature. This cycle was repeated 15 times. Finally, an integrated analysis of data gathered during the study was achieved by statistical and graphical methods. An attempt is made to establish some relationships between durability tests, pore size, geometry and structure of the pore network, and the properties related to fluid transport, as well as the petrographical features of the stones.

Results: analysis and discussion

Macro- and optical micropetrography

The commercial varieties 'Semi-rijo' (SR) and 'Moca Creme' (MC) are both calciclastic limestones. The 'Semi-rijo' (SR) is a light beige pelletal, oolitic, calciclastic limestone, with rare rounded bioclasts (Fig. 2a). The allochemical components have, in general, a wider range of sizes, from 0.08 to 1.6 mm. The 'Moca Creme' (MC) is a beige pelletal, oolitic, calciclastic limestone, richer in bioclasts (Fig. 2b). These ones are usually less rounded and have a centimetric size. The 'pellets' show a better uniformity of size, ranging from 0.03 to 0.08 mm. In thin sections the 'Semi-rijo' (SR) could be classified as a fossiliferous intrapelmicrosparite/grainstone (Fig. 2a) and the 'Moca Creme' (MC) as a biopelintrasparite/grainstone (Fig. 2b) (Pettijohn 1975).

As the porosity is not clearly visible in thin section by optical petrography, it is likely that very fine pores exist in these stones, as corroborated by the pore entry sizes obtained by MIP and SEM (see later).

Scanning electron microscopy (SEM)

For both stones the pores may occur associated with sparry calcite cement, according to classifications proposed by Bissell & Chilingar (1967), Pettijohn (1975) and Moore (1989), or intergranular and a with fissure-like shape (Fig. 3a) or within the pellets, showing a more irregular shape (Fig. 3b).

Dry and saturated bulk density, open and free porosity, capillary imbibition and Hirschwald coefficient

Both varieties, SR and MC, are moderately porous stones and have an open porosity accessible to water ranging, respectively, from 11.2 to 12.8% and from 12.1 to 12.9% (Table 1). In general, this

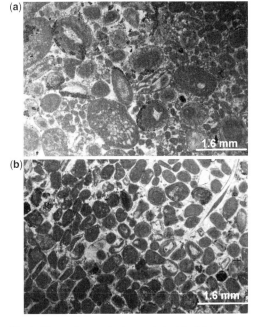

Fig. 2. Photomicrographs (crossed nicols) of thin sections of (a) 'Semi-rijo' (SR) (fossiliferous intrapelmicrosparite/grainstone) and (b) 'Moca Creme' (MC) (biopelintrasparite/ grainstone).

Fig. 3. SEM image of sparry calcite cement: (a) with intergranular fissure-like shape pores and of a pellet interior; and (b) with pores of irregular shape.

Table 1. Basic statistical parameters regarding stone fluid transport related properties and salt crystallization tests

	Stone types																
	'Semi-rijo' (SR)								'Moca Creme' (MC)								
	Samples ∥ to stratification				Samples ⊥ to stratification				Samples ∥ to stratification				Samples ⊥ to stratification			
	Mean	Max	Min	Std[‡]	Mean	Max	Min	Std[‡]	Mean	Max	Min	Std[‡]	Mean	Max	Min	Std[‡]
Dry bulk density (kg m^{-3})	2320.78	2339	2299	15.05	2322.28	2335	2305	11.70	2301.54	2312	2292	8.14	2308.92	2321	2294	9.85
Saturated bulk density (kg m^{-3})	2438.90	2451	2426	9.37	2439.76	2448	2431	6.57	2427.32	2436	2419	6.06	2432.08	2441	2419	8.11
Open porosity (%)	11.84	12.8	11.2	0.61	11.77	12.6	11.2	0.53	12.60	12.9	12.2	0.26	12.34	12.6	12.1	0.21
N48[*] (%)	11.58	12.4	11.0	0.52	11.40	12.3	10.6	0.66	12.24	12.6	11.8	0.33	11.95	12.4	11.5	0.37
S48[†] (%)	97.86	99.6	96.6	0.95	96.84	98.3	93.3	1.78	97.09	99.3	95.5	1.48	96.84	99.1	94.1	1.69
A[§] (g cm^{-2} h$^{-1/2}$)	0.223	0.28	0.20	0.023	0.218	0.26	0.18	0.027	0.216	0.26	0.20	0.020	0.173	0.20	0.14	0.017
B[§] (cm h$^{-1/2}$)	1.919	2.09	1.73	0.139	1.823	2.08	1.57	0.179	1.682	1.90	1.42	0.158	1.224	1.44	0.79	0.205
Weight loss (%)	11.42	13.03	9.88	–	11.23	12.90	9.67	–	20.82	22.64	18.62	–	17.06	19.98	11.34	–

Legend: [*]open porosity at 48 h; [†]Hirschwald coefficient; [‡]standard deviation; [§]capillary kinetic parameters; ∥, samples cut parallel to stratification; ⊥, samples cut perpendicularly to stratification.

compares well with the higher values of dry and saturated bulk density of the 'Semi-rijo' variety. Hirschwald coefficients are, in general, higher than 80%, suggesting freely interconnected and uniformly distributed pores (Mertz 1991; Hammecker 1993; Thomachot & Jeannette 2002). The 'Moca Creme' (MC) shows lower values for the two capillary kinetic parameters considered – A (mass increase per area per square root of time) and B (height of capillary rise per square root of time) – that range, respectively, from 0.143 to 0.256 g cm^{-2} h$^{-1/2}$ and from 0.7872 to 1.9029 cm h$^{-1/2}$, while 'Semi-rijo' samples have values of A ranging from 0.1812 to 0.2782 g cm^{-2} h$^{-1/2}$ and B ranging from 1.5721 to 2.0938 cm h$^{-1/2}$ (Fig. 4a, b). This could be related to the more heterogeneous texture and the presence of very thin (usually not larger than 0.5 cm) bedding planes in 'Moca Creme'. The lowest values were found for the samples cut perpendicular to the bedding

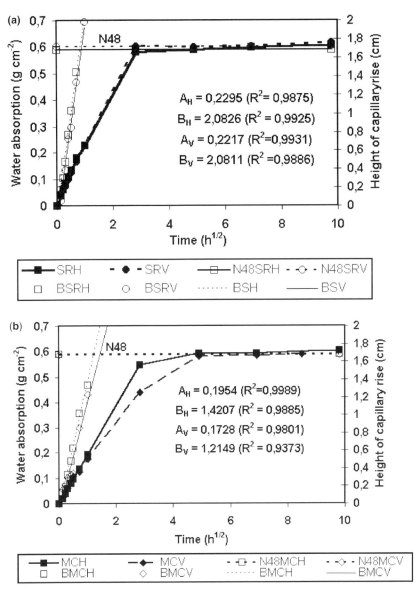

Fig. 4. Water absorption by capillary action on samples cut ∥ (H) and ⊥ (V) to the bedding planes: (**a**) 'Semi-rijo' (SR) (a) and (**b**) 'Moca Creme' (MC).

planes. The values of the capillary rise parameter, B, are relatively small when compared with the results from other carbonate stone types, such as those reported by Hammecker & Jeannette (1994): 2.4 cm h$^{-1/2}$ for 'Laspra limestone' (fine dolomicrite with an open porosity of 30.0%), 5.5 cm h$^{-1/2}$ for 'Lourdines limestone' (calcareous micrite with an open porosity of 25.8%) and 7.1 cm h$^{-1/2}$ for 'Hontoria limestone' (bioclastic limestone with an open porosity of 19.8%); or compared with results reported for less porous rocks, such as weathered granites (Alves *et al.* 1996; Begonha & Sequeira Braga 2002).

Mercury injection porosimetry (MIP)

According to MIP results (Fig. 5a, b and Table 2) the pore network structure for both stones, is composed essentially of micropores (pore radius ≤ 7.5 μm), comprising more than 90% of the total space invaded by mercury injection. However, the 'Semi-rijo' has, in general, a higher microporosity.

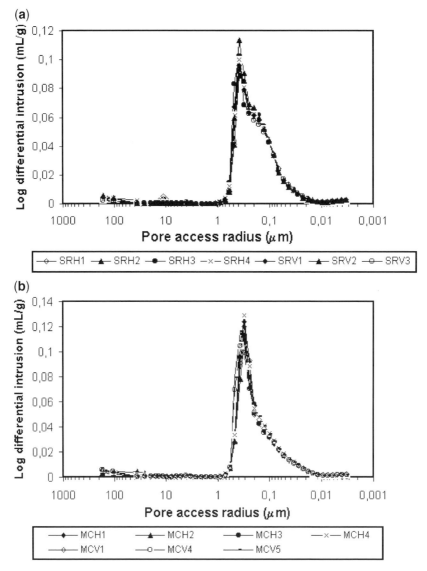

Fig. 5. Log differential mercury porosimetry curves for samples cut ∥ (H) and ⊥ (V) to the bedding planes: (**a**) 'Semi-rijo' (SR) and (**b**) 'Moca Creme' (MC).

Table 2. Basic statistical parameters for some characteristics of the pore system obtained by MIP

	Stone types																
	'Semi-rijo' (SR)								'Moca Creme' (MC)								
	Samples ∥ to stratification				Samples ⊥ to stratification				Samples ∥ to stratification				Samples ⊥ to stratification			
	Mean	Max	Min	Std[1]	Mean	Max	Min	Std[1]	Mean	Max	Min	Std[1]	Mean	Max	Min	Std[1]
NHg* (%)	14.20	14.6	13.7	0.38	14.18	15.1	13.5	0.75	14.39	15.2	13.9	0.70	14.62	15.2	14.2	0.43
NHg ($r^{\ddagger} \leq 7.5$ μm) (%)	13.65	14.2	13.3	0.39	12.69	14.2	9.7	2.07	12.40	14.2	8.2	2.39	12.92	14.4	11.4	1.27
NHg ($r > 7.5$ μm) (%)	0.55	0.6	0.4	0.10	1.48	3.9	0.4	1.64	1.98	7.0	0.5	2.79	1.69	3.2	0.7	1.20
NHg ($r \leq 7.5$ μm)/NHg (%)	96.13	97.0	95.5	0.74	89.33	97.0	71.1	12.19	86.66	96.3	54.1	18.24	88.41	95.2	78.7	8.21
$M_{prv}^{\ddagger}$ (μm)	0.193	0.20	0.19	0.007	0.195	0.21	0.19	0.009	0.260	0.42	0.21	0.089	0.235	0.26	0.22	0.014
$M_{pra}^{\S}$ (μm)	0.047	0.05	0.04	0.006	0.045	0.05	0.04	0.005	0.045	0.07	0.03	0.016	0.038	0.04	0.03	0.003
$A_{pr}^{\bullet}$ (μm)	0.090	0.10	0.08	0.006	0.091	0.10	0.09	0.004	0.099	0.12	0.09	0.016	0.094	0.10	0.09	0.007

Legend: *porosity by MIP; ‡ pore access radius (μm); ‡median pore radius by volume; §median pore radius by area; • average pore radius (Bear 1972; Micromeritics 1997); [1]standard deviation; ∥, samples cut parallel to stratification; ⊥, samples cut perpendicularly to stratification.

The curves obtained are typical of unimodal pore structures (Bear 1972; Mertz 1991; Hammecker 1993; Micromeritics 1997; Thomachot & Jeannette 2002), explaining the Hirschwald coefficient and the capillary imbibition results. Log differential curves are skewed towards low values of pore access radius (Fig. 5a, b).

The 'Moca Creme' with median (M_{prv}, median pore radius by volume) and average (A_{pr}) pore radii (for a definition see, for instance, Bear 1972 and Micromeritics 1997), ranging, respectively, from 0.2116 to 0.4185 μm and from 0.0851 to 0.1238 μm, has a larger pore access in comparison to 'Semi-rijo'. The values of porosity (NHg, %) estimated by MIP confirm the higher porosity values of 'Moca Creme' obtained by water imbibition.

Salt crystallization tests

Regarding the durability tests, the 'Moca Creme' seems to be less resistant to salt crystallization than 'Semi-rijo' (see Figs 6a–d and 7a–e, Table 1). All

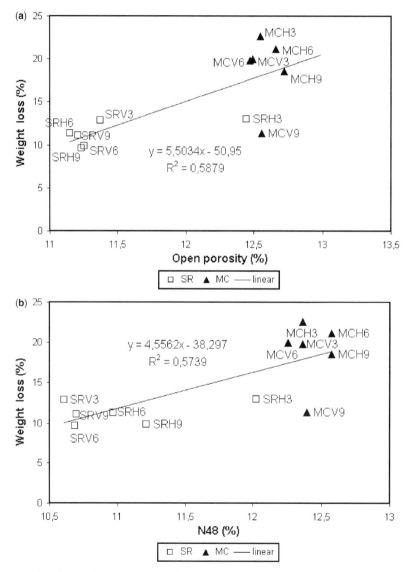

Fig. 6. Results of the salt crystallization test: plots of weight loss (%) v. (**a**) open porosity (%) and (**b**) open porosity at 48 h (N48, %); plots of weight change (%) against the number of loading cycles for samples cut ∥ (H) and ⊥ (V) to the bedding planes: (**c**) 'Semi-rijo' (SR) and (**d**) 'Moca Creme' (MC).

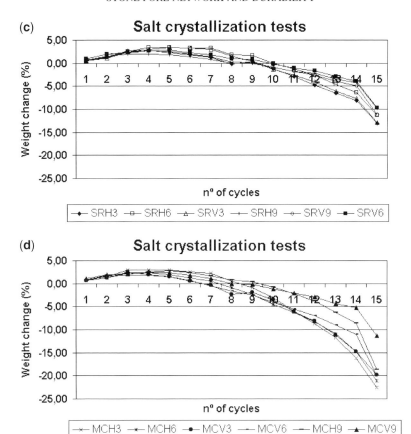

Fig. 6. *Continued.*

of the basic statistical parameters considered for weight loss (Table 1) are higher for samples parallel to the bedding planes than for samples normal to bedding planes in both stone types, the difference being greatest for the 'Moca Creme' samples.

Salt crystallization test results seem to be best correlated with the free (N48, %) and open porosity (%) accessible to water (Fig. 6a, b and Table 3). The positive correlation (correlation coefficients, r, respectively, equal to +0.7575 and +0.7667) found between the weight loss (%) and the free (N48, %) and open porosity (%) accessible to water suggests, in general, that the higher the pore space available to the salt solution the more deteriorated are the stones. A similar relationship was found by McGreevy (1996) for three varieties of Caen limestone, Calvados, France. Somehow, these results are different from those of Inkpen

Table 3. *Correlation coefficients between fluid related properties and weight loss by salt crystallization tests*

Physical properties	Weight loss by salt crystallization tests
Saturated bulk density (kg m^{-3})	-0.6942
Open porosity (%)	0.7667
Open porosity at 48 h (%) (N48)	0.7575
Hirschwald coefficient (%) (S48)	0.4381
Capillary kinetic parameter A (g cm^{-2} h$^{-1/2}$)	0.2557
Capillary kinetic parameter B (cm h$^{-1/2}$)	-0.0863

Fig. 7. Cubic samples (40 × 40 × 40 mm) before (**a**, **c**) and after (**b**, **d**, **e**) salt crystallization tests. 'Semi-rijo' (SR) (**a**, **b**, **e**) and 'Moca Creme' (MC) (**c**, **d**). In (**e**) some details are presented of salt decay features (fracturing and pelloid detachment) on 'Semi-rijo' specimens obtained by SEM observations at the end of the 15 cycles used in the salt crystallization tests.

et al. (2000) for Portland Stone. They obtained a stronger association between the saturation coefficient and weight loss. However, according to Lewin (1989), the susceptibility of a stone to salt decay is a function, among other factors, of porosity, i.e. the total void volume: a large porosity means relatively easy access of solution. Hence, the higher values of free (N48, %) and open porosity (%) for 'Moca Creme', combined also with its larger pore access, suggest for this stone a pore system better interconnected and more accessible to salt solution than that of 'Semi-rijo'. Based on this interpretation, the negative correlation ($r = -0.6942$) between the weight loss and the saturated bulk density is not surprising, considering the mineralogical homogeneity of the samples (Table 3).

Despite this observation, in some cases, of an incipient disintegration of some samples at the very beginning of the crystallization tests (third cycle, samples SRH9, MCV6 and MCV3), most of them only started losing weight after the fourth cycle (Fig. 6c, d). However, no evidence of break down of any sample was shown until a minimum weight increase of 2%, due to salt accumulation, had been achieved. Each stone sample behaves differently, however. The differences in response to salt crystallization observed within the set of samples from the 'Moca Creme' (MC) limestone are much more pronounced than those observed for the 'Semi-rijo' (SR) limestone samples. These differences are in agreement with the more heterogeneous structure/texture of the 'Moca Creme' (MC) limestone, as already pointed out.

The degradation of the stones is mainly due to pellet detachment, surface flaking and scaling, sparry calcite cement fracturing, and granular disintegration causing the rounding of edges and corners, and an unevenly distributed loss of mass and volume, and changing the shape of the cubic test samples (Fig. 7a–d) (patterns similar to those found by Langella et al. 2000; Kouzeli & Pavelis 2004; Rothert et al. 2007). As is shown in Figure 7e, the weathering forms caused by the mechanical breakdown by salt crystallization observed could mainly be related to the development of a secondary, inter- or transgranular microfracture network in the studied stones around the contacts between pelloids, bioclasts and sparry cement, as referred by Warke & Smith (2007).

By the end of the 15 test cycles no sample showed evidence of complete failure and a maximum weight loss of 22.6% was only registered in a sample (MCH3) of 'Moca Creme' (MC) limestone. In general, the samples that accumulated more salt during the tests were shown to be less resistant and experienced a greater loss in weight.

Final considerations

Results from salt crystallization tests point towards the importance of free (N48, %) and open porosity (%) on the susceptibility to salt weathering of these stones. Besides petrographical heterogeneities, like the presence of stratification, pellets, oolites and fossils, the imposed (cubic) geometry of the samples seems to control the morphological changes induced by the salt decay. Decay is clearly concentrated on the edges and corners of specimens.

Both 'Semi-rijo' (median pore radius ranging from 0.1864 to 0.2064 μm) and 'Moca Creme' (median pore radius ranging from 0.2116 to 0.4185 μm) stone types have freely interconnected pore networks with uniformly distributed small pore radius. These characteristics can explain the high Hirschwald coefficients (McGreevy 1996) as well as the relatively small values of the capillary rise parameter. However, the higher values of free (N48, %), open porosity (%) and pore access radius for the 'Moca Creme', suggest that this stone has a pore system better interconnected and more accessible to salt solution than the 'Semi-rijo'. This seems to explain its lower resistance to salt decay processes.

Taking into account the data already published regarding these dimension stones, the results presented in this paper are a valuable contribution to their technical characterization. They could be relevant to the assessment of stone susceptibility to weathering, conservation treatments (namely consolidation) and definition of specifications. However, as already suggested by several authors (see, for instance, Thomachot & Jeannette 2002; Andriani & Walsh 2007; Rothert et al. 2007; Warke & Smith 2007), a better understanding of the role of pore structure characteristics in controlling the durability of this type of material could be obtained through an improvement in the methodological approach used so far. This could be achieved by introducing some modifications to the standard laboratory-based simulation (*durability*) tests regarding the geometry of test specimens, climatic conditions and chemical characteristics of solutions, in order to reflect 'real-world' conditions and allow the development of stone-specific decay characteristics.

This study received funding from Portuguese Fundação para a Ciência e a Tecnologia (FCT), Project POCTI/CTA/44940/2002_PORENET, with financial support from European Union FEDER and the state budget of the Portuguese Republic and Centro de Petrologia e Geoquímica/IST/TULisbon subproject DECASTONE. We also would like to thank K. Malaga and P. Warke for their reviews, which improved the quality of this paper.

References

ALVES, C., SEQUEIRA BRAGA, M. A. & HAMMECKER, C. 1996. Water transfer and decay of granitic stones in monuments. *Comptes Rendus de l'Académie des Sciences, Paris*, **323**, 397–402.

AMOROSO, G. G. & FASSINA, V. 1983. *Stone Decay and Conservation: Atmospheric Pollution, Cleaning, Consolidation and Protection.* Materials Science Monographs, **11**.

ANDRIANI, G. F. & WALSH, N. 2007. The effects of wetting and drying, and marine salt crystallization on calcarenite rocks used as building material in historic monuments. *In*: PŘIKRYL, R. & SMITH, B. J. (eds) *Building Stone Decay: from Diagnosis to Conservation.* Geological Society, London, Special Publications, **271**, 179–188.

ANGELI, M., BIGAS, J. P., BENAVENTE, D., MENÉNDEZ, B., HÉBERT, R. & DAVID, C. 2007. Salt crystallization in pore: quantification and estimation of damage. *In*: SIEGESMUND, S. & STEIGER, M. (eds) *Salt Decay.* Environmental Geology, **52**, 187–195.

ARNOLD, A. & KUENG, A. 1985. Crystallization and habits of salt efflorescences on walls. Part I: Methods of investigations and habits. *In*: FÉLIX, G. (ed.) *Proceedings of the 5th International Congress on Deterioration and Conservation of Stone.* Volume 1. Presses Polytechniques Romandes, Lausanne, 255–267.

ARNOLD, A. & ZEHNDER, K. 1985. Crystallization and habits of salt efflorescences on walls. Part II: Conditions of crystallization. *In*: FÉLIX, G. (ed.) *Proceedings of the 5th International Congress on Deterioration and Conservation of Stone.* Volume 1. Presses Polytechniques Romandes, Lausanne, 269–277.

ARNOLD, A. & ZEHNDER, K. 1989. Salt weathering on monuments. *In*: ZEZZA, F. (ed.) *The Conservation of Monuments in the Mediterranean Basin. Proceedings of the 1st International Symposium on the Influence of Coastal Environment and Salt Spray on Limestone and Marble.* Grafo Edizioni, Bari, 31–58.

BEAR, J. 1972. *Dynamics of Fluids in Porous Media.* Dover Publications, New York.

BEGONHA, A. & SEQUEIRA BRAGA, M. A. 2002. Weathering of the Oporto granite: Geotechnical and physical properties. *Catena*, **49**, 57–76.

BENAVENTE, D., CUETO, N., MARTÍNEZ-MARTÍNEZ, J., CURA, M. A. G. & CAÑAVERAS, J. C. 2007. The influence of petrophysical properties on the SALT weathering of porous building rocks. *In*: SIEGESMUND, S. & STEIGER, M. (eds) *Salt Decay.* Environmental Geology, **52**, 197–206.

BISSELL, H. J. & CHILINGAR, G. V. 1967. Classification of sedimentary carbonate rocks. *In*: CHILINGAR, G. V., BISSELL, H. J. & FAIRBRIDGE, R. W. (eds) *Carbonate Rocks. Origin, Occurrence and Classification.* Developments in Sedimentology, **9A**, 87–168.

CARMICHAEL, R. S. 1989. *Practical Handbook of Physical Properties of Rocks and Mineral.* CRC Press, Boca Raton, FL.

CARÒ, F. & GIULIO, A. D. 2007. Rock petrophysics v. performance of protective and consolidation treatments: the case of Mt Arzolo Sandstone. *In*: PŘIKRYL, R. & SMITH, B. J. (eds) *Building Stone Decay: from Diagnosis to Conservation.* Geological Society, London, Special Publications, **271**, 287–294.

CARVALHO, J. 1995. Calcários Ornamentais e Industriais na Área de Pé da Pedreira (Maciço Calcário Estremenho). *Boletim de Minas, Lisboa*, **32**(1), 25–39.

CARVALHO, J., MANUPPELLA, G. & MOURA, A. C. 1998. Contribution to the geological knowledge of the Portuguese Ornamental Limestone. *Actas do V Congresso Nacional de Geologia*, **84**(2), 74–77.

CARVALHO, J., MANUPPELLA, G. & MOURA, A. C. 2000. Calcários Ornamentais Portugueses. *Boletim de Minas, Lisboa*, **37**(4), 223–232.

CATÁLOGO DE ROCHAS ORNAMENTAIS PORTUGUESAS (C. R. O. P.). 1983, 1984, 1985. *Volumes I, II, III, Direcção-Geral de Geologia e Minas (eds) Ministério da Indústria, Energia e Exportação.* Lisbon, Portugal.

COSTA, J. R. G., MOREIRA, J. C. B. & MANUPPELLA, G. 1988. Calcários Ornamentais do Maciço Calcário Estremenho. *Estudos, Notas e Trabalhos, D. G. G. M.*, **30**, 51–88.

DOEHNE, E. 2002. Salt weathering: a selective review. *In*: SIEGESMUND, S., WEISS, T. & VOLLBRECHT, A. (eds) *Natural Stone, Weathering Phenomena, Conservation Strategies and Case Studies.* Geological Society, London, Special Publications, **205**, 51–64.

EN 1925:1999. *Methods of Test for Natural Stone Units – Determination of Water Absorption Coefficient due to Capillary Action.* European Committee for Standardization, Brussels.

EN 1936:1999. *Methods of Test for Natural Stone Units – Determination of Real Density and Apparent Density and of Total and Open Porosity.* European Committee for Standardization, Brussels.

EN 12370:1999. *Natural Stone Test Methods: Determination of Resistance to Salt Crystallisation.* European Committee for Standardization, Brussels.

FIGUEIREDO, C. 1999. *Alteração, Alterabilidade e Património Cultural Construído: o caso da Basílica da Estrela.* PhD thesis, Technical University of Lisbon, IST, Portugal.

FITZNER, B. 1993. Porosity properties and weathering behaviour of natural stones. Methodology and examples. *In*: ZEZZA, F. (ed.) *Stone Material in Monuments: Diagnosis and Conservation (Second Course), Heraklion, Crete.* C. U. M. University School of Monument Conservation, Mario Adda Editore, Bari, 43–54.

FLATT, R. J., STEIGER, M. & SCHERER, G. W. 2007. A commented translation of the paper by C. W. Correns and W. Steinborn on crystallization pressure. *In*: SIEGESMUND, S. & STEIGER, M. (eds) *Salt Decay.* Environmental Geology, **52**, 221–237.

HAMMECKER, C. 1993. *Importance des transferts d'eau dans la dégradation des pierres en oeuvre.* PhD thesis, University Louis Pasteur, Strasbourg.

HAMMECKER, C. & JEANNETTE, D. 1994. Modelling the capillary imbibition kinetics in sedimentary rocks: role of petrographical features. *Transport in Porous Media*, **17**, 285–303.

IÑIGO, A. C., GARCÍA-TALEGÓN, J., TRUJILLANO, R. & VICENTE-TAVERA, S. 1996. Influence of consolidation and hydrofugation treatments on the physical properties of a Vila natural and artificially aged granite: a statistical approach. *In*: VICENTE, M. A.,

DELGADO-RODRIGUES, J. & ACEVEDO, J. (eds) *Degradation and Conservation of Granitic Rocks in Monuments*. Protection and Conservation of European Cultural Heritage, Research Report, **5**, 145–151.

INKPEN, R. J., PETLEY, D. & MURPHY, W. 2000. Durability and rock properties. *In*: SMITH, B. J. & TURKINGTON, A. V. (eds) *Stone Decay. Its Causes and Controls. Proceedings of Weathering 2000. An International Symposium held in Belfast*. Donhead, Shaftesbury, Dorset, 33–52.

KOUZELI, K. & PAVELIS, C. 2004. Study of the building materials of the 'Demossion Sema' monuments in Athens. *In*: KWIATKOWSKI, D. & LÖFVENDAHL, R. (eds) *Proceedings of the 10th International Congress on Deterioration and Conservation of Stone, Stockholm*. Volume 1. COMOS Sweden, Stockholm, 407–414.

LANGELLA, A., CALCATERRA, D., CAPPELLETTI, P., COLELLA, A., GENNARO, M. & GENNARO, R. 2000. Preliminary contribution on durability of some macroporous monumental stones in historical towns of Campania region, Southern Italy. *In*: FASSINA, V. (ed.) *Proceedings of the 9th International Congress on Deterioration and Conservation of Stone, Venice*. Volume 1. Elsevier Science, Amsterdam, 59–67.

LEWIN, S. 1989. The susceptibility of calcareous stones to salt decay. *In*: ZEZZA, F. (ed.) *The Conservation of Monuments in the Mediterranean Basin. Proceedings of the 1st International Symposium on the Influence of Coastal Environment and Salt Spray on Limestone and Marble*. Bari, Grafo Edizioni, 59–63.

MANUPPELLA, G., MOREIRA, J. C. B., COSTA, J. R. G. & CRISPIM, J. A. 1985. Calcários e dolomitos do Maciço Calcário Estremenho. *Estudos, Notas e Trabalhos, D. G. G. M.*, **27**, 3–48.

MASON, B. & MOORE, C. B. 1982. *Principles of Geochemistry*. 4th edn. Wiley, New York.

MCGREEVY, J. P. 1996. Pore properties of limestones as controls on salt weathering susceptibility: a case study. *In*: SMITH, B. J. & WARKE, P. A. (eds) *Processes of Urban Stone Decay*. Donhead, Shaftesbury, Dorset, 150–167.

MERTZ, J. D. 1991. *Structures de porosité et propriétés de transport dans les grès*. Sciences géologiques, **90**. PhD thesis, University Louis Pasteur, Strasbourg.

Micromeritics. 1997. *AutoPore III Operator's Manual V1.xx*. Micromeritics, Norcross, GA.

MOORE, C. H. 1989. Carbonate Diagenesis and Porosity. *Developments in Sedimentology*, **46**.

MOURA, A. C. 2001. A pedra natural ornamental em Portugal – Nota Breve. *Boletim de Minas, Lisboa*, **38**(3), 161–177.

N FB 10-504:1973. *Pierres calcaires: mesure du coefficient d'absorption d'eau*. Association Française de Normalisation (AFNOR), Paris.

PÉREZ-BERNAL, J. L. & BELLO, M. A. 2002. Weathering effects on stone pore size distributions. *In*: GALÁN, E. & ZEZZA, F. (eds) *Protection and Conservation of the Cultural Heritage of the Mediterranean Cities*. Swets & Zeitlinger, Lisse, 203–207.

PETTIJOHN, F. J. 1975. *Sedimentary Rocks*. 3rd edn. Harper & Row, New York.

Portuguese Natural Stones (P. N. S.). 1995. *CEVALOR (Centro Tecnológico para o Aproveitamento e Valorização das Rochas Ornamentais e Industriais)*. Borba, Portugal.

ROTHERT, E., EGGERS, T., CASSAR, J., RUEDRICH, J., FITZNER, B. & SIEGESMUND, S. 2007. Stones properties and weathering induced by salt crystallization of Maltese Globigerina Limestone. *In*: PŘIKRYL, R. & SMITH, B. J. (eds) *Building Stone Decay: From Diagnosis to Conservation*. Geological Society, London, Special Publications, **271**, 189–198.

SIEGESMUND, S., WEISS, T. & VOLLBRECHT, A. 2002. Natural stone, weathering phenomena, conservation strategies and case studies: introduction. *In*: SIEGESMUND, S., WEISS, T. & VOLLBRECHT, A. (eds) *Natural Stone, Weathering Phenomena, Conservation Strategies and Case Studies*. Geological Society, London, Special Publications, **205**, 1–7.

SMITH, B. J. & KENNEDY, E. M. 1998. Moisture loss from stone influenced by salt accumulation. *In*: JONES, M. S. & WAKEFIELD, R. D. (eds) *Stone Weathering and Atmospheric Pollution Network '97: Aspects of Stone Weathering, Decay and Conservation*. Imperial College Press, London, 55–64.

THOMACHOT, C. & JEANNETTE, D. 2002. Evolution of the petrophysical properties of two types of Alsatian sandstone subjected to simulated freeze–thaw conditions. *In*: SIEGESMUND, S., WEISS, T. & VOLLBRECHT, A. (eds) *Natural Stone, Weathering Phenomena, Conservation Strategies and Case Studies*. Geological Society, London, Special Publications, **205**, 19–32.

TRUJILLANO, R., IÑIGO, A. C., RIVES, V. & VICENTE, M. A. 1996. Behaviour of three different types of granite under forced alteration. *In*: VICENTE, M. A., DELGADO-RODRIGUES, J. & ACEVEDO, J. (eds) *Degradation and Conservation of Granitic Rocks in Monuments*. Protection and Conservation of European Cultural Heritage, Research Report, **5**, 89–93.

WARKE, P. A. & SMITH, B. J. 2007. Complex weathering effects on durability characteristics of building stone. *In*: PŘIKRYL, R. & SMITH, B. J. (eds) *Building Stone Decay: From Diagnosis to Conservation*. Geological Society, London, Special Publications, **271**, 211–224.

WINKLER, E. M. 1997. *Stone in Architecture. Properties, Durability*. 3rd edn. Springer, Berlin.

Limestone on the 'Don Pedro I' facade in the Real Alcázar compound, Seville, Spain

C. VAZQUEZ-CALVO[1]*, M. J. VARAS[1,2], M. ALVAREZ DE BUERGO[1] & R. FORT[1]

[1]*Instituto de Geología Económica, CSIC-UCM, Facultad de Ciencias Geológicas, Universidad Complutense de Madrid, C/José Antonio Nováis 2, 28040 Madrid, Spain*

[2]*Departamento de Petrología y Geoquímica, Facultad de Ciencias Geológicas, Universidad Complutense de Madrid, C/José Antonio Nováis 2, 28040 Madrid, Spain*

*Corresponding author (e-mail: carmenvazquez@geo.ucm.es)

Abstract: This paper discusses the research conducted prior to restoring the 'Don Pedro I' facade on the Real Alcázar or royal palace at Seville, Spain. The different types of stone on the facade were located and characterized, and their state of decay mapped. Although other materials (brick, rendering, ceramics, marble) are present on the facade, its main elements are made from two types of limestone: *palomera* and *tosca*, each in a different state of conservation and exhibiting distinct behaviour. Colour parameters, real and bulk densities, compactness, open porosity, water saturation coefficient and total porosity were determined to characterize the two varieties. In addition, ultrasonic techniques were used to map the various levels of decay on the facade, stone by stone, for future interventions. The findings show that owing to its petrographical and petrophysical properties, *palomera* stone is of a lower quality than *tosca* stone, and has undergone more intense deterioration.

The *Real Alcázar* or royal palace at Seville was built in Mudejar style in 913 by Muslim emir Abd Ar Rahman III (Chávez González 2004; Tabales 2005). The Don Pedro I facade (Fig. 1), which dates from the fourteenth century, was built in the same style. The royal compound underwent a number of transformations to meet the needs and preferences of successive rulers (Almagro 2005, 2006). The palace was substantially overhauled in the nineteenth century in light of its very poor state of repair at the time (Chávez González 2004). This so-called Romantic Restoration, which took place during the long reign of Isabel II, and particularly the works performed in the first half of the century were harshly criticized as they distorted and marred the original. Among others, in the restoration the original finishes on the facade were removed and whitewashed. Many of the most severely damaged stone ornaments and ashlars were replaced in the mid-nineteenth century. The work carried out in the second half of the century was much more respectful of the history of the building, tending to conserve and recover rather than restore. Further work was carried out on the facade in the twentieth century, primarily in 1937, again including the replacement of dimension stones.

The facade is divided from top to bottom into three storeys (upper, middle and lower), flanked on both sides by galleries. The wall itself consists in ashlars and ornate arabesques. Two types of limestone, namely *tosca* and *palomera*, were used in both the dimension stones and ornamental elements (Chávez González 2004). *Tosca* was used on the lower storey, while both ashlars and arabesques on the middle and upper storeys were made with *palomera*. The upper third of the facade has marble columns and elements made of other materials, such as ceramic tile, mortar and bricks.

The map of building materials used in the facade is given in Figure 2.

Methodology

Both stones were characterized and their state of deterioration determined; a petrographical analysis was conducted and the respective petrophysical properties were measured. While the samples taken consisted primarily of limestone scales and chips, core specimens (20 mm in diameter, 100 mm long) were also extracted. The resulting voids were filled with restoration mortar. The forms of decay identified on the surface of the facade were mapped on an elevation drawing. The equipment used for characterization and analysis was as follows.

- Olympus BX51 polarized light optical microscope (OM) coupled to an Olympus DP12 digital camera, using Olympus DP-Soft (version 3.2) software. Alizarin staining was

Fig. 1. Don Pedro I facade. The area studied is framed.

used to differentiate the carbonates. The Folk scheme was used to classify the rocks in this study (Folk 1962).
- JEOL JSM 6400 scanning electron microscope (SEM): acceleration voltage 0.2–40 kV; beam current, 6×10^{-10} A; vacuum conditions, 10^{-5} mbar; resolution, 35 Å, working distance, 8 mm at 35 kV. The images were recorded at an acceleration voltage of 20 kV. An Oxford-INCA X-ray energy dispersive spectrometer (EDS) with a nominal resolution of 133 eV at 5.39 kV was coupled to the SEM. Samples were graphite sputtered (15 mm thick cover) with a Balzers Med010 sputter coater (Balzers, Liechtenstein).
- Micromeritics Autopore IV 9500 mercury intrusion porosimeter (MIP). This instrument measures pore diameters from 0.001 to 1000 μm, from atmospheric pressure up to 60 000 psi (228 MPa). Core specimens 10 mm in diameter and at least 30 mm high were used. The cut-off defining micro- and macroporosity was 5 μm; the parameters recorded were the mean and median pore diameter, porosity accessible to mercury and specific surface area.
- CNS Electronics Pundit for measuring ultrasonic wave propagation velocity (V_p) on the facade using the indirect transmission method. The transducer frequency was 54 kHz and the contact area flat and circular, measuring 50 mm in diameter. A Plasticine-based couplant was used to secure the transducers to the surface. The measurement grid consisted of squares with sides 15 cm long. A total of around 4400 values were obtained. The petrological characteristics of each variety of stone influences the ultrasounds velocity, the lower the velocity, the more severe was the stone decay for each kind of stone.
- Chromatic parameters (CIE 1986) were likewise measured using a Minolta CM-2002 spectrophotometer with CM-1 colour data software and illuminant D65 at an observation angle of 10°. The chromatic co-ordinates ($a*$ and $b*$) were recorded, along with luminosity ($L*$), white index (WI) and yellow index (YI) (Materials, A.S.f.T.a. 2000). Chroma ($C* = \sqrt{[(\Delta a*)^2 + (\Delta b*)^2]}$) and overall colour variation ($\Delta E* = \sqrt{[(\Delta L*)^2 + (\Delta a*)^2 + (\Delta b*)^2]}$) were measured to characterize the two types of stones, as well as to determine the degree of soiling on each compared to the original colour.
- Surface hardness was measured using the sclerometric method (Original-Schmidt Type N Proceq test hammer) *in situ*, directly on the facade. Surface hardness is yet another parameter indicative of material deterioration.
- The other petrophysical properties determined included capillary water absorption, real and bulk densities, compactness, open porosity or porosity accessible to water and water saturation (RILEM 1980).

Geological setting

The building rocks studied were extracted from Tertiary outcrops (Upper Miocene) in the sub-Betic area, i.e. the southern border of the Guadalquivir River basin. *Palomera* or *franca* stone is Tortonian and comprises primarily white limestone (biomicrite according to the Folk 1962 classification). *Tosca*, stratigraphically found above *palomera*, is a coarse-grained bioclastic limestone (biosparrudite according to Folk 1962) (IGME 1959, 1977, 1986, 1987).

These stones were most probably drawn from the quarries located around what are today Utrera and Morón de la Frontera in the province of Seville, and El Puerto de Santa María in Cadiz, which were extensively mined for construction until the eighteenth century. Utrera and Morón are 35 and 65 km SE of Seville, respectively, while El Puerto de Santa María is 120 km nearly due south.

These two stones were used to build a number of Sevillian heritage monuments, such as the cathedral, El Salvador Church and the city hall (Herrero Fernández 2007; Barrios Padura *et al.* 2006).

Results

As noted earlier, *tosca*, a medium- to coarse-grained carbonate rock, was chiefly used in the lower storey ashlars; while *palomera*, a fine-grained limestone, is present in the upper storey ashlars, as well as on the ornaments (arabesques) in the middle and lower thirds of the facade, and as a support for ceramic tiles, paint, lime rendering and so forth. The ochre-coloured *tosca* stone (Table 1), while not severely decayed, was found to be highly soiled (Table 1), with material missing on the socle. The fine-grained,

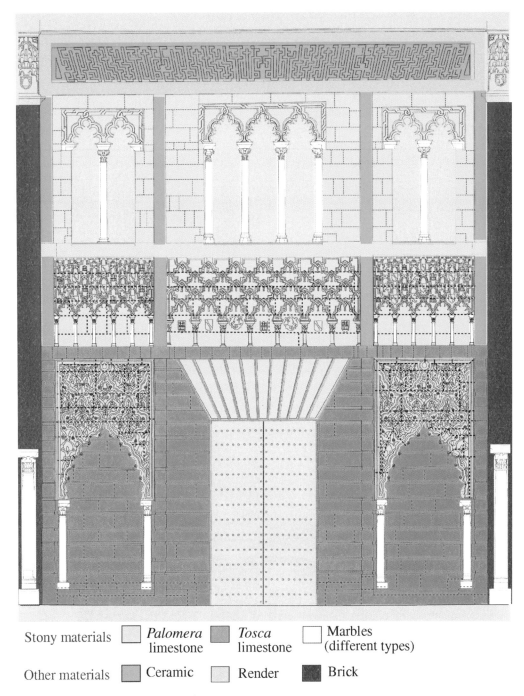

Fig. 2. Map of building materials on the facade.

low-cohesion *palomera*, also known as *franca* stone, is very soft and massive (Fig. 3a). Originally whitish-ochre (Table 1) it, like *tosca*, was observed to be badly soiled (Table 1).

Petrographically speaking, the *tosca* samples contained a high percentage of seashells – bioclasts – along with quartz grains, and has a rough surface. While at first glance they appeared to be compact,

Table 1. *Fresh and soiled* tosca *and* palomera *stone colour parameters*

Stone	L^*	a^*	b^*	C^*	YI	WI	–
Tosca	78.8	2.6	16.4	16.6	27.7	−5.5	–
Palomera	83.5	1.3	14.5	14.6	23.6	3.6	–
	ΔL^*	Δa^*	Δb^*	ΔC^*	ΔYI	ΔWI	ΔE^*
Tosca (soiled)	−30.0	0.7	−4.0	−3.8	2.3	1.3	30.4
Palomera (soiled)	−20.6	3.1	1.7	2.3	9.0	−9.5	22.0

L^*, Luminosity; $+a^*$, red hue; $-a^*$, green hue; $+b^*$, yellow hue; $-b^*$, blue hue; C^*, colour saturation; YI, yellow index; WI, white index.

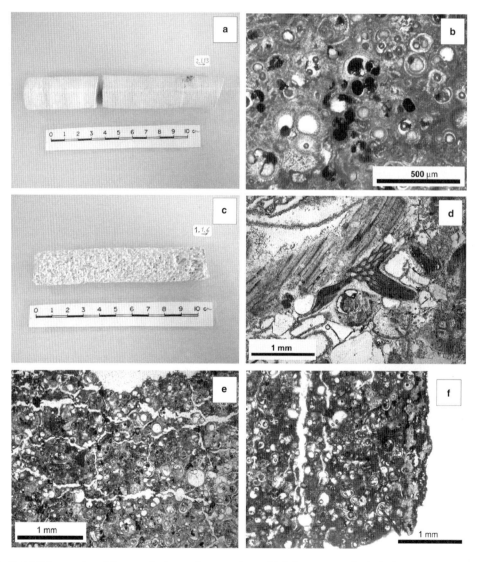

Fig. 3. (**a**) *Palomera* stone. Macroscopic picture of core specimen; (**b**) *Palomera* stone. Optical microscopic image; (**c**) *Tosca* stone. Macroscopic picture of core specimen; (**d**) *Tosca* stone. Optical microscopic image; (**e**) Fissures following every space direction in *palomera* stone from arabesques; (**f**) the fissural porosity is scarce and not deep (1–2 mm below the surface) in the *palomera* stone placed at the back of the arabesques. Fissures follow a horizontal direction.

they were in fact fairly porous. Bioclasts accounted for nearly 40% of the entire rock content, with particle sizes over 2 mm in many of the samples studied (Fig. 3c). The petrographical study (Fig. 3d) revealed that bryozoa, echinoderms, bivalves and some foraminifers can be distinguished in the bioclasts, all greatly fragmented. The bryozoan and bivalve shells contained micrite (10%) and pseudosparite (recrystallization of micrite). The cement overgrowth visible around the echinoderm plates probably formed during burial diagenesis, while the acicular cement around the bryozoan and bivalve fragments observed are typical of phreatic marine areas.

In addition to the bioclasts, the samples of these limestones had around 30% round, monocrystalline, quartz grains whose average size ranged from 500 μm to 1 mm. Like the bioclasts, they were extensively cracked. Rock porosity was mainly interparticle. Over 10% of the pores had been generated as a result of matrix dissolution processes. In the Folk (1962) classification, this is a biosparrudite rock.

Petrographically, *palomera* is a massive, grain-supported stone (Fig. 3b). The main skeletal grains found in the samples were globigerina fossils (around 75%), with a mean size of 150–200 μm. The greenish-brown marly interparticle matrix (15%) was a result of dissolution. Some intraparticle and some mouldic porosity were likewise present. Some of the mouldic pores were partially cemented with sparry cement. The bioclasts were surrounded by an acicular cement, typical of marine phreatic zones. These limestone samples also had a quartz content of around 5%. These monocrystalline grains, with an average size ranging from 65 to 125 μm, were angular in shape, contrary to the quartz particles in *tosca* limestone, and extensively

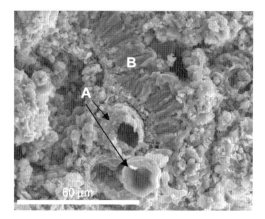

Fig. 4. SEM-SE mode image in which the organic material filling the interior of globigerina chamber (A) and the fibrous cement that surround them can be observed.

cracked. Glauconite was likewise present as an authigenic mineral. The globigerina chambers were very often seen to be filled with an organic material (carbon was the only element detected by EDS) (Fig. 4). This rock can be classified as a biomicrite (Folk 1962).

The prevalent forms of decay observed in *palomera* stone were flaking, granular disintegration and cracking (Fig. 5). The outcome of the surface instability resulting from these pathologies will be detachment and the concomitant distortion of the shape of the arabesques.

The organic material inside the shells in the stone was observed to oxidize and decompose very rapidly on exposed surfaces, increasing intraparticle porosity in an outward direction and favouring the

Fig. 5. (a) Granular disintegration and loss of material and (b) fissures, both from *Palomera* stone samples.

appearance of fissure porosity. The degree of stone decay was found to depend on its location on the facade.

A – Fissure porosity was high in the reliefs and arabesques, over 20% in many cases. These fissures were multidirectional, causing the material to crack and eventually to flake, with intergranular disintegration at surface level. In addition, intraparticle porosity rose, probably as a result of the decomposition of the material mentioned above (Fig. 3e).

B – Fissure porosity was rare and fairly shallow (1–2 mm below the surface) in the stone at the rear of the arabesques (Fig. 3f). Here the fissures were horizontal and intraparticle porosity was likewise rare, inferring that the high-relief arabesques protected the stone from weathering.

C – The stone in the pilaster areas (middle and upper thirds) was very poorly conserved, exhibiting high fissure porosity. Moreover, the decayed material had not been removed before applying the restoration mortar used to rebuild the pilasters, increasing the risk of detachment of the present mortar and tile cover.

D – The original ashlars on the upper storey showed fissure porosity. Although the fissures were no deeper than 300–400 μm, they caused flaking on this part of the facade. Microcrystalline gypsum was also detected in the inter- and intraparticle pores to a depth of 1 mm from the surface.

Table 2 shows the mean values for real and bulk densities, compactness (real to bulk density ratio), open porosity and the water saturation coefficients for the two materials. The real density in both stones was around 2.6 g cm^{-3}. The mean real density was slightly lower in *palomera* than in *tosca* stone. The difference between real and bulk density was smaller for *tosca* than for *palomera* stone. The low compactness index in both stones, 0.66 for *tosca* and 0.56 for *palomera*, was related to their high porosity (33.5 and 43.7% for *tosca* and *palomera*, respectively).

The water saturation coefficient, in turn, was higher in the *palomera* stone, with a mean value of $29.4 \pm 2.3\%$ compared to $18.9 \pm 0.6\%$ in *tosca*. Open porosity was also slightly higher in *palomera* ($43.7 \pm 1.9\%$) than in *tosca* ($33.6 \pm 0.7\%$). The higher standard deviation value for open porosity in *palomera* stone is indicative of the heterogeneity typical of an advanced stage of decay.

Water absorption was observed to progress very rapidly in the building, as shown in Figure 6, which depicts capillarity-induced absorption in the stone. Capillarity-induced absorption tests were conducted on core specimens 7 cm in length. Water was absorbed particularly rapidly in *tosca* limestone, which took up from 84 to 90% of the total water it could absorb in the first 3 min. Thereafter, absorption stabilized, rising to 93–96% in the following 2 days. Water absorption was more gradual in *palomera* stone, which absorbed from 8 to 14% of the total absorbed in the first few minutes, and 29–45% after the first hour. These values rose to 85–91% after the first day. This behaviour was corroborated by the capillarity test results, with coefficients of $12.0-15.7 \text{ kg m}^{-2} \text{ h}^{-1/2}$ for *tosca* and $4.3-8.5 \text{ kg m}^{-2} \text{ h}^{-1/2}$ for *palomera* stone.

The mercury intrusion test results showed that total porosity or porosity accessible to Hg (Table 3) was lower in *tosca* than the *palomera* stone. The porosity values for *tosca* stone ranged from 27 to 32%, while the *palomera* stone percentages were 35–45%. Sample porosity varied depending on the location of the sample (on the facade) and the degree of stone decay. Porosity was higher in the *tosca* stone on the socle (32%), for instance, than in the middle storey (27%).

The protected *palomera* stone at the rear of the arabesques, which was only slightly decayed, exhibited lower porosity (35%). Higher porosity was found in the *palomera* stone on the arabesques reliefs, ranging from 38.9 to 39.7% on the lower and from 44.7 to 45.2% on the upper third of the facade.

The prevalent pore size in *palomera* stone was smaller than 5 μm (65–96% micropores), whereas *tosca* stone showed a prevalence of macropores (84–86%).

The mean pore diameter was much higher in *tosca* (32.4–57.5 μm) than in *palomera* (1.4–3.1 μm). At the same time, the specific surface was smaller in *tosca* ($0.3 \text{ m}^2 \text{ g}^{-1}$) than in *palomera* stone ($0.6 \text{ m}^2 \text{ g}^{-1}$) pores. Consequently, *palomera* is more susceptible to decay and soiling.

Table 2. *Petrophysical characteristics of* tosca *and* palomera *stone*

Stone	Real density (g cm^{-3})	Bulk density (g cm^{-3})	Compactness	Open porosity (%)	Water saturation coefficient (%)
Tosca	2.69 ± 0.01	1.78 ± 0.02	0.661 ± 0.01	33.6 ± 0.7	18.9 ± 0.6
Palomera	2.65 ± 0.03	1.49 ± 0.06	0.56 ± 0.02	43.7 ± 1.9	29.4 ± 2.3

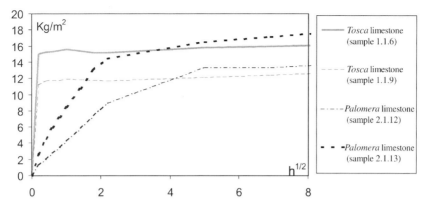

Fig. 6. The graph shows two examples of capillary water absorption for each stone: *tosca*, samples 1.1.6 and 1.1.9, and *palomera*, samples 2.1.12 and 2.1.13.

Pore size distribution was modified by stone weathering (Fig. 7). In intact *palomera* stone the pore diameter, deep into the specimen, was 1.6 μm, compared to 0.3 μm (both are average values) in the most severely weathered samples taken from the arabesques. The concomitant rise in the specific surface made the weathered stone more susceptible to further weathering. Moreover, the number of 10–30 μm diameter pores grew, thereby increasing macroporosity. While microporosity scarcely changed in *tosca* (increasing slightly but maintaining the same pore size distribution), its macroporosity pore size distribution climbed from 10–80 to 60–200 μm. The mean values ranged from 32.4 to 57.5 μm in diameter.

The mean sclerometric surface hardness results obtained for the materials are given in Table 4. The values shown for *palomera* stone refer to the measurements taken on ashlars and the uncarved areas at the rear of the arabesques, as the low consistency of the carved areas precluded any such testing. The results showed that the *tosca* ashlars had lower sclerometric indices than found for the *palomera* dimension stone.

Ultrasonic analysis was conducted to map ultrasonic velocities (m s^{-1}) across the entire facade (Fig. 8). The mean, standard deviation, and maximum and minimum velocity for each stone are given in Table 5. Because, as noted above, the behaviour of the two stones varies, the same ultrasonic velocity value in one and the other may not indicate the same rate of deterioration. The ultrasonic velocity values were therefore divided into four classes, obtained by representing the measured values and their occurrence in percentage (Fig. 9a, b). A statistical study was performed on the 4400 measurements taken to determine different degrees of stone deterioration and establish the guidelines for future conservation measures. The first step was to separate the ultrasonic velocity results obtained for the two stone varieties. The values obtained were plotted on a 'normal probability curve'. As a result, sectors with different slopes could be distinguished, with each deterioration interval being delimited by points of inflection. The segment with the highest values denoted stone that was in the best condition, while the lowest values corresponded to the areas where deterioration was most severe.

In other words, Class I identifies the areas where velocities were lowest and decay most advanced, while the highest velocity, least decayed zones were in Class IV. When the iso-velocity maps were replotted (Fig. 9c) in accordance with these classes the following intervals were obtained:

- Class I: under 1000 m s^{-1} for *tosca* and under 1200 m s^{-1} for *palomera*;
- Class II: 1000–1500 m s^{-1} for *tosca* and 1200–1750 m s^{-1} for *palomera*;

Table 3. *Petrophysical characteristics (Hg-porosimetry) of less decayed* tosca *and* palomera *stone*

Stone	Pore diameter average (μm)	Pore diameter median (Vol.) (μm)	Specific surface area (m^2 g^{-1})	Porosity accessible to Hg (%)	Micro < 5 μm (%)	Macro > 5 μm (%)
Tosca	1.5	32.4	0.36	26.9	14	86.0
Palomera	2.0	3.1	0.58	41.0	95.6	4.4

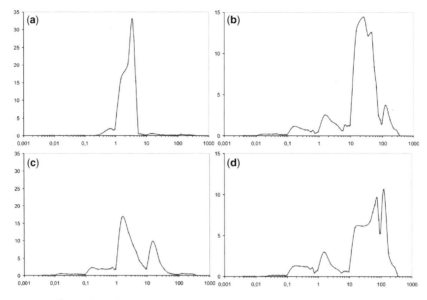

Fig. 7. MIP curves (Δ intrusion volume (%) v. pore diameter (μm)): results for (**a**) unaltered *palomera* stone; (**b**) unaltered *tosca* stone; (**c**) altered *palomera* stone; and (**d**) altered *tosca* stone.

- Class III: 1500–2100 m s^{-1} for *tosca* and 1750–2200 m s^{-1} for *palomera*;
- Class IV: over 2100 m s^{-1} for *tosca* and over 2200 m s^{-1} for *palomera*.

These were the intervals used to map the degree of stone deterioration on the facade, sectioned off according to the Kriging gridding method; Kriging being a group of geostatistical techniques to interpolate the value of a random field at an unobserved location from observations of its value at nearby locations. Surfer 8.0 Software was used.

The resulting maps identify the weakest parts of the facade, with account taken of the properties of each variety of stone.

Discussion: causes and effects of facade decay

The study showed that the structural stability of the facade is not at risk. The general state of conservation of the stone depends on its position on the facade. The ashlars, which are in good condition, show no breaking, cracking or significant loss of material that might generate loads or pressure able to destabilize the structural system or cause imminent collapse.

The ashlars present only slight signs of wear, mainly around the edges and on the surface due to the passing of the time. Some cracks tend to follow the masonry joints. The lack of mortar in the joints also facilitates water seepage. Aesthetically speaking, the facade is extremely decayed owing to soiling on the lower and middle storeys, and weathering on the upper third (reliefs and arabesques).

Deterioration is particularly severe in the arabesques on the lower and middle storeys, the pilasters on the middle and upper thirds, and the foliated arches over the windows in the top storey. The most common pathologies are weathering-induced flaking and granular disintegration. The agents that

Table 4. *Limestone surface hardness: sclerometric index*

	Tosca		Palomera	
	Lower-level ashlars	Lower-level imposts	Lower- and middle-level arabesques	Upper-level ashlars
Mean	15.6	12.9	20.2	22.7
Standard deviation	6.1	2.4	5.5	3.4

Fig. 8. (Top) Don Pedro I facade; and (bottom) ultrasonic analysis map.

in all likelihood contribute most to the decay observed are: the daily and yearly variations in temperature and humidity that characterize the region; absorption of rainwater that pools in shaded areas; and wear caused by wind-driven rain and soil. These agents may cause the flaking and granular disintegration through humidity variations (Fort et al. 2002; Gomez-Heras 2006). Another agent is the biological activity of small organisms, such as spiders that nest in fissures and holes.

Table 5. *Ultrasonic velocity (V_p) results for* tosca *and* palomera *stone*

Stone	Mean (m s^{-1})	Standard deviation (m s^{-1})	Maximum (m s^{-1})	Minimum (m s^{-1})
Tosca	2056	556	3947	765
Palomera	1542	529	3261	252

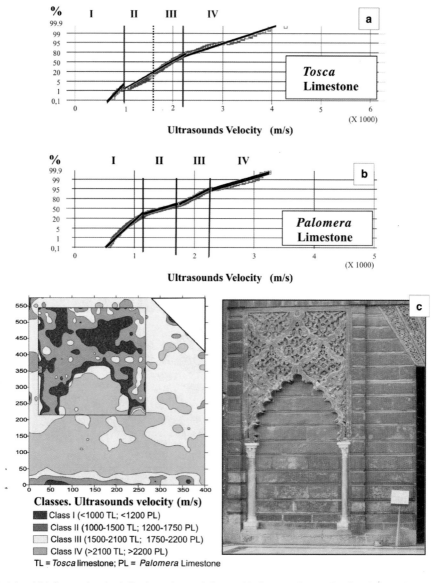

Fig. 9. (**a**) and (**b**) Curves showing inflection points and classes; (**c**) picture and maps showing the most decayed areas, based on the ultrasonic velocity classes established and stone characteristics.

The end result is the loss of cohesion of the carved elements, rendering the original shape of the figures unidentifiable in some cases.

Water pooling around the imposts exposes the *palomera* stone to the capillary rise of water, with the concomitant decay in the lower sections of stone. In some places, the resulting loss of material has been repaired with pinkish mortars, which have now worked loose as capillary rise continues.

A highly decayed calcareous material was identified under surface roughcasting in many of the pilasters on the middle and upper storeys, and in the foliated arches over the windows in the top third of the facade. This generates instability underneath the rendering used to rebuild these areas, with the risk of detachment owing to the low strength of the underlying stone support.

The different types of rendering applied to the facade have caused not only aesthetic but tangible damage, due at times to the chemistry of the materials used (i.e. cement and gypsum in contact with limestone) and at others to their physical properties (Varas *et al.* 2007, 2008). The restoration mortars at the base of the arabesques on the middle storey exemplify the latter. These repair mortars have a very low porosity, with a predominance of micropores (diameter 0.75–0.95 μm). Their resulting water repellence makes them incompatible with all the other materials and the cause of the loss of material in the arabesques. The reason for this physical incompatibility is the inability of water to circulate freely through the pores of the two materials, which is more evident in the lower storey where it pools more readily, favouring absorption, such as around the imposts.

Final remarks

The study conducted can be used as grounds for planning interventions in the short term for the determination and mapping of stone decay, along with the characterization of the two varieties used in the facade, and will contribute to the properly designed and engineered restoration and conservation of Spanish cultural heritage. Of the various analytical and instrumental techniques used, ultrasonic analysis proved to be particularly well suited to determining the state of material decay, particularly on the surface. Establishing four classes of decay for each stone variety on the grounds of ultrasonic velocities will help identify the areas requiring urgent intervention.

A few remarks and recommendations stemming from the study follow.

The *palomera* stone forming the arabesques and surrounding area is intensely deteriorated. Because the stone in these areas is liable to work loose, intervention is believed to be needed in the Class I and II areas. The *tosca* stone (lower storey) is in an acceptable state of conservation, intervention being required in the Class I areas only, and perhaps in the Class II zones.

The *palomera* stone in the middle corpus also exhibits significant decay. Attention is drawn in particular to the arabesque area on the extreme right of the facade, where Class I stone, the poorest quality, prevails. Decay is concentrated in the carved parts of the arabesques and the tiled pilasters, which were inappropriately reconstructed using gypsum mortar.

Palomera stone decay is less severe in the upper third of the facade, probably due to the fact that dimension stone was used to build this section; some of these ashlars were replaced around 1937. The elements in the poorest state of conservation in this storey are the cusped arches. The *tosca* stone ashlars are in good condition. In fact, the only Class I ashlars are some of the ones that were replaced.

Consequently, intervention is recommended for the *palomera* stone, the more deteriorated of the two varieties.

This study was financed by Patronato del Real Alcázar de Sevilla in the framework of an agreement with the Spanish National Research Council. We would also like to thank MATERNAS (S-0505/MAT/94) program, financed by Madrid Regional Government and the Consolider-Ingenio 2007 (CSD2007-0058) program, financed by Spanish Science and Innovation Ministry.

References

ALMAGRO, A. 2005. La recuperación del jardín medieval del Patio de las Doncellas. *Apuntes del Alcázar de Sevilla*, **6**, 44–67.

ALMAGRO, A. 2006. *Estudio Histórico-arqueológico de la Fachada del Palacio del Rey D. Pedro en el Alcázar de Sevilla. Informe Preliminar*. Escuela de Estudios Árabes. CSIC. Inédito.

BARRIOS PADURA, A., BARRIOS SEVILLA, J. & GARCÍA NAVARRO, J. 2006. Alterations in the stone masonry of the Capitol Room of Sevilla City Hall. *Materiales de Construccion*, **56**(284), 87–94.

CHÁVEZ GONZÁLEZ, M. R. 2004. *El Alcázar de Sevilla en el siglo XIX*. Patronato del Real Alcázar de Sevilla.

CIE. 1986. *Colorimetry*. 2nd edn. Publication CIE 15.2, Bureau central de la Commission Internationale de l'Eclairage (CIE), Paris.

FOLK, R. L. 1962. Spectral subdivision of limestone types. *In*: HAM, W. E. (ed.) *Classification of Carbonate Rocks*. AAPG Memoirs, **1**, 62–84.

FORT, R., BERNABÉU, A., GARCÍA DEL CURA, M. A., LÓPEZ DE AZCONA, M. C., ORDÓÑEZ, S. & MINGARRO, F. 2002. Novelda stone: widely used withing the Spanish architectural heritage. *Materiales de Construcción*, **52**(266), 19–32.

GÓMEZ-HERAS, M. 2006. *Procesos y formas de deterioro térmico en piedra natural del patrimonio arquitectónico*. UCM, Servicio de Publicaciones, Madrid. World wide web address: http://www.ucm.es/BUCM/tesis/geo/ucm-t28551.pdf (last accessed 05/09/2008).

HERRERO FERNÁNDEZ, H. 2007. Estudios sobre la alteración de la piedra empleada en la iglesia de El Salvador de Sevilla. World wide web address: http://www.rtphc.csic.es/Marie Curie Research Training/18_Herrero.pdf (last accessed 04/09/2008).

IGME. 1959. *Hoja geológica 1:50.000 n° 1061 (n°282H). Cádiz*. Instituto Geológico y Minero de España, Madrid.

IGME. 1977. *Hoja geológica 1:50.000 N° 1003. Utrera*. Instituto Geológico y Minero de España, Madrid.

IGME. 1986. *Hoja geológica 1:50.000 n° 1021. Morón de la Frontera*. Instituto Geológico y Minero de España, Madrid.

IGME. 1987. *Hoja geológica 1:50.000 n° 1061. Cádiz*. Instituto Geológico y Minero de España, Madrid.

MATERIALS, A. S. F. T. A. 2000. *E313-00. Standard Practice for Calculating Yellowness and Whiteness Indices from Instrumentally Measured Color Coordinates*. American Society for Testing of Materials, West Conshohocken, PA.

RILEM. 1980. Recommended tests to measure the deterioration of stone and to assess the effectiveness of treatment methods. *Materials and Structures*, **75**, 175–253.

TABALES, M. A. 2005. El Patio de las Doncellas del Palacio de Don Pedro I de Castilla: Génesis y Transformación. *Apuntes del Alcázar de Sevilla*, **6**, 6–43.

VARAS, M. J., ALVAREZ DE BUERGO, M. & FORT, R. 2006. The origin and development of natural cements: the Spanish experience. *Construction and Building Materials*, **21**, 436–445.

VARAS, M. J., ALVAREZ DE BUERGO, M., PEREZ-MONSERRAT, E. & FORT, R. 2008. Decay of the restoration render mortar of the church of San Manuel and San Benito, Madrid, Spain: results from optical and electron microscopy. *Materials Characterization*, **59**, 1531–1540; doi:10.1016/j.matchar.2007.11.008.

The church of Santa Engrácia (the National Pantheon, Lisbon, Portugal): building campaigns, conservation works, stones and pathologies

C. FIGUEIREDO[1]*, L. AIRES-BARROS[1] & M. J. NETO[2]

[1]*Centre of Petrology and Geochemistry, CEPGIST, IST, Av. Rovisco Pais, 1049-001, Lisbon, Portugal*

[2]*Faculdade de Letras, Lisbon, Portugal*

*Corresponding author (e-mail: carlos.m.figueiredo@ist.utl.pt)

Abstract: A study of the different building campaigns, restoration works, pathologies and stones used through time in the Church of Santa Engrácia (Portuguese National Pantheon) is presented. The changing fortunes of the monument make it a useful case study in terms of building stone use and decay. This bold baroque project, begun in the seventeenth century, remained without a roof until the 1960s. At that time the leader of the conservative regime of Estado Novo ('New State'), António de Oliveira Salazar, decided to complete the building as a national pantheon to provide a memorial to some of the personalities of Portuguese history. A complementary approach between the geosciences and history of art identified the ancient and more recent quarries, and the main stone types, used during its different building campaigns and restoration/conservation works. The local stone variety 'lioz', extracted in the area around Lisbon, was the main type used. An overview of the current knowledge of the chemical, physical and mechanical characteristics of these stones is now available. Besides new data on the typology, causes and processes of the major weathering forms observed inside and outside the building, the building campaigns and restoration works are also presented.

The preservation of architectural heritage requires an interdisciplinary perspective; gathering knowledge from subjects as different as archaeology, history of art, geology, materials science, engineering and architecture (Aires-Barros *et al.* 1998). Several projects have examined historical quarries and exploitation methods, such as Portuguese projects on the quarries of the Batalha Monastery (Neto *et al.* 2000) or a similar study on Lisbon's cathedral and Jerónimos Monastery.

The need for a multidisciplinary approach to conservation or restoration has been clearly demonstrated in several international studies, such as the cases of the Tower of Pisa and the Basilica of Assisi in Italy (Croci 2000), the Madara Horseman monument in Bulgaria (Delalieux *et al.* 2001), and by Smith & Turkington (2000) in the studies of the causes and controls of stone decay. Data on provenance and technical properties of building material are required to evaluate the weathering processes, and to successfully preserve and reconstruct historical and monumental buildings (Amoroso & Fassina 1983; Michalski *et al.* 2002).

The Santa Engrácia National Pantheon (Fig. 1a) is located on a hill in the eastern part of the historical centre of Lisbon (Santa Engrácia parish) and in the vicinity of the River Tagus (280 m early). The pantheon presents some peculiar aspects: its location, its baroque architecture (a early example in Portugal) and its history (mainly the lack of a roof until the 1960s). In addition, there is no account of the building and restoration campaigns undertaken, and even less study of the stones used or the conservation state of the monument (Fig. 1b).

The objectives of this complementary approach between geosciences and history of art are: the establishment of the building periods of the Santa Engrácia church (Portuguese National Pantheon); the characterization of stones used; and the location and identification of recent and ancient quarries exploited for its several building campaigns and restoration works. Furthermore, a detailed survey of stone decay phenomena was carried out on the interior and exterior of the monument to establish the typology, causes and processes of deterioration.

Methodological approach

A complementary study, involving geosciences and history of art, was developed to look into: historical and architectonic aspects of the monument (building campaigns, restoration/conservation works and stone provenance – that is, recent and ancient

Fig. 1. (a) General view of the main façade of the Santa Engrácia National Pantheon. (b) General view of the paving showing the play of 'colours' created by the use of different ornamental stones: LZ ('Lioz'), EN ('Encarnadão'), AN ('Amarelo de Negrais'), AS ('Azul de Sintra'), NM ('Negro de Mem Martins'), BA ('Breccia of Arrábida') and RV ('Ruivina').

quarries); visual inspection and photographic recording of stones and weathering forms; environmental monitoring; technological characterization of stones; and laboratory study of stone decay products. A through bibliographical and documental survey of historical archives and recent paper and electronic databases was also been performed. Extensive literature regarding the technological characterization of the stones used is, in general, available today (Martins 1982, 1991; *Catálogo de Rochas Ornamentais Portuguesas* (C.R.O.P.) 1983, 1984, 1985; Ramos 1985; Moura 1991, 2001; Reynaud & Vintém 1994; Figueiredo 1999; Carvalho *et al.* 2000). Hence, an overview of the current knowledge of their provenance, geological, chemical, physical and mechanical characteristics is presented. According to C.R.O.P. (1983, 1984, 1985), Portuguese National Stones (P.N.S.) (1995), Figueiredo (1999), Carvalho *et al.* (2000) and Moura (2001), the physical and mechanical tests were carried out following Portuguese and International Standards: apparent density (DIN Standard 52102); open porosity (LNEC Specification E 216-1968); water absorption coefficient (DIN Standard 52103 and 52106); compressive strength (DIN Standard 52105; LNEC Specification E226); compressive strength after freeze–thaw tests (DIN Standard 52104); bending strength (DIN Standard 52112); and abrasion test (Portuguese Standard NP-309). According to C.R.O.P. (1983, 1984, 1985) and Figueiredo (1999), the chemical composition analysis involved the use of several chemical analytical techniques such as gravimetric analysis, colorimetry and atomic absorption spectrometry (AAS/AES).

Macro- and micropetrography, Fourier transform infrared spectroscopy (FTIR) and X-ray fluorescence analysis (XRF) were used to study the restoration works, stone decay material and salt efflorescences collected in several field campaigns. Infrared thermography was also used as a complementary, non-invasive tool in field studies for monitoring and diagnosing the conditions of the building and materials. A thermographic camera, FLIR B20, was used in several thermographic analyses carried out both outside and inside the Pantheon. Several thermograms (thermographic images, Figs 2a & 3a) in colour or grey levels were then obtained. In these images every colour/grey level corresponds to a different temperature. In addition, some significant profiles giving the variation in temperature along a chosen carefully given line were also prepared (Figs 2c & 3c). As the intensity of emitted infrared radiation depends on the temperature and the ability of each material to radiate (Mendonça 2005), it was possible, based on thermographic analysis, to detect a change in the external architecture and the presence of some pathologies (see, for instance, Figs 2a–d & 3a–c).

Finally, an integrated systematization and analysis has been applied to all of the collected elements.

The baroque church of Santa Engrácia

Building campaigns

The foundation of the Santa Engrácia parish church is due to Princess Maria (1521–1577), daughter of

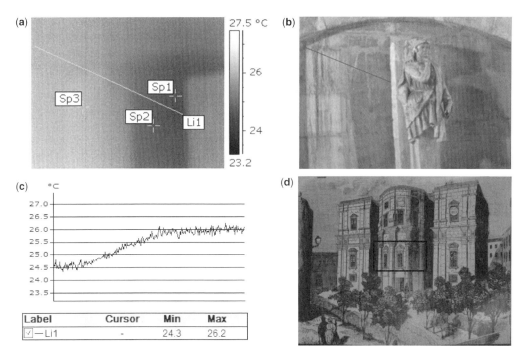

Fig. 2. Infrared thermography: (**a**) thermographic image of an internal wall (**b**); (**c**) a plot of temperature along the profile selected on the real image (b); and (**d**) an old painting showing the primitive architecture: the rectangle in black outlines the area changed on the external architecture and which shows the lower temperature values.

Manuel I, and took place in 1568 under the plan of Nicolau de Frias. However, nothing remains that dates to this time. At the end of the seventeenth century, the architect João Antunes conceived the baroque project for the Church of Santa Engrácia, in an audacious Roman-style design previously unknown in Portugal, to be built in Lisbon. The centralized plan, the dynamism of the concave and convex sides, and the play of colours in the interior decoration were achieved by the use of various types of stone that characterize João Antunes' work, which was halted by his death in 1712 (Carvalho 1971). It seems that the façades had been built up to the cyma, whilst the side chapels, the upper choir and the galleries had been completed inside. The crossing was still to be covered, probably by a cupola, but the plan is unrecorded.

From then the construction process became more like a saga and the monument started to be recognized as 'the never ending works of Santa Engrácia'. More than two centuries were needed to finish the church and the building remained, unfortunately, without a roof until the 1960s. It was only in 1956, during the commemorations of the 30 years of the Estado Novo ('New State') regime, that the decision was taken to open a competition for a programme to conclude and adapt the building as the national pantheon. However, it was only in 1964 that António de Oliveira Salazar, the leader of the conservative regime of Estado Novo, decided to finish the works in 2 years, programming the inauguration as part of the commemorations for the 40th anniversary of the regime. The possibility of smashing the popular myth about the 'works of Santa Engrácia' was used by the dictator as a final propaganda manoeuvre during a difficult time both at home and abroad.

Under the influence of the spirit of the Venice Congress, the project directed by the architect João Vaz Martins had tried a simplified solution for finishing the church, but without abandoning the intention of using a traditional language. He created an ample terrace marked out by a balustrade and planned a sober-lined double cupola, in harmony with the original style of the monument (Fig. 1a). Nevertheless, a change in the external architecture was discovered by thermographic analysis. There were some windows (Fig. 2d, outlined by a rectangle in black) that were closed, allowing the placing of statues (Fig. 2b). The internal side of the external wall presents a different radiation (lower values) profile on these sections (Fig. 2a, c). This was confirmed by carrying out research on the archive, where some old paintings

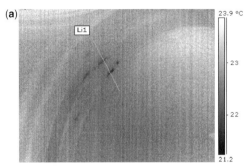

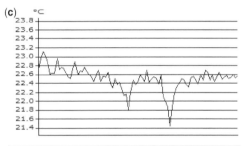

Fig. 3. Detection of fissures and leakage points (that show the lower values of temperature) by infrared thermography: (**a**) thermogram of a selected area on an inside wall (**b**) presenting different radiation along the profile (**c**) analysed.

were found showing the primitive architecture (see Figs 1a & 2d).

Conservation works

During the building works performed between 1964 and 1966 to conclude the monument, a huge conservation campaign of the existent interior stonework was carried out. It meant a thorough cleaning, which was carried out with brushes and detergents (*Hacker Steinbrize*) on the worked surfaces, such as the capitals and cymas, and with a stiff brush on the flat surfaces. Polishing machines were used on the vaulting and arches, because most of the joins had 'stalactites coming from the dissolution and recrystallisation of the limestone' (Mantas 2002). The washing was followed by sealing, using clear wax. A lot of the stone had to be replaced or repaired to fill gaps. 'Azul de Sintra', breccia from Arrábida, 'ruivina' from Estremoz, 'amarelo de Negrais', and 'encarnadão' from Pêro Pinheiro and Mem Martins, as well as Mem Martins 'negro', were rigorously carved, following moulds, to fit the gaps left by the missing stone, once the 'old glues' had been cleared away.

From 1999 to 2000 a new cleaning and repair work project was carried out on the exterior and interior of the monument. The main weathering forms were removed using different cleaning methods and techniques ['water spray' by high-pressure jet and 'nebulizers', scrubbing (with nylon brush), microblasting appliance (Air-brasive^{TM}), absorbent powders and special clays ('poultice'), herbicide (*Krovac I* from Sapec) and biocide (*R80* from Bayer) for removal of 'biological colonization']. Some interior walls and opened joints and fissures were repaired as well by rendering, surface protection and impregnation with impermeable material (*Gumasil*^{TM}), respectively.

Building materials: the stones

Geological and geographical setting

Several stone types (commercial and local varieties) used as structural, dimension and ornamental stone in the National Pantheon were identified: 'Lioz' (LZ), 'Encarnadão' (EN), 'Amarelo de Negrais' (AN), 'Azul de Sintra' (AS), 'Negro de Mem Martins' (NM), 'Breccia of Arrábida' (BA) and 'Ruivina' (RV). Most of them are regional stones that are, in general, still extracted today from quarries located in the surrounding of Lisbon, the capital of the country (Fig. 4).

The Jurassic ('Azul de Sintra' and 'Negro de Mem Martins' stones) and the Cretaceous ('Lioz', 'Encarnadão' and 'Amarelo de Negrais' stones) limestones came from quarries located in the western Lisbon-Sintra (Pêro Pinheiro) area (Martins 1982, 1991; C.R.O.P. 1983, 1984, 1985; P.N.S. 1995; Figueiredo 1999). Regarding the ancient quarries, although they are not active today, some of them are, however, still available for sampling.

In the region of Setúbal (south of Lisbon), the 'Breccia of Arrábida', a local variety of ornamental stone, was largely extracted in the past from thickest banks mainly composed of calcareous rocks forming the important Jurassic massif of 'Serra da Arrábida' (Sousa 1898; Ribeiro *et al.* 1979; Teixeira & Gonçalves 1980).

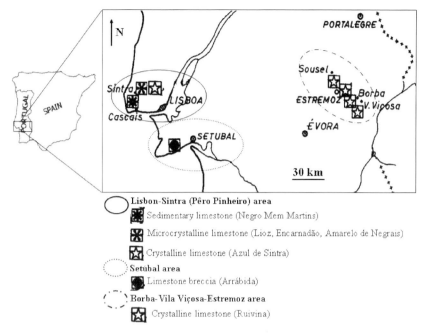

Fig. 4. Map of Portugal showing the origins of the stone used in the building of the Church of Santa Engrácia. Modified after C.R.O.P. (1983, 1984, 1985).

The 'Ruivina' marble is exploited from some Lower Cambrian metamorphic carbonate rock formations that crop out in the anticlinorium of Estremoz–Borba–Vila Viçosa (south of Portugal) (C.R.O.P. 1983, 1984, 1985; Moura 1991, 2001).

Petrography

The limestones locally known as 'Lioz' (beige–white limestone), 'Encarnadão de Negrais' (red–pink limestone) and 'Amarelo de Negrais' (yellowish limestone) are colour varieties of the Cretaceous microcrystalline limestone composed essentially of calcite (Martins 1991; Figueiredo 1999). While the 'Lioz', the whitish beige, bioclastic and calciclastic variety, is in thin section and by Folk's classification (Pettijohn 1975) a biomicrosparite–biomicrosparrudite, the 'Encarnadão' and 'Amarelo de Negrais' are, respectively, biopelsparite and biomicrudite.

The 'Negro de Mem Martins' limestone is a blackish, very-fine calciclastic and fossiliferous limestone. By Folk's classification, it is a fossiliferous pelmicrite. Calcite is the essential mineral, and the accessory ones are dolomite, quartz, micas and opaque minerals.

The 'Azul de Sintra' is a greyish-blue limestone with a granoblastic texture. Calcite is the essential mineral, and the accessory ones are quartz, muscovite and opaque minerals.

The 'Breccia of Arrábida' is a poligenic, polychrome limestone conglomerate used as a dimensional stone with a polished surface finish (Sousa 1898; Ribeiro et al. 1979; Teixeira & Gonçalves 1980).

The 'Ruivina' marble is a dark-grey marble with granoblastic texture, uniformly grained (medium–fine), showing slightly deformed zones. Calcite is the essential mineral, and the accessory ones are quartz, muscovite and opaque minerals (C.R.O.P. 1983, 1984, 1985; Moura 1991, 2001).

Chemical composition

Some basic statistical parameters (average, maximum and minimum values) concerning the major chemical elements forming these stones are presented in Table 1. These parameters were estimated from data published by C.R.O.P. (1983, 1984, 1985) and Figueiredo (1999). Based on these results, the limestones 'Lioz', 'Encarnadão de Negrais' and 'Azul de Sintra', and the marble 'Ruivina', are more or less pure, having, in general, more than 55% of calcium oxide and less than 3% of silica (Mason & Moore 1982; Martins 1991; Klein & Hurlbut 1993). However, the yellowish and the blackish varieties, taking into account the largest values of their content in SiO_2, Al_2O_3, Fe_2O_3 and MgO, are slightly magnesian and clayey (Pettijohn 1975; Mason & Moore 1982; Klein & Hurlbut 1993).

Table 1. *Basic statistical parameters regarding the chemical composition, in weight per cent, of the main stone types used in the Church of Santa Engrácia*

	Stone types ('local and commercial terms')																				
	'Lioz'			'Encarnadão'			'Amarelo de Negrais'			'Negro Mem Martins'*	'Azul de Sintra'*			'Ruivina'		
	Mean	Max	Min	Mean	Max	Min	Mean	Max	Min	Mean	Mean			Mean	Max	Min
SiO_2	0.50	1.41	0.09	1.15	3.50	0.10	5.42	11.21	1.59	8.77	2.48				1.07	0.36
TiO_2	0.00	0.00	0.00	0.00	0.00	0.00	0.03	0.08	0.00	0.12	0.02				0.00	0.00
Al_2O_3	0.33	0.48	0.09	0.30	0.77	0.00	1.46	3.00	0.25	1.90	0.61				0.69	0.36
Fe_2O_3 (total)	0.03	0.05	0.00	0.12	0.48	0.02	0.68	1.00	0.37	0.17	0.31				0.11	0.09
MnO	0.00	0.00	0.00	0.00	0.01	0.00	0.01	0.01	0.01	0.01	0.01				0.00	0.00
MgO	0.34	0.61	0.16	0.60	2.49	0.12	2.36	5.39	0.41	2.96	0.84				0.99	0.18
CaO	55.38	55.86	54.65	54.34	55.60	50.37	48.66	53.02	45.30	46.46	52.74				55.07	53.89
Na_2O	0.03	0.05	0.00	0.04	0.18	0.00	0.04	0.07	0.02	0.06	0.02				0.05	0.04
K_2O	0.03	0.07	0.00	0.04	0.10	0.00	0.14	0.23	0.02	0.58	0.03				0.12	0.06
L.O.I.	43.29	43.95	42.39	43.19	43.80	42.01	40.99	43.22	36.64	39.69	42.28				43.47	43.42
CO_2	44.00	44.07	43.95	43.76	43.77	43.75	42.02	–	–	–	–				43.42	43.37

Legend: *average values from only one source, Figueiredo (1999); –, no data; L.O.I., loss on ignition.

Physical and mechanical properties

Basic statistical parameters of some physical and mechanical properties are summarized in Table 2. These parameters were estimated from data already published by several authors (C.R.O.P. 1983, 1984, 1985; P. N. S. 1995; Figueiredo 1999; Carvalho *et al.* 2000; Moura 2001).

The selected properties referred to in Table 2 are supposed to be appropriate in order to show the particularities of each stone, and to help in understanding their particular use and behaviour under the conditions to which they are applied (C.R.O.P. 1983, 1984, 1985; Bradley 1998; Weber & Lepper 2002).

The results obtained indicate that the stones used for the Santa Engrácia church are, in general, highly resistant to mechanical stress (with compressive strength values, in general, larger than 105 MPa) (Carmichael 1989; Weber & Lepper 2002) and to decay processes due to water–rock interaction (open porosity and water absorption values, in general, less than 1% and 0.5%, respectively) (C.R.O.P. 1983, 1984, 1985; Bajare & Svinka 2000; Carvalho *et al.* 2000). After the freeze–thaw tests they also show a change in the compressive strength; in general, of less than 20%. These results compare well to those derived from thermal ageing accelerated tests indicating that the yellow limestone shows a susceptibility to weathering greater than the other limestones (Figueiredo 1999). However, the white variety presents the best durability of all these limestones.

Applications

The 'Lioz' limestone was the stone type mainly used as structural, as well as ornamental, stone. All of the façades of the Santa Engrácia church are built in 'Lioz' limestone. The other stone types had a very particular application, as it can be seen on the pavement and in specific architectural features, such as interior vaultings, socles and frames, producing a very dynamic decoration (see Fig. 1a, b).

Based on extant documentation, both in the monument itself and in the Directorate-General of National Buildings and Monuments archives (DGEMN), Table 3 was compiled. It presents the essential information regarding the type of stones, their origin and their main application in the Santa Engrácia church.

To complete the work on Santa Engrácia, modern construction techniques were employed in raising the dome and much stonework was required to finish the building. Large quantities of stone were then needed to finish the construction process and move on to restoring the existing stonework, which had deteriorated with the passage of time. All of

Table 2. Basic statistical parameters for some physical and mechanical properties of the main stone types used in the Church of Santa Engrácia

	Stone types ('local and commercial terms')													
	'Lioz'			'Encarnadão'			'Amarelo de Negrais'			'Negro Mem Martins'[‡]	'Azul de Sintra'[‡]	'Ruivina'		
	Mean	Max	Min	Mean	Max	Min	Mean	Max	Min	Mean	Mean	Mean	Max	Min
Compressive strength (MPa)	104.23	138.96	87.97	112.32	139.94	90.22	143.72	146.02	141.41	157.00	45.01	91.20	84.63
Compressive strength* (MPa)	117.68	135.33	101.69	110.03	129.94	97.77	109.83	–	–	–	–	95.81	93.36
Bending strength (MPa)	17.09	20.50	11.96	18.32	21.67	12.16	19.91	–	–	–	–	26.48	25.11
Apparent density (kg m^{-3})	2678	2703	2600	2683	2708	2603	2623	2674	2597	2582	2583	2715	2703
Open porosity (%)	0.50	0.90	0.21	0.38	0.71	0.17	1.08	1.36	0.71	0.38	0.38	0.28	0.14
Water absorption (%)	0.15	0.21	0.11	0.14	0.27	0.07	0.41	0.53	0.27	0.15	0.15	0.10	0.05
Abrasion test (mm)	2.35	2.50	2.20	2.60	3.40	1.80	2.80	–	–	–	–	2.60	2.00
Impact test (cm): minimum fall height	45	45	45	46	55	40	–	55	50	–	–	55	55
Linear thermal expansion coefficient[†]	4.33	5.90	3.30	4.72	5.90	4.10	4.40	–	–	–	–	14.80	5.40

Legend: –, no data; *after 25 cycles of freeze–thaw tests; [†] ×10^{-6} per °C; [‡] average values from only one source, Figueiredo (1999).

Table 3. *Origin and main applications of the stone of the Church of Santa Engrácia*

Stones ('local and commercial terms')	Origin	Applications	Observations
'Lioz'	Pêro Pinheiro	Paving, columns, cornices, socles, lining pieces, pilasters, angles, decorative stonework, balustrades, ornamentation, vaulting, statues in the main façade.	Light-coloured unblemished stone, uniform, without stains, streaks, rust-coloured veins and geode-like cavities
'Encarnadão'	Pêro Pinheiro	Interior vaulting socle, frames.	Stone 0.15 or 0.10 m thick, respectively, for the socle frames and curved ornamentation
'Negro de Mem Martins'	Mem Martins	Panel lining	Stone 0.06 m thick
'Breccia'	Arrábida	Vaulting panels	
'Ruivina'	Estremoz	Tombs	Polished finish
Several coloured marble	Estremoz and Sintra	Worked stonework pieces, Paving and statues.	White, blue uniformly coloured stone
Several coloured stones	Pêro Pinheiro	Decorative stonework in the chapels and the church	Stones between 0.08 and 0.12 m thick

the existing documentation shows the care taken in the provision of stone for the works. The stonework suppliers were limited in number and chosen from among those who regularly worked with the DGEMN. The explanation for not opening this aspect up to public tender was based on the 'responsibility involved in the work which requires great care in the uniformity of the stones, in terms of both quality and size' (Mantas 2002). This had to be 'identical to that existing in the work' (Mantas 2002). It had to be the 'lioz type', with a 'uniform colour' and be without 'rust-coloured veins' (Mantas 2002). Particular attention was paid to the provision of the 16 columns for the exterior decoration of the cupola, as they were intended to be large, separate pieces. The DGEMN specialists went to the Pêro Pinheiro quarry to confirm that there were lioz blocks 'already being removed, with a length of 10 m and a 1.90 m × 1.10 m section, in sufficient quantity for the sixteen columns' (Mantas 2002).

There is, however, also a reference to 'encarnadão' from Mem Martins for the interior lining of the cupola. The stone had to be worked according to the similar moulds and pieces already in the church. The large blocks, which needed to be hoisted, would have had cavities opened to receive the 'iron glove', to avoid the use of bands that could damage the corners during the lifting.

The memorial stones were made in Estremoz marble and black 'ruivina', from the same region of the Alentejo. It was these stones, along with white and pink lioz, that were used in the magnificent paving inside the church, one of the most remarkable examples of stonework from the period (Fig. 1b).

Stone pathologies

A detailed survey of exterior and interior stone decay phenomena was carried out on the National Pantheon. Several weathering forms have been identified. On the exterior (Fig. 5a) the presence of vegetation in specific areas of the façades should be noted: a brownish surface deposit covering protected zones; fissuring and cracking of stones of some columns, balconies and balustrades. On the interior (Fig. 5b), physical–mechanical weathering forms (deformation, fissuring, cracking, differential degradation, flaking, scaling and spalling) prevail. These are mainly related to the differential settlement of the building, rising damp, rainwater infiltration and the local occurrence of soluble salts [gypsum ($CaSO_4 \cdot 2H_2O$), thenardite (Na_2SO_4) and niter (KNO_3)]. These soluble salts were, however, only found in very localized areas on the interior wall of the western façade, in the 'show room', at the elevated choir (thenardite, niter and gypsum) and also on the walls in the tomb rooms (gypsum and thenardite).

As the monument is located in a moderately polluted area (Aires-Barros 2001) only very particular reasons could help explain such occurrences. A local source and/or enrichment of salt solution in alkaline elements might be singled out as a possible explanation for such small and confined occurrences. While the occurrence of gypsum could be mainly

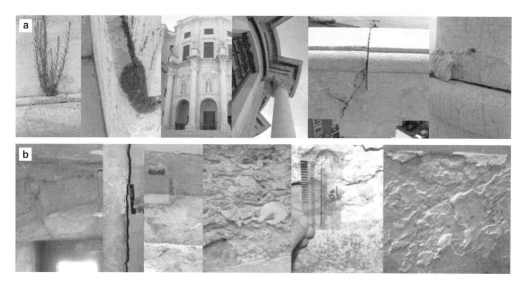

Fig. 5. Major weathering forms identified on the Santa Engrácia church. (a) On the exterior, vegetation in specific areas of the façades; a brownish surface deposit covering protected zones; fissuring and cracking of stones of some columns, balconies and balustrades. (b) On the interior, deformation, fissuring, cracking, differential degradation, flaking, scaling and spalling are visible.

related to repair works performed from 1999 to 2000, the main source for thenardite and niter could be the interaction between rain and seepage waters and stone, joint mortar and the concrete used in the construction of the double cupola (Arnold & Zehnder 1989). Some fissures and leakage points not visible to the naked eye have been detected using infrared thermography (Fig. 3a–c). A possible contribution to the enrichment of the salt solution in alkaline elements from cleaning activities should also be taken into account (Arnold & Zehnder 1989). The presence of pigeons and the effects of their faeces should also be taken into account as well (Gómez-Heras et al. 2004).

As seepage waters are usually undersaturated with respect to gypsum, thenardite and niter, these salts could only be precipitated through seepage evaporation (Arnold & Zehnder 1989; Winkler 1997). When combined with the presence of rainwater percolating through the monument stone structures and the dissolution of stone and joint materials, the microclimate brings about changes in water composition mainly due to the evaporation and precipitation of some water components (Figueiredo 1999; Figueiredo et al. 2000). However, taking into account that the interior microclimate of the National Pantheon, controlled by an air-conditioning system, is mostly characterized over the year by an air temperature and relative humidity ranging, respectively, from around 15.8 to 19 °C and from 35 to 63.3%, not many wetting–drying cycles should be expected. In spite of this fact, however, large dark stain zones related to the presence of moisture in the walls and the pavement could be observed.

Final remarks

Despite the huge cleaning and repair project carried out in 1999 and 2000, several stone decay phenomena can still been identified on the exterior and interior of the National Pantheon. Also, thanks to this campaign, the impact of these pathologies on the overall evaluation of the conservation state of the monument has been significantly minimized. Most of them only have a very local significance. Nowadays, the exterior appearance much improved owing to the removal of the major signs of weathering and deterioration that existed prior to the application of the cleaning methods.

The major concern is still that of the interior mainly due to the prevalence of physical pathologies such as fractures and fissures related to structural problems. Mostly due to the differential settlement of the building, the fractures and fissures are largely related to the presence of water in the interior of the monument caused by rainwater infiltration. Other sources of water are rising damp and condensation of the air humidity, mainly on the glasses of windows of the dome, despite controlling the inside microclimate using an air-conditioning system.

Wetting and drying cycles, combined with the local presence of soluble salts such as niter, thenardite and gypsum, could promote the hydration and dehydration, as well as the crystallization and dissolution cycles, of these salts, which could help explain the other physical degradation forms (granular disintegration, scaling and so on). Water–rock/stone interaction could also promote the enrichment of the solution carrying these salts.

Following its complex, extended building programme, that lasted from the sixteenth to the twentieth century, only the signs of differential deterioration observed on the interior wall of the western façade of the building, at the elevated choir, still remains clearly visible. This could be considered as the main perennial imprint of the period when the monument was left uncovered.

The stones used on Santa Engrácia church are, in general, of very good quality and are resistant to weathering. Only the less pure stones ('Negro de Mem Martins' and 'Amarelo de Negrais') with a low clay matrix content represent moderate- to good-quality dimension stones that are more prone to weathering.

This work was partially funded by FCT project POCTI/EAT/35063/2000 and by Centro de Petrologia e Geoquímica do Instituto Superior Técnico.

References

AIRES-BARROS, L., NETO, M. J. & DIONÍSO, M. A. 1998. Methodological approach to the study of the cathedral of Lisbon stones and their pathologies. In: MATRÁN, M. A. F. (ed.) Proceedings of the 4th Congreso Internacional de Rehabilitación del Património Arquitectónico y Edificación, La Habana. CICOP, Spain, 381–383.

AIRES-BARROS, L. 2001. As rochas dos monumentos Portugueses: tipologias e patologias. Cadernos do Instituto Português do Património Arquitectónico (IPPAR), Ministério da Cultura, **I**, **II**.

AMOROSO, G. G. & FASSINA, V. 1983. Stone Decay and Conservation: Atmospheric Pollution, Cleaning, Consolidation and Protection. Materials Science Monographs, **11**.

ARNOLD, A. & ZEHNDER, K. 1989. Salt weathering on monuments. In: ZEZZA, F. (ed.) The Conservation of Monuments in the Mediterranean Basin. Proceedings of the 1st International Symposium on the Influence of Coastal Environment and Salt Spray on Limestone and Marble. Grafo Edizioni, Bari, 31–58.

BAJARE, D. & SVINKA, V. 2000. Restoration of the historical brick masonry. In: FASSINA, V. (ed.) Proceedings of the 9th International Congress on Deterioration and Conservation of Stone, Venice, Volume **1**. Elsevies Science, Amsterdam, 3–11.

BRADLEY, F. 1998. Natural Stone: A Guide to Selection. W. W. Norton, New York.

CARMICHAEL, R. S. 1989. Practical Handbook of Physical Properties of Rocks and Mineral. CRC Press, Boca Raton, FL.

CARVALHO, A. 1971. As obras de Santa Engrácia e os seus artistas. Academia Nacional de Belas-Artes, Lisboa.

CARVALHO, J., MANUPPELLA, G. & MOURA, A. C. 2000. Calcários ornamentais portugueses. Boletim de Minas, Lisboa, **37**(4), 223–232.

CATÁLOGO DE ROCHAS ORNAMENTAIS PORTUGUESAS (C.R.O.P.). 1983, 1984, 1985. Volumes I, II, III, Direcção-Geral de Geologia e Minas (eds), Ministério da Indústria, Energia e Exportação, Lisbon, Portugal.

CROCI, G. 2000. General methodology for the structural restoration of historic buildings: the case of the Tower of Pisa and the Basilica of Assisi. Journal of Cultural Heritage, **1**, 7–18.

DELALIEUX, F., CARDELL, C., TODOROV, V., DEKOV, V. & VAN GRIEKEN, R. 2001. Environmental conditions controlling the chemical weathering of the Madara Horseman monument, NE Bulgaria. Journal of Cultural Heritage, **2**, 43–54.

FIGUEIREDO, C. 1999. Alteração, Alterabilidade e Património Cultural Construído: o caso da Basílica da Estrela, PhD thesis, Technical University of Lisbon, IST, Portugal.

FIGUEIREDO, C., MARQUES, J. M., MAURÍCIO, A. & AIRES-BARROS, L. 2000. Water–rock interaction and monument stone decay: the case of Basilica da Estrela, Portugal. In: FASSINA, V. (ed.) Proceedings of the 9th International Congress on Deterioration and Conservation of Stone, Venice, Volume **1**. Elsevier Science, Amsterdam, 79–87.

GÓMEZ-HERAS, M., BENAVENTE, D., ÁLVAREZ DE BUERGO, M. & FORT, R. 2004. Soluble salt minerals from pigeon droppings as potential contributors to the decay of stone based cultural heritage. European Journal of Mineralogy, **16**, 505–509.

KLEIN, C. & HURLBUT, C. S. 1993. Manual of Mineralogy after James D. Dana. Wiley, New York.

MANTAS, H. 2002. O Panteão Nacional – Memória e afirmação de um ideário en decadência. A intervenção da Direcção Geral dos Edifícios e Monumentos Nacionais na igreja de Santa Engrácia (1956–1966). MSc thesis, Faculdade de Letras da Universidade de Lisboa, Portugal.

MARTINS, R. O. 1982. Rochas ornamentais calcárias. Geonovas, Associação Portuguesa de Geólogos, **1**(3), 85–98.

MARTINS, O. R. 1991. Estudo dos calcários ornamentais da região de Pêro Pinheiro. Estudos Notas e Trabalhos, D.G.G.M. **33**, 105–163.

MASON, B. & MOORE, C. B. 1982. Principles of Geochemistry, 4th edn. Wiley, New York.

MENDONÇA, L. V. 2005. Termografia por Infravermelhos Inspecção de Betão. Engenharia & Vida, **16**, 53–57.

MICHALSKI, S., GOTZE, J., SIEDEL, H., MAGNUS, M. & HEIMANN, R. B. 2002. Investigations into the provenance and properties of ancient building sandstones of the Zittau/Görlitz region (Upper Lusatia, Eastern saxony, Germany). In: SIEGESMUND, S., WEISS, T. & VOLLBRECHT, A. (eds) Natural Stone, Weathering Phenomena, Conservation Strategies and Case Studies. Geological Society, London, Special Publications, **205**, 283–297.

MOURA, A. C. 1991. Rochas ornamentais carbonatadas de Portugal. Os mármores. Definição e características gerais. *Boletim de Minas, Lisboa*, **28**(1), 3–15.

MOURA, A. C. 2001. A pedra natural ornamental em Portugal. Nota breve. *Boletim de Minas, Lisboa*, **38**(3), 161–177.

NETO, M. J., SOARES, C. & AIRES-BARROS, L. 2000. Batalha Monastery (Portugal): interdisciplinary survey of historic quarries and restoration works. *In*: CALVI, G. & ZEZZA, U. (eds) *Proceedings of the International Congress on Quarry–Laboratory–Monument, Pavia, Italy*. Università Degli Studi Di Pavia, La Goliardica, Pavese srl, 449–454.

PETTIJOHN, F. J. 1975. *Sedimentary Rocks*. 3rd edn. Harper & Row, New York.

PORTUGUESE NATURAL STONES (P.N.S.). 1995. *CEVALOR (Centro Tecnológico para o Aproveitamento e Valorização das Rochas Ornamentais e Industriais)*. Borba, Portugal.

RAMOS, J. M. F. 1985. Matérias-primas não metálicas de Portugal (Contribuição para o seu conhecimento). *Geonovas*, Associação Portuguesa de Geólogos, **8/9**, 69–91.

REYNAUD, R. & VINTÉM, C. 1994. Estudo da jazida de calcários cristalinos de Estremoz-Borba-Vila Viçosa. *Boletim de Minas, Lisboa*, **31**(4), 355–473.

RIBEIRO, A., ANTUNES, M. T. ET AL. 1979. *Introduction à la geologie générale du Portugal*. Serviços Geológicos de Portugal, Lisbon.

SMITH, B. J. & TURKINGTON, A. V. 2000. Introduction: the need for interdisciplinary thinking in stone decay and conservation studies. *In*: SMITH, B. J. & TURKINGTON, A. V. (eds) *Stone Decay. Its Causes and Controls. Proceedings of Weathering 2000. An International Symposium, Belfast*. Donhead Publishing Ltd, UK, 1–11.

SOUSA, F. P. 1898. Subsidios para o estudo dos calcareos do distrito de Lisboa. *Revista Engenharia Militar*, **1**, **3**, **5**, **6**.

TEIXEIRA, C. & GONÇSALVES, F. 1980. *Introdução à geologia de Portugal*. INIC, Lisbon.

WEBER, J. & LEPPER, J. 2002. Depositional environment and diagenesis as controlling factors for petro-physical properties and weathering resistance of siliciclastic dimension stones: integrative case study on the 'Wesersandstein' (northern Germany, Middle Buntsandstein). *In*: SIEGESMUND, S., WEISS, T. & VOLLBRECHT, A. (eds) *Natural Stone, Weathering Phenomena, Conservation Strategies and Case Studies*. Geological Society, London, Special Publications, **205**, 103–114.

WINKLER, E. M. 1997. *Stone in Architecture. Properties, Durability*. 3rd edn. Springer, Berlin.

Decay of the Campanile limestone used as building material in Tudela Cathedral (Navarra, Spain)

O. BUJ*, J. GISBERT, B. FRANCO, N. MATEOS & B. BAULUZ

Facultad de ciencias de la Tierra, Dpto. de Petrología y Geoquímica, Universidad de Zaragoza, C/ Pedro Cerbuna 12, 50009 Zaragoza, Spain

*Corresponding author (e-mail: oscarbuj@unizar.es)

Abstract: A characterization is presented of the building materials used in the Cathedral of Tudela, as well as of the different forms of decay, with the aim of establishing the cause and mechanisms of this decay. The Cathedral of Tudela was built mainly with Campanile limestone from the upper Miocene. The campanile limestone is a wackestone, with a terrigenous content of 2.6% and 1.5–2.5% of organic matter.

After a detailed investigation of all the different forms of stone decay, our conclusion is that the main type of damage affecting Campanile limestone has morphologies similar to a mechanical fracture with breakages of convex surfaces and resulting very sharp edges. The process of decay is caused by the expansion of the rock during the drying process, which has a very rapid and aggressive effect on the rock.

Laboratory tests showed that through extreme drying in the presence of a magnesium sulphate solution, the salt crystallization inside the stone generates a strength greater than the tensile strength of the stone, thus causing a fracture and the loosening of rock fragments. The materials introduced in recent restorations (sandstone and Portland cement) provide the necessary magnesium for the development of this weathering in Campanile limestone.

The original building of the cathedral dates from 1180 to 1190, with various extensions in the fifteenth and sixteenth centuries. A more thorough analyses of the documentation regarding the building history of the cathedral, which was provided by the Principe de Viana Institute, indicates that the cathedral was built with Campanile limestone. Sandstone was introduced as a building material in 1682 when the tower was rebuilt. The use of Portland cement began in 1930, although its more generalized use dates from the restoration of the cloister in 1949–1956 as well as from later interventions.

The current damage in the cathedral is the result of humidity and salt crystallization, which are relatively common processes in other historic buildings in the Ebro Basin. Less common is a very severe and rapid form of decay that presents morphology similar to a mechanical fracture causing breakages of convex surfaces resulting and sharp edges. This pathology causes the ashlars to crack and generates the loosening of numerous fragments from the capitals, especially in the carved elements that have a greater artistic value. This process of weathering, which affects the Campanile limestone, has always existed, although with moderate manifestations. However, this decay has become more widespread and has increased in intensity since the beginning of the 1970s and it still ongoing today.

The main morphological expressions of this deterioration are conditioned by the bedding planes and the location of the stone (Fig. 1). The fractures are similar to those produced in the compressive strength test. These fractures occur in spite of the weight they hold being well below the maximum strength of Campanile limestone. In the pilaster in Figure 1b, the architect from Principe de Viana calculated that the weight was 70–80 kg cm^{-2} and the strength of the rock is 470 kg cm^{-2}. This calculation ruled out a weight fracture, which is what the morphology of the fissure suggested.

When the ashlar is positioned with the bedding perpendicular, a loosening of the flakes parallel to the ashlar's surface takes place (Fig. 1.1). However, when it is positioned with the bedding parallel, the ashlar shows fissures throughout its thickness with planes that often are at 60° from the horizontal (Fig. 1.2). The process is selective and it affects with greater intensity the more exposed parts (Fig. 1.3) like edge, columns and carved elements (capital volutes, noses, ears, etc.) in contrast with horizontal construction elements or walls. In the sculptural stone decoration there is a tendency towards the loosening of carved elements with significant protuberances.

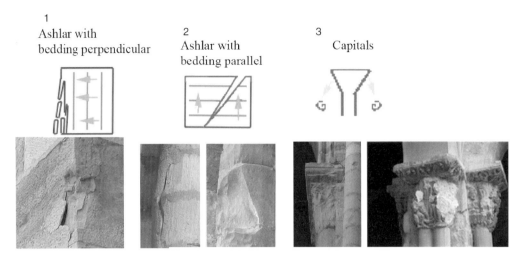

Fig. 1. Morphological expressions of this decay mechanism.

The initial hypothesis, in order to investigate the origins of this pathology, was that the deterioration was caused by rock expansion – a possibility that seemed feasible once the weight excess had been ruled out. Nevertheless, the possible hydric expansion and expansion while drying with the different saline species present in the cathedral were investigated.

Distribution of pathologies

Façades and exterior perimeters (lower section)

In the whole of the exterior perimeter there are capillary rises (more moderate in the south perimeter) that have damaged the construction elements in the north and west façades by salt crystallization and expansion. In the south façade (and the construction elements with this orientation) the damage is minor, although there is some loosening of fragments, as a results of expansion. There are black patinas with irregular development in the three façades caused by atmospheric pollution and, at times, by pigeon droppings.

Inside the cathedral

There are important problems with historic damp through capillary action. (Inactive after the last restoration). This affects the whole of the inside perimeter. In addition to this problem, construction elements have absorbed salts that form efflorescence in many parts. There are also significant damp areas caused by filtrations in coverings (inactive after the last intervention), particularly at the intersection of the transept with the nave, in the north side of this nave and in the base of some of the window frames.

The cloister

The damage is due to expansion, antrophic fractures, pulverization, flaking off, pitting and black patinas. The majority of the capitals present a calcitic patina, which is the result of the application of a restoration product during restoration work in 1980.

In various interventions of reintegration and replacement numerous new materials, different from the original composition of the cloister, have been introduced. We can highlight among them epoxy resins, Portland mortar, restoration mortars, ungalvanized iron bars and huge volumes of sandstone rocks. Of the original capitals, there are 15 in a severe condition of deterioration, another 26 present significant damage and 44 are in an acceptable condition. There is comprehensive photographic documentation available for the cloister from the years 1917–1949. Detailed comparisons with the current situation allow us to assert that 75% of the damage caused by drying expansion occurred after 1949.

Experimental procedure

Samples of the stone were taken from the quarry extraction sites as well as from the cathedral itself, making use of the ashlars that were dismantled in restoration works during the 1990s.

The petrographic characterization of these materials was performed by studying them through an optical microscope. The mineral characterization of the rocks, as well as of the salts present in them, was performed using X-ray diffraction (XRD) (Phillips 1710 diffractometer). The insoluble residue, the organic matter and the CO_2 of the stone were determined by calcinations. We prepared a documental study comparing historical photographs of the years 1917–1949, provided by the Principe de Viana Institute, with current photographs.

The chemical analyses of the major elements in the samples was performed by means of X-ray fluorescence (XRF) in the Analysis Department of Granada University. The equipment used was a Phillips sequential spectrometer, model PW2440.

The distribution of pore sizes was characterized by mercury intrusion porosimetry (MIP) (Van Braker et al. 1981). The equipment used was a PASCAL 140 (macroporosity) and a PASCAL 240 (microporosity) porosimeter. The test samples used were cylinders 3 cm in height and 0.5 cm in diameter.

The physical and mechanical characterization was performed following the UNE (EN 1925, EN 13755, 22950-2), ISRM and RILEM (comissions 25-PEM 1980) recommendations. The equipment used to measure expansion was a comparator (Digico) with micron precision, which was connected to a computer to enable constant data recording. This was then assembled onto a device, as described in ISRM 1979, that allows for the introduction of solutions and the forced drying of material without manipulating the test samples.

The expansion tests were performed with distilled water as well as with 5% solutions of NaCl, $Na_2SO_4 \cdot 10H_2O$, $MgSO_4 \cdot 7H_2O$, $Mg(NO_3)_2$, $MgCl_2$ and NH_4NO_3. The expansion test was performed with prismatic test samples of $4 \times 4 \times 10 \pm 0.5$ cm once the faces were rectified and polished.

Each one of the cycles consisted of saturation by immersion of the sample for 24 h, and its further drying under temperature conditions of 22 ± 3 °C and a relative humidity of $20 \pm 5\%$. The drying phase was considered to be over once the expansion of the test samples was over and lasted for 12–17 days.

Materials

The limestones studied date from the Miocene age and are excavated from quarries in the Tertiary Ebro Basin. This rock is interpreted as a carbonate deposit from a lacustrine palustrine environment with normal salinity. The limestone is classified as wackestone and biomicrite wackestone following the classifications of Dunham (1962) and Folk (1962).

The chemical composition comprises mainly calcite with small quantities of clay minerals (80–85% micrite, 15% bioclastic fragments, 2% sparite cement, and 1–2% clay mineral and grain quartz). The stone is characterized by a micritic texture and by a clear preferred orientation of bioclastic components parallel to the bedding. The composition of the bioclasts is that of calcite and they represent 26% of the structure. The maximum size of the bioclasts is 1 mm and the average size is 0.3 mm. The larger size bioclasts (bivalves and carofites) present a preferred orientation parallel to the bedding. The cement is composed of 3% sparite calcite, which has an average size of 0.08 mm, and is located preferentially in the moldic porosity. The open porosity (10%) is the result of bioturbation, with an average pore size of 0.2 mm.

In Tables 1 and 2 the chemical analyses of the major elements of the complete rock obtained from XRF are presented, as well as that of the insoluble residue, water content, organic matter and CO_2 calculated using weight loss after ignition.

The values for the insoluble residue indicate that we are dealing with very pure limestone, presenting a maximum value of 2.44%, of which 1.5% corresponds to organic matter and the other 1% to the the silicate fraction. Within the silicate fraction we have 25% philosilicates against 75% quartz and feldspars. The philosilicates are, in order of abundance, illite, chlorite and smectite.

In order to ascertain the presence of salts in the Campanile limestone used in the Cathedral, chemical analyses were carried out in areas where this

Table 1. *Values of water content, organic matter (M.O.) CO_2 and insoluble residue (R.I.) for Campanil limestone*

H_2O	M.O.	CO_2	R.I.
0.614%	1.114%	41.601%	1.011%

Table 2. *Concentrations of major elements for samples analysed by X-ray fluorescence. All elements expressed in per cent*

SiO_2	TiO_2	Al_2O_3	Fe_2O_3	MnO	MgO	CaO	Na_2O	K_2O	P_2O_5	LOI	TOTAL
0.511	<LLD	0.101	0.094	0.096	0.859	54.543	0.148	0.023	0.027	43.290	99.692

LLD, lower limit on detection; LOI, loss on ignition.

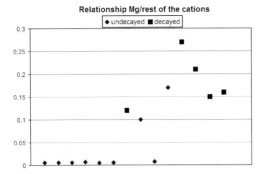

Fig. 2. Relationship between Mg and the rest of the cations in areas in which this decay mechanism is present.

pathology was present. These were performed on the extracted salts using arbocell compresses. The results of the analyses indicated that in all areas where this pathology is found, and regardless of the absolute quantity of the salts present, there was a high proportion of magnesium salts (Fig. 2). Thus, we were also able to determine the relationship between the magnesium present and the rest of the cations. This was determined for the Campanile limestone as well as for the replacement sandstone and the Portland cement. The ratio of Mg to the rest of the cations is 0.08 in Campanile Limestone, and 0.16 and 0.17 in sandstone and Portland mortar, respectively. The establishment of this correlation of different cations in relation to Mg produced very variable results.

The sandstone is classified as a medium-grain litharenitic rock of greyish-orange colour. The grain size, shape and sorting of this rocks is homogeneous, with subrounded grains of an average size of 0.25–0.35 mm and maximum size of 0.7–0.9 mm, and sorting values more or less constant at between 1.2 and 1.4 (Buj *et al.* 2007). The skeleton is grain-supported with heterogeneous fragments of carbonated rocks (calcite and dolomite-, mono- and polycrystalline), fragments of metamorphic and volcanic rocks, quartz (mono- and polycrystalline) and feldspars. Its accessory minerals are zircons, sphene, tourmalines, pyroxene, iron oxides and muscovite. The cement is made up of 90–95% sparite and microsparite carbonate (calcite and dolomite) and 5–10% silica. Carbonated sparry cement is occasionally of ferric calcite. Amongst the cement and the carbonated clasts there is 5% dolomite.

Results

The pore-size distribution of the Campanile limestone, calculated using mercury intrusion porosimetry, shows a unimodal distribution of pore sizes (Fig. 3d); with almost in its entire access to pores in a small pore radius range (Table 3).

The results of the hydric tests have been summarized in Table 4. The absorption, capillary suction and desorption graphics of this rock (Fig. 3a–c) present a well-connected porous system that allows a fast movement of water in its interior through large access points. The fast desorption of the material and the critical water content indicate a low permanence of humidity inside the stone.

The measure of the expansion not confined at room temperature during the absorption and drying processes was performed using distilled water and other saline solutions. Those carried out using 5% magnesium sulphate solution created the greatest expansion of the material during the drying process.

The results of the expansion tests with distilled water showed an expansion value of 1×10^{-4}, thus corroborating the view that the expansion which takes place in this rock is not a consequence of the presence of a small proportion of swelling clays. On the other hand, the expansion results while drying in the presence of magnesium sulphate salts show a residual average expansion of 44×10^{-4}, with values ranging from 30×10^{-4} to 60×10^{-4} after eight cycles.

The graphic in Figure 4 shows the linear expansion in relation to time for one of the characterized test samples. In this graphic there are eight different cycles. Each cycle has two stages: the absorption stage and the drying stage. The first cycle was performed with distilled water, and the others with 5% of magnesium sulphate solutions. In the first cycle, during the water uptake and with subsequent drying, the test sample does not show any changes in size. In the remaining cycles the rock shows an important expansion during the drying phase and a slight contraction during the rehydration phase.

The contraction of the material that takes place during the rehydration phase is lower than the expansion produced during the drying phase, thus generating a residual expansion that continues to increase as the cycles progresses and which reaches a maximum value of 35×10^{-4}. This contraction after rehydration of the material is equivalent to the sum of the effects caused by the loosening of the tensions generated during the crystallization of the salts, and the dissolution of the salts that might have been crystallized between the surface of the test samples and the equipment.

There was a constant register of the length of the test samples that helped in the understanding of the temporal evolution of the process. Figure 5 shows an outline of the process to which the test samples were subjected. In the first phase (A) the test sample is completely immersed in a 5% solution of

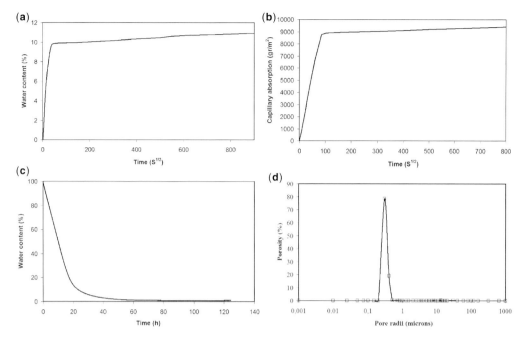

Fig. 3. (a) Absorption graph (water content), (b) capillary absorption graph, (c) desorption graph and (d) pore-radii distribution.

magnesium sulphate and showed no changes in length. In the second phase (B), with constant relative humidity and temperature, the drying of the material takes place, producing an expansion of 12×10^{-4}. In the third phase (C) the test sample is rehydrated again with a 50 g l^{-1} solution of magnesium sulphate, which causes part of the expansion generated during the drying phase to disappear, leaving a residual expansion of 4.5×10^{-4}.

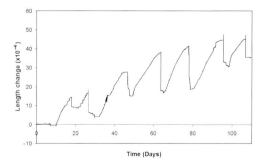

Fig. 4. Length change behaviour of one limestone sample. The first cycle was performed with distilled water, and the others with 5% magnesium sulphate solutions.

Discussion

The comparison of historic photographs from 1917 to 1949 with current ones shows clearly that this pathology was already present, but with a lesser intensity than it is today. The documentary data available, as well as the need for interventions to clean the fractures, show a recent development and indicate that there has been a spread of this pathology since the 1970s. This spread exists in correlation with the magnesium sulphate additions that were caused by the restorations, in which sandstone (combined with Campanile limestone construction elements) and Portland cement were introduced.

The crystallization of soluble salts inside the porous system of limestone is regarded as the main deterioration mechanism in cultural heritage monuments (Correns 1949; Wely 1959; Benavente et al. 1999; Rodríguez-Navarro & Dohene 1999; Scherer 1999). Different experimental studies have shown that porosity and the distribution of pore sizes play an important role in the susceptibility of stone to weathering by salts (McGreevy 1996; Ordoñez et al. 1997). It seems probable that one of the determinant factors in the development of this pathology in Campanile limestone is its pore-space distribution. One of the distinctive features of this pore system is its high porosity, which allows an

Table 3. *Pore-radii distribution and porosity of the Campanile limestone*

Average size (μm)	0.001–0.01 μm	0.01–0.1 μm	0.1–1 μm	1–10 μm	>10 μm	Maximum	Macroporosity (%)	Microporosity (%)
1.660	0.000	0.000	98.909	0.332	0.803	0.2–0.15	0.12	12.59

Table 4. *Petrophysical properties of the Campanile limestone*

W	P_o	P	S_1	W_d	W_{cri}	CAC	S_c
11.23	22.02	23.28	90.32	1.06	17.35	96.18	85.9

W, humidity content (%); P_o, open porosity (%); P, total porosity (%); S_1, saturation rate at 1 h (%); W_d, water retention after drying (%); W_{ar}, critical water content (%); CAC, capillary absorption rate (g m^{-2}s$^{-0.5}$); S_c, capillary saturation (%).

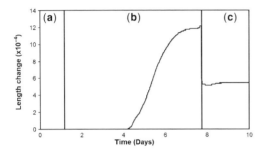

Fig. 5. Outline of the length change behaviour. (**a**) The test sample is completely immersed in a solution of magnesium sulphate, (**b**) drying of the material takes place and (**c**) the test sample is rehydrated.

intense crystallization of salts as well as a distribution of practically the entirety of pore sizes (98.91%) in the range of 0.1 μm. Porous materials that present a distribution of pore sizes in a small range have an increased susceptibility to expansive processes (Ruedrich & Siegesmund 2006).

The results of the expansion tests carried out with distilled water show an expansion value smaller than 1×10^{-4}, thus confirming that the expansion that occurs in this rock is not a consequence of the expansion of the clay minerals.

The mean expansion that we have calculated for this rock is 4 mm m^{-1} after eight cycles. The final change in length is the result of progressive microfracturation caused by the crystallization pressure during the drying cycles. The cycles of saturation and drying processes create a stress that generates the breakdown of the stone. There are no data available in the References on absolute expansion in systems contaminated by salts, but there are data available on hydric expansions. Poschlod, in 1990, calculated the maximum expansion in marble of 0.1 mm/m. Grimm (1999) calculated a maximum expansion of 0.1 mm m^{-1} in sandstones rich in clays, and Snethlage & Wendler (1997) calculated maximum expansions in sandstones of 0.5 mm m^{-1}.

The internal stress generated during these cycles by the crystallization of salts can cause damage to the material if the strength generated is greater that of the tensile strength of the stone. The tensile strength for this rock is 7.08 MPa, while the stress caused by the swelling of the stone (generated by the crystallization of salts in the rock) is 19 MPa. These stresses have been calculated by direct analogy to the stresses caused by thermal stress. The stress caused by the swelling of the stone is similar to the stress calculated by Pusch in 2006 (he indicated a maximum stress of up to 20 MPa) for the hydration of smectites in sandstone.

Nevertheless, the expansion phenomenon has a significant variability, with a coefficient of variation of 0.31 in the same groups of samples in which the coefficient of variation is 0.02 for the absorption test. This low coefficient of variation, obtained from the absorption test, points to a very homogeneous porous system, which leaves the exclusive attribution of expansion to the porous system with little credibility. The remaining factors that play a role in the expansion of this material have not been established with certainty yet; therefore, further investigation is necessary. The most suitable hypothesis is the presence of organic matter, as it provides an explanation for associated phenomena that would otherwise remain unexplained. Organic compounds, particularly the alginates (bearing in mind that we are dealing with Carofites limestone), are able to drive the salt crystals; this would explain the phenomenon of expansion during the drying process when all of the crystals grow in the same direction. However, the gelification by cationic bridge is a phenomenon that may explain the greater expansion associated with saline solutions as it causes the organic gel to swell through hydration. In addition to this, the complex degradation of organic matter and its irregular distribution on the rock would explain the increased coefficient of variation in the expansion test in relation to the coefficient of variation in the absorption test.

Conclusions

The decay mechanism that causes this pathology is the stress originated in the rock during the drying of the material resulting from the expansion of Campanile limestone in extreme drying conditions and in the presence of magnesium salts. Other researchers have documented this particular phenomenon in systems contaminated by sodium chloride saline solutions, magnesium sulphate and calcium nitrate (Snethlage & Wendler *et al.* 1997; Ruedrich *et al.* 2007). The expansion during the drying process is not yet clear, but could be explained by salt crystallization in the pores of greater dimensions, as is the case in the model of freeze–thaw (Fitzner & Snethlage 1982) or by the formation of salt films that attach themselves to the grains pushing them while they grow (Pühringer *et al.* 1985).

The deterioration mechanism of expansion through drying has always existed in the cathedral, but with moderate dimensions and manifestations. However, since the beginning of the 1970s a widespread and an increased deterioration has taken its toll and this continues to this day.

Portland cement and sandstone, introduced in restorations that were undertaken between 1950 and 1980, have had an important role in the alteration of the stone as they have been responsible for the appearance of magnesium salts.

Within the cathedral we find various construction elements showing areas of damp and salt crystallization. This damage developed in the uppermost section of the capillary rise, with no point of connection with the very important saline efflorescence. This upper section is also the driest of all of the sections where the presence of salts occurs. Generally, all of the carved forms or those with a large exposed area present greater damage because the deterioration takes place at the point of maximum drying.

Campanile limestone shows a larger expansion during the drying stage generated by the crystallization of magnesium sulphate solutions. The cyclic repetition of saturation and drying with this saline solution generates fissures and fractures of the stone in a very short space of time. The force generated by this salt during crystallization in this rock is clearly greater than the tensile strength of the material, thus being responsible for the fracture and loosening of fragments.

The nature of this process is not perfectly clear and further investigation is necessary. It seems probable that the main factors affecting this pathology are the crystallization pressure of magnesium salts, the structure of the pore system and the organic matter present in this rock.

We would like to thank the CADIA Institute and the Principe de Viana Institute, both of them in Pamplona, for the permission to publish this data, which is part of a report commissioned by the above-mentioned Institutes. We would also like to thank Mr. Javier Sancho and Mr. Antonio Aretxabala for their comments, data contribution and personal support throughout the whole process. And finally, we also would like to thank the firm Zubillaga for the ashlars and tests provided.

References

BENAVENTE, D., GARCÍA DEL CURA, M. A., FORT, R. & ORDOÑEZ, S. 1999. Thermodynamic modelling of changes induced by salt pressure crystallization in porous media of stone. *Journal of Crystal Growth*, **204**, 168–178.

BUJ, O. & GISBERT, J. 2007. Petrophysical characterization of three commercial varieties of miocene sandstones from the Ebro valley. *Materiales de construcción*, **57**, 62–73.

CORRENS, C. W. 1949. Growth and dissolution of crystals under linear pressure. *Discussions of the Faraday Society*, **5**, 267–271.

DUNHAM, R. J. 1962. Classification of carbonate rocks according to depositional texture. *In*: HAM, W. E. (ed.) *Classification of Carbonate Rocks*. AAPG Memoirs, **1**, 108–121.

FITZNER, B. & SNETHLAGE, R. 1982. Zum Einfluß der Porenradienverti Eilung auf das Verwitterungsverhalten ausgewählter Sandsteine. *Bautenschult Bausan*, **82**, 97–102.

FOLK, R. L. 1962. Spectral subdivision of limestones types. *In*: HAM, W. E. (ed.) *Classification of Carbonate Rocks*. AAPG Memoirs, **1**, 62–84.

GRIMM, W. D. 1999. Beobachtungen und Ueberlegungen zur Verformung von Marmorobjekten durch Gefuegeauflockerung. *Zeitschrift der Deutschen Geologischen Gesellschaft*, **150**(2), 195–235.

MCGREEVY, J. P. 1996. Pore properties of limestone as controls on salt weathering susceptibility: a case study. *In*: SMITH, B. J. & WARKE, P. A. (eds) *Processes of Urban Stone Decay*. Donhead, Shaftesbury, Dorset, 150–167.

NORMA UNE-EN 1925:1999. *Métodos de ensayo para piedra natural. Determinación del coeficiente de absorción de agua por capilaridad*. Spanish Association for Standardization and Certification (AENOR), Madrid.

NORMA UNE-EN 13755:2002. *Métodos de ensayo para piedra natural. Determinación de la absorción de agua a presión atmosférica*. Spanish Association for Standardization and Certification (AENOR), Madrid.

NORMA UNE 22950-2:2003. *Resistencia a la tracción. Determinación indirecta.* Spanish Association for Standardization and Certification (AENOR), Madrid.

ORDÓÑEZ, S., FORT, R. & GARCIA DEL CURA, M. A. 1997. Pore size distribution and the durability of a porous limestone. *Quarterly Journal of Engineering Geology and Hydrogeology*, **30**, 221–230.

POSCHLOD, K. 1990. Das Washer im Porenraum kristalliner Naturwerksteine und sein Einfluss aufdie Verwitterung. *Muencher Geowissenschaftliche Abhandlungen*, **7**, 1–62.

PÜHRINGER, J., BERNTSSON, L. & HEDBERG, B. 1985. Hydrate salts and degradation of materials. *In*: *Proceedings of the Fifth International Congress on Deterioration and Conservation of Stone, Lausanne*. Presses Polytechniques Romandes, 231–240.

PUSCH, R. 2006. Mechanical properties of clays and clay minerals. *In*: BERGAYA, F., THENG, B. K. G. & LAGALY, G. (eds) *Handbook of Clay Science*. Developments in Clay Science, **1**, 247–260.

RILEM 25-PEM PROTECCIÓN Y EROSIÓN DE MONUMENTOS. 1980. Ensayo N° II. *Dilatación lineal por absorción de agua*. RILEM, Bagneux, France.

RODRÍGUEZ-NAVARRO, C. & DOHENE, E. 1999. Salt weathering: influence of evaporation rate, supersaturation and crystallisation pattern. *Earth Surface Processes and Landforms*, **24**, 191–209.

RUEDRICH, J. & SIEGESMUND, S. 2006. Fabric dependence of length change behaviour induced by ice crystallisation in the pore space of natural building stone. *Heritage Weathering and Conservation*, **I**, 497–505.

RUEDRICH, J., SEIDEL, M., ROTHERT, E. & SIEGESMUND, S. 2007. Length changes of sandstones caused by salt crystallization. *In*: PŘIKRYL, R. & SMITH, B. J. (eds) *Building Stone Decay: From Diagnosis to Conservation*. Geological Society, London, Special Publications, **271**, 199–209.

SCHERER, G. W. 1999. Crystallisation in pores. *Cement and Concrete Research*, **29**, 1347–1358.

SNETHLAGE, R. & WENDLER, E. 1997. Moisture cycles and sandstone Degradation. *In*: BAER, N. S. & SNETHALAGE, R. (eds) *Saving our Architectural Heritage. The conservation of historic stone structures*. John Wiley and Sons, Chichester, 7–24.

VAN BRAKEL, J., MODRÝ, S. & SVATÁ, M. 1981. Mercury porosimetry: state of the art. *Powder Technology*, **29**, 1–12.

WELY, P. K. 1959. Pressure solution and the force of crystallisation. A phenomenological theory. *Journal of Geophysical Research*, **64**, 2001–2025.

On-site evaluation of the 'mechanical' properties of Maastricht limestone and their relationship with the physical characteristics

S. RESCIC*, F. FRATINI & P. TIANO

CNR – Istituto per la Conservazione e Valorizzazione dei Beni Culturali (ICVBC), Via Madonna del Piano 10, 50019 Sesto Fiorentino (FI), Italy

*Corresponding author (e-mail: s.rescic@icvbc.cnr.it)

Abstract: Maastricht limestone is a soft bioclastic calcarenite of the Upper Cretaceous period cropping out in southern Limburg between Belgium and The Netherlands. This material was widely used from the Middle Ages to the Renaissance. Four different varieties can be distinguished according to fossil content and petrographic characteristics, which determine slight differences in compressive strength. Despite its poor mechanical characteristics, the material is very durable with remarkable frost resistance. This is mainly due to the pore dimensions (the most frequent pore radius class is 16–64 μm) but also to the particular kind of weathering that causes the formation of a protective 'skin' through a process of dissolution of unstable aragonite from serpulids and calcite precipitation in the pores of the external layer. The physical characteristics and the mechanical properties (using the drilling resistance measurement system (DRMS) method) of the hard layer that developed on the surface of Tongeren Cathedral, constructed using the Sibbe variety of Maastricht limestone, were investigated and compared with those of the quarry material. This comparison made it possible to emphasize the particular hardness of this surface in contrast to the outer layer of the quarry material. Moreover, it was possible to determine its thickness and to infer that this hard layer was formed after only 15 years of exposure.

In this paper the physical characteristics (determined in laboratory tests) and the cohesion (*in situ* tests) of weathered Maastricht limestone (Tongeren Cathedral, Belgium) were studied in comparison to the unweathered material from the Sibbe quarry (in Limburg, The Netherlands). This material was widely used from the Middle Ages to the Renaissance period and it characterizes the architecture of Limburg on both sides of the Meuse river, as well as the sixteenth century architecture of some cities in central Holland (Utrecht and Zaltbommel). At present it is mainly used for restoration purposes. The only location where the Maastricht limestone can be quarried today is in Sibbe, The Netherlands. Other sites exist but are used principally for the production of cement additives. The quarry is located exactly beneath the village of Sibbe, and extends over an area of approximately 100 hectares (1 × 1 km). The underground facility consists of an extended labyrinth of tunnels down to a depth as great as 1 km.

Maastricht limestone is very homogeneous, and layering is rarely observed. It is extracted using ordinary motor-operated chainsaws (commonly used to cut timber).

The Maastricht limestone

Maastricht limestone (Maastricht stone, Tuffeau de Maastricht, Mergel, Maastrichtien) is a soft bioclastic calcarenite of the Upper Cretaceous period belonging to the Maastricht Formation cropping out in southern Limburg between Belgium and The Netherlands. Four main varieties can be distinguished according to fossil content and petrographic characteristics, which determine slight differences in the compressive strength. Despite its poor mechanical characteristics, the material displays a good durability with remarkable frost resistance (Dreesen & Dusar 2004; Dubelaar *et al.* 2006). This is mainly due to the type of porosity but also to the formation, particularly in the Sibbe variety and in open air conditions, of a protective 'skin' through a process of dissolution of unstable aragonite from serpulids and calcite precipitation in the pores of the external layer (Dubelaar *et al.* 2006). The decay develops mainly through detachment of the crust (Fig. 1).

This kind of weathering, which affects soft bioclastic limestones with the formation of a hard superficial layer, is widespread and it has been studied by many authors who distinguished different genesis mechanisms (Rossi Manaresi & Tucci 1989; Vannucci *et al.* 1994; Smith *et al.* 2003; Siegesmund *et al.* 2007; Török *et al.* 2007).

Under an optical microscope the Sibbe variety shows a good sorting, with an average grain size of approximately 100 μm (Fig. 2).

The grains, subangular in shape, consist mainly of sparitic calcite (shell fragments and skeletons of sea organisms) and secondarily of micritic calcite.

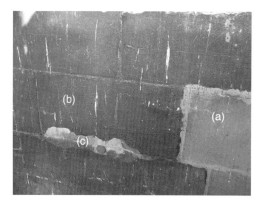

Fig. 1. External wall of Tongeren Cathedral where it is possible to observe the development of a crust due to dissolution and reprecipitation of aragonite/calcite: (a) new ashlar; (b) old ashlar with dark crust; (c) loss of crust.

Fig. 2. Thin section image (plane polarized light) that shows a structure consisting mainly of shell fragments, along with micritic calcite grains with a scarce binder of sparitic calcite.

There are also rare silicatic grains. The binder is scarce and comprises sparitic calcite. Porosity is high (>50%) and mainly macroporosity. The petrographical classification is grainstone (Dunham 1962) and intrasparite (Folk 1959, 1962).

Methods

Three different porosity determinations (total open porosity, pore-size distribution and porosity accessible to water) were carried out on samples taken from the quarry material, Tongeren Cathedral original (old zone) and recently replaced (new zone, 15 years old) ashlars, respectively. Each type of porosity determination was performed on five samples measuring $2 \times 2 \times 2$ cm. The samples from the cathedral included the outer hard layer, which, therefore, was measured together with the softer internal portion.

Total open porosity (P%)

The total open porosity was determined through the helium and mercury pycnometer method (UNI 9724/2-90:1990; Barsottelli et al. 1998). This parameter was obtained measuring the following data: dried weight (W_d) measured with an analytical balance (precision of ± 0.0001 g); apparent volume (V_b) measured with a mercury pycnometer (Chandler Engineering Co.); real volume (V_r) measured with a helium Penta-Pycnometer (Quanta Chrome). The bulk (γ_b) and real (γ) density, and the total open porosity ($P\%$), were calculated according to the following equations:

$$\gamma = W_d/V_r \quad (1)$$
$$\gamma_b = W_d/V_b \quad (2)$$
$$P\% = [(V_b - V_r)/V_b] \times 100. \quad (3)$$

Pore-size distribution (MesoP%)

Pore-size distribution was determined using the Hg porosimeter method (Normal 4/80:1980; Barsottelli et al. 1998).

A Thermofinnigan mercury porosimeter (Pascal 140 and 240 units) was utilized to investigate the pore-size distribution and the amount of porosity in the 0.0037–150 μm (pore radius) range. This pore range has been defined as mesoporosity (Barsottelli et al. 1998).

Porosity accessible to water (WP%)

Water porosity was determined through the hydrostatic balance method (ISO 6783:1982). In order to determine this parameter, the dry weight (W_d), hydrostatic weight (W_i) and wet weight (W_w) were measured. For the dry weight the stone samples were put in a oven at 60 °C until the weight difference after two consecutive weight controls, over a 24 h period, was less than 0.1%. For the hydrostatic weight the specimens were put under vacuum for 24 h to eliminate air and were then soaked in water. After a further 24 h without vacuum, the hydrostatic weight was determined by the hydrostatic balance (Mettler Toledo AG204). The wet weight was determined after lightly wiping the liquid water from the sample surface with a chammy. The porosity accessible to water ($WP\%$) was calculated according to the following equation:

$$WP\% = (W_w - W_d/W_w - W_i) \times 100. \quad (4)$$

Capillary water absorption (CIW%)

The capillary water absorption was determined using the UNI 10859:2000 method. Five samples (5 × 5 × 2 cm) for each typology (quarry, *in situ* old and new ashlar) were dried at 60 °C until constant weight (W_d) (within 0.1% of the sample mass). These were then placed on a pack of filter paper completely saturated with distilled water (the liquid water should not directly touch the stone material). The samples were weighed (W_w) at given time intervals up to saturation (in this case 10 days). Before weighing, the stone surface was gently wiped with a chammy. The surface of the samples in contact with the filter paper was the one exposed to weathering both for the quarry samples and the samples from the cathedral. The imbibition coefficient expressed in weight (*CIW%*) is calculated according to the following equation:

$$CIW\% = (W_w - W_d/W_d) \times 100. \quad (5)$$

Drilling resistance measurement system (DRMS)

Mechanical 'cohesion' was determined *in situ* in the quarry material, Tongeren Cathedral original (old zone) and recently replaced (new zone) ashlars using the drilling resistance measurement system (DRMS) (Tiano *et al.* 2000; Tiano 2001; Fratini *et al.* 2006).

The DRMS is a new portable instrument developed and validated by the authors in collaboration with Sint Technology (Calenzano, Florence, Italy), who produce and market it. The DRMS is a device designed to perform a simple but precise 'drilling resistance' (DR is expressed as a force in Newton) test by continuously measuring the force required to drill a hole in stone material during drill-bit penetration. Both rotational speed and penetration rate are kept constant during the test.

The device was developed both for laboratory tests and outdoor applications.

Results

The petro-physical data are shown in Table 1.

The material from the cathedral always displays lower porosity values when compared to the quarry material. This is due to the formation of the hard layer through a process of dissolution of unstable aragonite and precipitation of calcite into the external layer's pores (Dreesen & Dusar 2004; Dubelaar *et al.* 2006). This hard layer is characterized by porosity values lower than the values reported in Table 1, which are related to a thickness of 2 cm while the thickness of the hard layer is about 2 mm (as determined by DRMS). This discrepancy depends on the fact that the method used to determine the porosity does not permit the use of samples of less than 1 cm^3. The pore size distribution (Fig. 3) and the capillary absorption diagrams (Fig. 4) both show a variation.

The porosity distribution shifts towards pores of smaller dimensions in those samples where the formation of the hard layer occurred (the cathedral). The capillary absorption curves show that the formation of this layer causes a delay in water uptake (lower absorption gradient) during the first minutes and lower absorption in general. Moreover, no important differences can be noted between the samples from the replacement ashlars and from the original ones.

The petro-physical characterization data concur with those of the drilling resistance findings shown in Table 2.

The mean values of the force calculated at different depths are, indeed, higher in the cathedral ashlars than in the material from the quarry. An interesting datum is the mean force in the depth range 0–2 mm from the surface. The cohesion of the crust of the original ashlars is, indeed, markedly higher than the cohesion of the quarry material. This is even more evident looking at the drilling curves (Fig. 5).

These curves show that the crust also begins to form in the case of the replacement ashlar (15 years old). Furthermore, from the drilling curves it is possible to see that at a depth of 6–10 mm the old ashlar has cohesion values similar to that of the quarry material at the same depth. This is also confirmed by the mean values of the force computed for this depth range (Table 2).

This datum apparently does not agree with an extraction of material through dissolution of

Table 1. *Petro-physical characteristics*

	γ_b (g cm^{-3})	γ (g cm^{-3})	P (%)	MesoP* (%)	WP (%)
Sibbe quarry	1.34 ± 0.07	2.68 ± 0.01	50 ± 2	49 ± 2	49 ± 2
Tongeren Cathedral new zone	1.45 ± 0.04	2.77 ± 0.02	48 ± 2	45 ± 2	43 ± 2
Tongeren Cathedral old zone	1.49 ± 0.04	2.83 ± 0.02	47 ± 2	42 ± 2	41 ± 2

*Pore radius range 0.0037–150 μm, mean values on five samples

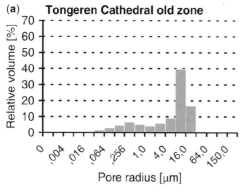

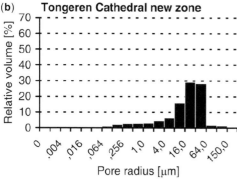

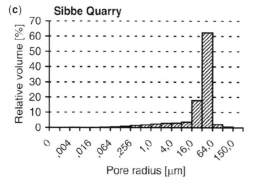

Fig. 3. Pore-size distribution showing a shift of the pore radius towards smaller size pores in the cathedral samples (old (**a**) and new (**b**) zones) compared with the quarry material (**c**).

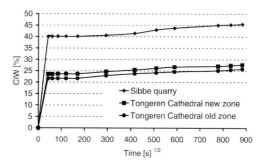

Fig. 4. Capillary water absorption curves showing the delay in water uptake during the first minutes and lower absorption in the cathedral samples (old and new zones) compared with the quarry material.

calcite and aragonite from the rock interior (McAlister *et al.* 2003). However, this dissolution should be considered as spread through a large volume of rock and, therefore, the weakening effects on the inside of the rock are minimized.

As mentioned earlier, the data on the 'hardness' and thickness of the superficial crust determined with the DRMS, together with the porosity data obtained from the $2 \times 2 \times 2$ cm samples, allowed us to calculate the approximate value of crust porosity. This was done because there is a correlation between drilling resistance (DR in Newtons) and total open porosity (Fratini *et al.* 2006). Thus, a porosity–drilling resistance correlation curve relative to the 0–20 mm depth range was developed. The relative equation:

$$\text{porosity} = -2.9174 \times \text{DR} + 56.487 \quad (6)$$

enabled us to find the drilling resistance of the crust (Table 3).

Regarding the frost resistance of the material, as reported by Dreesen & Dusar (2004) and Dubelaar *et al.* (2006), the data of pore-size distribution give reasons for this good behaviour. As a matter of fact, the more represented class has a pore radius of 16–64 μm, significantly higher than the pore radius threshold reported by many authors (Walker *et al.* 1969; Vincenzini 1974; Brownell 1976; Davison 1980; Maage 1984; Robinson 1984) that falls in the 1–5 μm range.

Table 2. *DFMS results*

	Force (N)		
	0–10 mm	0–2 mm	6–10 mm
Sibbe quarry	1.80 ± 0.93	0.62 ± 0.33	2.82 ± 1.47
Tongeren Cathedral new zone	2.27 ± 1.49	3.27 ± 1.41	2.28 ± 1.76
Tongeren Cathedral old zone	3.76 ± 0.59	8.50 ± 2.21	2.79 ± 0.48

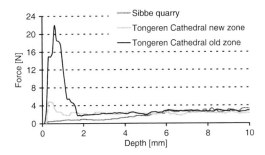

Fig. 5. DRMS drilling outlines showing the presence of a more or less hard layer evidenced by an initial peak of the drilling profile in the case of the cathedral (old and new zones) v. the quarry material.

Table 3. *Total open porosity*

	P (%) 0–20 mm*	P (%) 0–2 mm†
Sibbe quarry	50	54
Tongeren Cathedral new zone	48	47
Tongeren Cathedral old zone	47	32

*Whole samples; †crust.

Conclusions

Despite its poor mechanical characteristics, the good durability of the Maastricht tuffeau depends mainly on two factors: the low number of pores with a radius of around 1 μm that justifies the good frost resistance; and the weathering that develops through the formation of a protective 'skin'. This protective layer shows the following characteristics:

- thickness of about 2 mm;
- 'cohesion' 20 times higher than the underlying material and the quarry material;
- decrease in capability to absorb water through capillarity;
- total open porosity of 32% with respect to 48% of the underlying material;
- shifts in pore distribution towards pores of smaller size.

The strong physical and mechanical differences between the 'hard layer' and the substrate give rise to the formation of a discontinuity that favours its detachment and may expose a weaker substrate now susceptible to retreading with a new crust.

The authors are grateful for the financial support from the EU 5th Framework Project 'Integrated tool for *in situ* characterization of effectiveness and durability of conservation techniques in historical structures' (DIAS) with contract no. DIAS EVK4-CT-2002-00080.

References

BARSOTTELLI, M., FRATINI, F., GIORGETTI, G., MANGANELLI DEL FÀ, C. & MOLLI, G. 1998. Microfabric and alteration in Carrara Marble: a preliminary study. *In: Science and Technology for Cultural Heritage. Istituti Editoriali e Poligrafici Internazionali Pisa*, **7**(2), 115–126.

BROWNELL, W. E. 1976. Structural clay products. *Applied Mineralogy*, **57**, 201–210.

DAVISON, J. I. 1980. Linear expansion due to freezing and other properties of bricks, *Proceedings of the second Canadian Masonry Symposium*. Carleton University, Ottawa, Ontario, Canada, 13–14.

DREESEN, R. & DUSAR, M. 2004. Historical Building stones in the province of Limburg (NE Belgium): role of petrography in provenance and durability assessment. *Materials Characterization*, **53**, 273–287.

DUBELAAR, C. W., DUSAR, M., DREESEN, R., FELDER, W. M. & NIJLAND, T. G. 2006. Maastricht limestone: A regionally significant building stone in Belgium and The Netherlands. Extremely weak, yet time resistant. *In*: FORT, R., ALVAREZ DE BUERGO, M., GOMEZ HERAZ, M. & VAZQUEZ CALVO, C. (eds) *Proceedings of the International Conference on Heritage, Weathering and Conservation*. HWC 2006, 21–24 June 2006, Madrid, Spain. Taylor and Francis Group, UK, Volume 1, 9–14.

DUNHAM, R. J. 1962. Classification of carbonate rocks according to depositional textures. *In*: HAM, W. E. (ed.) *Classifications of Carbonate Rocks*. AAPG Memoirs, **1**, 108–121.

FOLK, R. L. 1959. Practical petrographic classification of limestones. *Bulletin AAPG*, **43**, 1–38.

FOLK, R. L. 1962. Spectral subdivision of limestone types. *In*: HAM, W. E. (ed.) *Classifications of Carbonate Rocks*, AAPG Memoirs, **1**, 64–84.

FRATINI, F., RESCIC, S. & TIANO, P. 2006. A new portable system for determining the state of conservation of monumental stones. *Materials and Structures*, **39**(2), 139–147.

ISO 6783:1982. *Determination of the Water Porosity (Hydrostatic Balance Method)*. International Organization for Standardization, Geneva.

MAAGE, M. 1984. Frost resistance and pore size distribution in bricks. *Materials and Structures*, **17**(101), 345–350.

MCALISTER, J. J., SMITH, B. J. & CURRAN, J. A. 2003. The use of sequential extraction to examine iron trace metal mobilisation and the case-hardening of building sandstone: a preliminary investigation. *Microchemical Journal*, **74**, 5–18.

Normal 4/80:1980. *Distribution of Pore Volume as a Function of Pore Diameter*. C.N.R.-I.C.R., Rome.

ROBINSON, G. C. 1984. The relationship between pore structure and durability of brick. *Ceramic Bulletin*, **63**, 295–299.

ROSSI MANARESI, R. & TUCCI, A. 1989. Pore structure and salt crystallization: 'salt decay' of Agrigento biocalcarenite and 'case hardening' in sandstone. *In*: ZEZZA, F. (ed.) *Proceeding of the 1st Internaional Symposium of the Conservation of Monuments in the Mediterranean Basin*. Bari, 7–10 June, Grafo Edizioni, Brescia, Italy, 97–100.

SIEGESMUND, S., TÖRÖK, A., HÜPERS, A., MÜLLER, C. & KLEMM, W. 2007. Mineralogical, geochemical and microfabric evidences of gypsum crusts: case study from Budapest (Hungary). *Environmental Geology*, **52**, 385–397.

SMITH, B. J., TÖRÖK, A., MCALISTER, J. J. & MEGARRY, Y. 2003. Observations on the factor influencing stability of building stones following contour scaling: a case study of the oolitic sandstone from Budapest (Hungary). *Building and Environment*, **38**, 1173–1183.

TIANO, P. 2001. The use of microdrilling techniques for the characterization of stone materials. *In*: BINDA, L. & DE VEKEY, R. C. (eds) *Site Control and Non Destructive Evaluation of Masonry Structures and Material; Proceedings of the Rilem tc177 mdt International Workshop, Mantova (I), 12–13 November 2001, RILEM PRO26*. RILEM, Paris, 203–214.

TIANO, P., FILARETO, C., PONTICELLI, S., FERRARI, M. & VALENTINI, E. 2000. Drilling force measurement system, a new standardisable methodology to determine the 'superficial hardness' of monument stones: prototype design and validation. *International Journal for Restoration of Buildings and Monuments*, **6**(2), 115–132.

TÖRÖK, A., SIEGESMUND, S., MÜLLER, C., HÜPERS, A., HOPPERT, M. & WEISS, T. 2007. Differences in textures, physical properties and microbiology of weatherinfg crust and host rock: a case study of the porous limestone of Budapest (Hungary). *In*: PŘIKRYL, R. & SMITH, B. J. (eds) *Building Stone Decay: From Diagnosis to Conservation*. Geological Society, London, Special Publications, **271**, 261–276.

UNI 9724/2–90:1990. *Materiali lapidei, Determinazione della massa volumica apparente e del coefficiente d'imbibizione*. Ente Italiano di Unificazione, Milan.

UNI 10859:2000. *Determination of Water Absorption by Capillarity*. Ente Italiano di Unificazione, Milan.

VANNUCCI, S., ALESSANDRINI, G., CASSAR, J., TAMPONE, G. & VANNUCCI, M. L. 1994. Prehistoric megalithic temples of the maltese arcipelago: causes and mode of the deterioration of Globigerina Limestone. *In*: FASSINA, V., OTT, H. & ZEZZA, F. (eds) *Proceedings of the 3rd International Symposium of the Conservation of Monuments in the Mediterranean Basin*. Venezia 22–25 June 1994, 555–565.

VINCENZINI, P. 1974. Le prove di laboratorio nella previsione del comportamento al gelo dei materiali ceramici per l' edilizia. *Ceramurgia*, **4**, 176–182.

WALKER, R. D., PENCE, H. J., HAZLETT, W. H. & ONG, W. J. 1969. One cycle slow freeze test for evaluation aggregate performances in frozen concrete. *National Cooperative Highway Research Program Report*, **65**, 21.

The effect of combustion-derived particulates on the short-term modification of temperature and moisture loss from Portland Limestone

D. E. SEARLE[1]* & D. J. MITCHELL[2]

[1]Room MI144, Construction & Infrastructure, School of Engineering and the Built Environment, University of Wolverhampton, Wolverhampton, West Midlands WV1 1SB, UK

[2]School of Applied Sciences, University of Wolverhampton, Wolverhampton, West Midlands WV1 1SB, UK

*Corresponding author: (e-mail: d.searle@wlv.ac.uk)

Abstract: It is known that cyclic heating–cooling and wetting–drying can play a significant role in the long-term deterioration of building stone. These cycles can be modified by the deposition of atmospheric particulates, which darken surfaces, resulting in changes in the absorptivity and emissivity characteristics of the stone. The capacity of diesel and coal particulates to modify the moisture and temperature regime of Portland Limestone and Hollington Sandstone was investigated. Through a greater capacity to lower the albedo of the stone and enhance the absorption of radiant energy, diesel particulate was shown to significantly increase the rate of moisture loss, temperature, and rates of heating and cooling of Portland Limestone. With particulates from diesel combustion now becoming one of the dominant particulate types found in urban centres, potential implications for future stone conservation are discussed.

Urban particulate pollution derived from the use of fossil fuels has a long history of affecting stone buildings and monuments in the UK. Where the nature of the particulate itself has evolved from being coal-derived industrial and domestic sources to being dominated by those resulting from vehicular sources, primarily diesel, they both have negative effects, perceived and actual, on the UK's built stone heritage.

The role of particulates in stone decay has been identified has having an effect on the aesthetics (Grossi & Brimblecombe 2004; Brimblecombe & Grossi 2005) in the form of the soiling of surfaces, a potential catalytic role in the formation of gypsum black damage crusts (Rodriguez-Navarro & Sebastian 1996; Simão et al. 2006) and modification to other physical factors that contribute to accelerated weathering. While the role of coal particulates in stone degradation has been the subject of many studies and investigations, the role of diesel particulates requires further investigation.

Movement of moisture and variation of temperature have an important role in weathering processes either directly, in the form of insolation weathering (Warke et al. 1996) and wetting–drying cycles, or indirectly by facilitating the movement of salts and enhancing chemical processes.

Although stone weathering is the result of many varying and interactive factors, the development of equipment to sample diesel particulate and apply equal amounts of different particulate types to stone surfaces presented an opportunity to directly study how the presence of particulate on stone surfaces may affect surface temperature and moisture loss.

This was undertaken by a comparative experimental design, which compared the mass of water loss and temperature variations, over time, between stone samples with different particulate treatments. The experiment was carried out under controlled environmental conditions designed to provide a combination of meteorological conditions conducive to drying; i.e. low relative humidity (RH) and high ambient temperature. In particular, an attempt was made to explore any changes in the rate of moisture loss that might occur from the change in surface albedo owing to the presence of particulate.

Methods and Materials

Samples

Two stone types were used, Portland Limestone and Hollington Sandstone (for comparison purposes), which were chosen for their ubiquitousness in culturally important buildings and research studies into urban stone decay. For each stone type 12 stone tablets of equal size (nominally 52 × 52 × 17 mm) were cut, placed into a Perspex surround and sealed

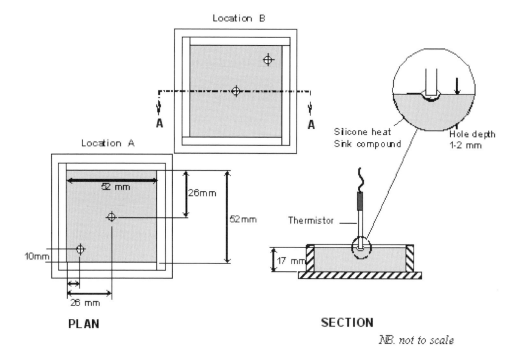

Fig. 1. Stone sample configuration showing thermistor coupling and location.

along the top edge with silicon sealant to form a sample unit. As a result of this any moisture movement would only occur through the top faces of the samples.

The same 24 samples were used in the surface temperature study but with the addition of two indentations in the surface to accommodate thermistors (Fig. 1).

Particulate impregnation of samples

All samples were coated with coal and diesel particulates both separately and in combination, as shown in Figure 2. Coal particulates were obtained from a coal-burning power station and diesel particulates from the 'live' exhaust lines of large passenger transport vehicles (Searle & Mitchell 2006).

Although both particulate types result from the combustion of fossil fuels, they differ significantly in chemical and composition and size (Table 1).

A known mass of particulates in the form of a suspension with deionized water was 'ponded' onto the top of the sample and then a vacuum of 500 mmHg applied. The particulate loading of 0.38 mg cm^{-2} was used to represent 'real-life' deposition levels of approximately 11 years in a typical urban UK environment. The purpose of the combined treatment, where there are successive layers of coal and diesel particulates, was to examine any synergistic effects that may occur where 'historic' coal particulate pollution is overlain by more contemporary traffic-derived diesel particulates. While it is acknowledged that the particulates may have undergone some modification when in suspension in water, chemical analysis of the particles themselves (Table 1), the liquid component of the suspension and examination of particle morphologies using scanning electron microscopy (SEM) showed them to be in broad agreement with the literature.

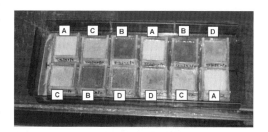

Fig. 2. Treated samples used for the rate of drying study, showing colour differences resulting from the treatment. A, untreated Portland Limestone; B, Portland Limestone + diesel particulate (coverage 0.38 mg cm^{-2}); C, Portland Limestone + coal particulate (coverage 0.38 mg cm^{-2}); and D, Portland Limestone + coal/diesel combination, 50:50 ratio (coverage 0.38 mg cm^{-2}).

Table 1. *Chemical composition and size range of diesel and coal particulates*

Major elements	Diesel particulate	Coal particulate
	Elemental composition (% by mass)	
Carbon (total)	80.0 (elemental: 29.5*) (organic/carbonate: 50.5*)	5–10*
Aluminium	0.2	11.8
Silicon	2.5	19.5
Sulphur	2.5	0.6
Potassium	<0.1	3.5
Calcium	1.2	2.3
Iron	<0.1	15.7
Other elements	13.5	37.0*
Particulate diameter range	2.5 μm – 30 nm	1–50 μm

*Estimated values.

Experimental set-up

As it was expected that the particulate-induced modification of the stone albedo would play a significant role, it was important to replicate the role of sunlight in providing the thermal regime for the experiment. To facilitate this, a radiant source of heat was used to deliver thermal energy to the stone surface. The use of such direct heating has been identified as necessary to elucidate differences between the thermal properties of various rock lithologies (Warke & Smith 1998). To maintain constant conditions for the experiments a Sanyo SG C097.CPX 'Fitotron' environmental chamber was used with an internal volume of approximately 1.4 m^3. Within the chamber, ambient temperature was kept at a constant 26 °C and 46% relative humidity, these levels were based on average daily maximum temperatures and minimum relative humidity's recorded in a UK urban centre between June and July in 1999. General Electric HE 060.10 250 W clear infrared bulbs were suspended 0.510 m above the surface of the samples (Fig. 3). The sample holder and samples were placed so that the geometric centre of the holder was aligned directly under the axis of the infrared bulb. In preliminary runs, it was shown that there was a temperature variation of less than 1 °C under the infrared lamp across the samples. In addition one-way ANOVA (analysis of variance) showed that there was no significant difference ($P < 0.05$, $n = 5$) between moisture loss from the 24 samples prior to particulate impregnation. Five complete 24 h cycles of wetting and drying were undertaken.

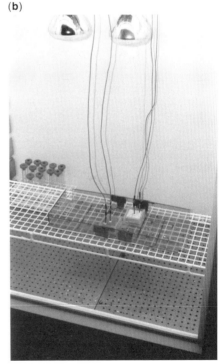

Fig. 3. Experimental set-up for: (**a**) the drying study; and (**b**) the surface temperature study.

All of the limestone samples, the untreated and diesel-treated sandstone samples, and the experimental set-up used in the rate of drying experiments were re-used for the albedo study. Two samples at a time were monitored in the cabinet during an experimental run (Fig. 4b). Four Unidata stainless steel thermistors, type 6507 (15k), were inserted into the indentations with a thermal heat sink compound to ensure a good thermal contact (communication with the manufacturer confirmed that the actual temperature measurement occurred from the tip to approximately 1–2 mm up the shaft of the thermistor body). These were connected to a Unidata 6040A 128k 'Starlogger' data logger, which read the thermistors at 5 min intervals. It was considered that the depth of the indentation combined with the heat sink compound would facilitate both a good thermal contact with the stone and quantitative measurement of the temperature in the surface layer.

Procedure

In the drying study samples were dried to constant weight at 40 °C and cooled to room temperature (approximately 18 °C) prior to commencement of the experiment. A known mass of deionized water was added to each sample at a mass ratio of 1:20 water:stone. When the water had been fully absorbed into the stone (indicated by a change in the surface reflectance of the sample from 'glossy' to 'dull') the samples were weighed and placed into the chamber. Samples were then removed at 60 min intervals and reweighed seven more times, with a final mass determination 24 h from the start of the experiment. The cumulative mass loss of water for each sample was calculated for each time interval.

Sample preparation for the surface temperature study was the same as the drying study. Temperatures from the four thermistors were logged at 5 min intervals for a minimum of 23 h. After this time the chamber door was opened, the infrared lamp was turned off and the chamber temperature reduced to 18 °C. The samples were removed after a minimum of 30 min after removal of the heat source; temperatures continued to be logged during this period.

Results and analysis

Statistical analysis

For the drying study a paired t-test was undertaken on the individual samples before and after treatment to help elucidate any differences that might exist. A one-way ANOVA was also carried out on the sample groups before treatment to ensure no significant differences existed between samples before treatment was applied.

The mean of the two readings from each sample was considered to represent the surface temperature and, as such, was used in the following analysis. To determine any significant differences at the 95% level between the treatments, a one-way ANOVA was undertaken on these mean temperatures for each stone type at hourly intervals. This was followed by Tukey and LSD (least significance difference) *post-hoc* tests to explore any significant difference that might be indicated by the ANOVA.

Drying study

Descriptive statistics (Table 2 and Fig. 4) show the mass of moisture loss at hourly intervals up to 7 h and after 24 h. These represent mean values from three samples, which individually constitute the mean of 5×24 h cycles of moisture loss. The lack of any significant differences ($P < 0.05$) between the untreated samples pretreatment indicate that the stones' response to moisture loss is the same across the four treatment groups for each stone type before any particulate treatment was applied.

The values obtained for moisture loss (Table 2) enabled a direct comparison to be undertaken of the moisture loss from the same samples before and after treatment. This was facilitated by a paired t-test on the three sample means in each treatment group (Tables 3 and 4).

Differences observed in the rate of moisture loss, after the different particulate treatments had been applied, were observed on the Portland Limestone samples. Only one value was significantly different after treatment on the Hollington Sandstone, it showed that less moisture was lost after 1 h of the drying cycle after treatment with coal particulate ($t = 10$, $P = 0.010$).

In contrast, there was evidence, some of it strong ($P < 0.010$), that the rate of moisture loss was increased on the diesel-treated limestone samples during the first 2 h of drying. The mass of moisture loss from the same treatment after 24 h was also observed to be significantly greater than the samples before treatment, with an increase from a mean value for the untreated stone of 4.51 g (SD = 0.06) to 4.64 g (SD = 0.05) after treatment. The first hour of moisture loss for the coal-treated samples was also significantly increased ($P = 0.024$) from that of the untreated limestone.

Surface temperature study

The mean of two measurements of surface temperature logged at 5 min intervals over an approximate 24 h period are shown in Figure 5. Comparative statistics (ANOVA) were calculated on the values obtained at 60 min intervals commencing from installation of the samples into the cabinet.

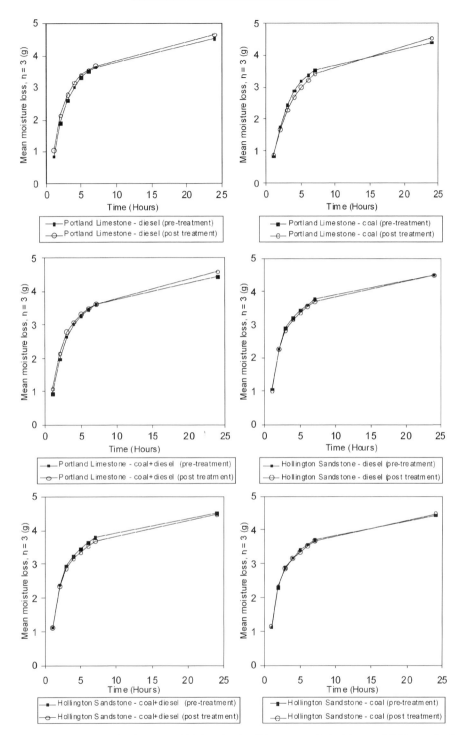

Fig. 4. Cumulative mass loss of moisture over 24 h from Portland Limestone (principal data) and Hollington Sandstone pre- and post-treatment.

Table 2. *Descriptive statistics for the cumulative mass loss of moisture over 24 h from Portland Limestone (principal data) and Hollington Sandstone pre- and post-treatment*

Time (h)	Diesel (pretreatment) CML (g)	SD	Diesel (post-treatment) CML (g)	SD	Coal (pretreatment) CML (g)	SD	Coal (post-treatment) CML (g)	SD	Coal–Diesel (pretreatment) CML (g)	SD	Coal–Diesel (post-treatment) CML (g)	SD
Portland Limestone (principle data)												
1	0.86	0.06	1.05	0.05	0.83	0.13	0.85	0.14	0.93	0.11	1.08	0.13
2	1.88	0.08	2.14	0.02	1.77	0.20	1.67	0.21	1.97	0.16	2.14	0.18
3	2.59	0.08	2.77	0.08	2.44	0.16	2.29	0.21	2.65	0.14	2.81	0.25
4	3.01	0.07	3.13	0.08	2.88	0.11	2.70	0.18	3.03	0.13	3.09	0.14
5	3.29	0.05	3.37	0.07	3.18	0.09	3.00	0.16	3.28	0.13	3.33	0.13
6	3.48	0.05	3.55	0.06	3.37	0.09	3.23	0.14	3.46	0.14	3.43	0.12
7	3.64	0.04	3.69	0.04	3.54	0.08	3.43	0.12	3.61	0.13	3.62	0.12
24	4.51	0.06	4.64	0.05	4.40	0.10	4.55	0.09	4.44	0.12	4.59	0.02
Hollington Sandstone												
1	1.05	0.06	1.01	0.03	1.13	0.25	1.17	0.24	1.13	0.15	1.13	0.12
2	2.27	0.12	2.27	0.17	2.30	0.40	2.33	0.31	2.38	0.23	2.35	0.18
3	2.88	0.09	2.83	0.14	2.88	0.15	2.87	0.11	2.95	0.11	2.87	0.12
4	3.20	0.06	3.14	0.12	3.17	0.09	3.16	0.07	3.25	0.08	3.16	0.10
5	3.44	0.09	3.36	0.12	3.40	0.08	3.36	0.06	3.47	0.07	3.36	0.10
6	3.60	0.05	3.55	0.11	3.57	0.08	3.54	0.06	3.64	0.07	3.55	0.10
7	3.76	0.05	3.71	0.11	3.70	0.08	3.68	0.07	3.79	0.06	3.70	0.10
24	4.50	0.03	4.49	0.02	4.45	0.04	4.48	0.05	4.50	0.03	4.46	0.07

CML, cumulative mean moisture loss ($n = 3$); SD, standard deviation.

Table 3. *Paired t-test of mean moisture loss from Portland Limestone samples pre- and post-treatment (principal data)*

Time (h)	Diesel treatment t-value	Sig	Dir	Coal treatment t-value	Sig	Dir	Coal–diesel treatment t-value	Sig	Dir
1	27.500	0.001	UT < D	2.646	0.118	PT < C	6.379	0.024	PT < C + D
2	4.914	0.039	UT < D	1.788	0.216	PT > C	2.396	0.139	PT < C + D
3	2.590	0.122	UT < D	2.110	0.169	PT > C	1.831	0.209	PT < C + D
4	1.998	0.184	UT < D	2.302	0.148	PT > C	3.062	0.360	PT < C + D
5	1.772	0.218	UT < D	2.048	0.177	PT > C	0.993	0.425	PT < C + D
6	1.377	0.302	UT < D	1.753	0.223	PT > C	0.846	0.486	PT < C + D
7	1.213	0.349	UT < D	1.973	0.187	PT > C	0.335	0.770	PT < C + D
24	7.181	0.019	UT < D	9.707	0.010	PT < C	2.135	0.165	PT < C + D

Sig, significance ($n = 3$); Dir, direction of relationship between treatments; PT, pretreatment; D, post-diesel-particulate treatment; C, post-coal-particulate treatment; C + D, post-coal–diesel-particulate treatment.

Post-hoc analysis of the ANOVAs for the Portland Limestone showed strong evidence ($P < 0.010$) that the diesel-coated samples have higher surface temperatures than both the control and the other treatments, under the same environmental conditions. Excess mean surface temperatures of approximately +2.2, +1.8 and +1 °C were recorded for the diesel treatment when compared to the control, coal and coal–diesel combination, respectively.

This experiment resulted in evidence, some of it strong ($P < 0.010$), that the coal–diesel treatment also had higher mean surface temperatures than the other two sample groups, although this was infrequently observed when compared to the coal treatment. The difference between the coal treatment and the control samples became consistently significant ($P < 0.05$) during the last half of the heating cycle.

The standard deviations recorded for the mean of the sample groups ($n = 3$) were low. This is reflected in the relatively small differences in temperature observed combined with the large numbers of significant differences found. For all the treatments any differences observed become significant after 2 h from the installation of the samples into the cabinet. This indicates that the rate of heat gain was similar for all sample groups. The

Table 4. *Paired t-test of mean loss of from Hollington Sandstone samples pre- and post-treatment*

Time (h)	Diesel treatment			Coal treatment			Coal–diesel treatment		
	t-value	Sig	Dir	t-value	Sig	Dir	t-value	Sig	Dir
1	0.813	0.502	PT > D	10.000	0.010	PT < C	0.180	0.874	PT = C + D
2	0.105	0.926	PT > D	0.596	0.612	PT < C	0.721	0.546	PT > C + D
3	1.639	0.243	PT > D	0.122	0.914	PT < C	2.667	0.117	PT > C + D
4	1.664	0.051	PT > D	0.378	0.742	PT < C	2.800	0.107	PT > C + D
5	3.965	0.243	PT > D	2.309	0.147	PT < C	3.024	0.094	PT > C + D
6	1.637	0.243	PT > D	1.427	0.287	PT < C	2.395	0.139	PT > C + D
7	1.533	0.265	PT > D	3.024	0.094	PT < C	2.682	0.115	PT > C + D
24	0.855	0.483	PT > D	2.774	0.109	PT < C	1.571	0.257	PT > C + D

Sig, significance ($n = 3$); Dir, direction of relationship between treatments; PT, pretreatment; D, post-diesel-particulate treatment; C, post-coal-particulate treatment; C + D, post-coal–diesel-particulate treatment.

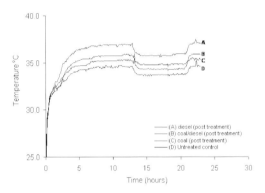

Fig. 5. Mean surface temperature of Portland Limestone samples under a radiant heat source ($n = 3$).

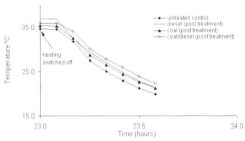

Fig. 6. Mean surface temperature of Portland Limestone samples after removal of the heat source ($n = 3$).

depression in surface temperature from all groups observed between 13 and 21 h from commencement of the experiment relates to night-time hours and the switching off of central heating circuits in the laboratory. As the experiments were undertaken during December–February it is likely that this was exacerbated by low ambient night-time temperatures. Untreated and diesel-treated Hollington Sandstone samples were tested for surface temperature but were found to have a mean temperature difference of 0.3 °C compared to 2–2.5 °C observed on similarly treated limestone samples.

After the 23 h heating period, the heat source was removed and surface temperatures continued to be monitored at 5 min intervals for approximately 40 min. The values obtained for each sample group and the rate of heat loss is shown in Figure 6.

With the exception of the diesel treatment, the pattern of heat loss is similar for the untreated control and the treatments. In general, the order of loss is the same as observed during steady-state conditions (i.e. coal–diesel > coal > untreated control). It appeared that the diesel-treated samples were losing heat at a greater rate than the other sample groups in the initial stages of cooling.

Discussion

When light-absorbing particles, such as diesel or coal particulate, accumulate on a surface a reduction occurs in the amount of solar electromagnetic radiation (short wavelengths 0.15–3.0 μm) reflected from the surface (Whittow 1984; Pesava *et al.* 1999). This can also be termed as a reduction in the reflection coefficient, solar reflectance or albedo of the surface, which will increase the thermal absorption of the surface (McGreevy 1985; Levinson & Akbari 2002). The role of albedo in the weathering of natural rock structures has been investigated in a range of studies (a summary of which can be found in Hall *et al.* 2005). These studies show that the surface temperature of rock under sunlight is mainly a function of rock thermal conductivity and specific heat capacity in addition to albedo (McGreevy *et al.* 2000). With the first two factors controlled in the experiment, it could be argued that any differences observed in temperature could solely be attributed to the reduction in albedo.

Diesel particulates, owing to their high elemental carbon content (Table 1), have been shown to increase soiling capacity when compared with particulates from coal combustion (Hamilton & Mansfield 1991). This can be clearly seen in Figure 3 where, for the same mass of particulate per unit area, the diesel-coated Portland Limestone samples are considerably darker than the other sample groups. Predictably, this resulted in a lower albedo and, hence, the significantly higher ($P < 0.05$) temperatures seen on these samples. This was also supported by the lack of any evidence for significant temperature differences between treatments for Hollington Sandstone samples. Compared to the light-grey colour of the Portland stone, the sandstone is a relatively dark reddish brown; therefore, there is not the same degree of reduction in albedo and, hence, resultant temperature increase. In addition, the relatively open texture of the sandstone may result in a fraction of the particulate treatment applied residing in the bottom of surface pores and therefore not fully contributing to the overall albedo reduction.

The temperature differences obtained between the untreated and the diesel-coated Portland Limestone samples were around 2–3 °C, which agreed broadly with two comparable studies undertaken by Warke et al. (1996) and Hall et al. (2005). Hall et al. (2005) examined how albedo variation exclusively affects temperature using paving bricks painted in 20% greyscale colour reflectance increments from white to black and exposed to the external environment daytime temperature ranges of approximately 6–22 °C. They found that for 20 and 80% reflectance values, arguably the closest approximation to the albedo of the limestone diesel-treated and untreated samples, temperature variations of 3–4 °C between these two reflectance values were recorded. However, this was at a lower overall temperature when compared to the current study. The results from the study by Warke et al. (1996), undertaken on oil- and coal-flyash-coated Portland Limestone also showed elevated temperatures for darker samples, albeit with a greater temperature difference (7–10 °C). This may be due in part to albedo differences, but could also be the result of evaporative cooling caused by the moisture loss from the samples in the current study.

The enhanced temperatures seen on the diesel-coated limestone samples correspond to the significantly increased ($P < 0.05$) moisture loss observed in the first few hours of heating (Table 3). In addition they heated up quicker, maintained higher surface temperatures and cooled down quicker than the other sample groups. This is consistent with the slightly higher emissivity levels and lower albedo associated with darker surfaces (Warke et al. 1996; Hall et al. 2005; Gomez-Heras et al. 2006).

Evaporation of moisture from saturated stone has been described as having a number of different phases (Amoroso & Fassina 1983; De Barquin & Dereppe 1996; Tournier et al. 2000). The first phase is characterized by a comparatively high loss of water from the surface of the stone, which is constant as long as the surface remains wetted (i.e. the rate of liquid water moving to the surface is equal to the evaporation loss). If this condition ceases and the evaporating surface is dry, then moisture loss, at a reduced rate, is by vapour diffusion through the pore network (phases II and III). It is likely that it is Phase 1 of the drying cycle on which the increased surface temperature resulting from the decrease in albedo is having an effect. This may explain the significant differences observed for the first 2 h of the drying experiments, as such an enhanced temperature would initially affect the upper layers of the stone.

It is proposed that such changes to the temperature regime and moisture flux may result in an increase in the intensity of heating and cooling cycles seen on soiled stone buildings and monuments. Through differential thermal expansion this would increase mechanical stresses between surface and subsurface layers, and to a lesser extent between adjacent areas of rain-washed 'clean' stone and sheltered 'soiled' areas (insolation weathering). In addition, it may enhance or modify other weathering mechanisms such as salt crystallization, chemical weathering and freeze–thaw cycles (Warke & Smith 1998; Ashurst & Dimes 1999).

It could be argued that compared to historical gradual deposition of coal particulates, stone structures close to heavily trafficked roads may undergo rapid resoiling after cleaning owing to diesel particulates. In addition, the adsorption forces exhibited by the diesel particulate due to their smaller particulate size (Table 1) may also result in a greater resistance to removal by rain wash than previous coal particulate deposition.

The consequence of this may be a change in the way soiling manifests itself over a building surface and how the visual appearance is altered. The large contrasts in shade traditionally observed between rain-washed 'white' and soiled areas on carbonate stone may reduce, and the prevalence of brown tones become more apparent (Bonazza et al. 2007). However, in the context of increasing pollution control and heightened government responses to the effects of climate change is soiling on urban stone surfaces a declining problem and, hence, research into its effects no longer worthy of attention?

Road traffic emissions are a major contributor to particulate pollution in the UK, contributing 27% in 2001 to total UK emissions of PM_{10} (particles measuring 10 μm or less) (AQEG 2005). However, the overall trend for PM_{10} emissions has been that of a decreasing one, with a 47% drop between 1970 and 2000, and this is projected to fall by a further 28% by 2010 (AQEG 2005). Diesel particulate, containing a high percentage of elemental carbon/black smoke (the primary darkening agent in relation to building soiling, Table 1), is a major proportion of this total. Although this would seem to present a case supporting cleaner buildings in the future, the following points should be taken into consideration.

- National PM_{10} emissions represent mean concentrations across a national network of monitoring stations, where deposition of particulates onto buildings is arguably dominated by point sources in their immediate vicinity (roadside measurements of PM_{10} are often double that of background levels: AQEG 2005).
- The strong downward trend observed in urban PM_{10} in the 1990s appears to be levelling off over the period 2000–2003. This could indicate that the predicted future reductions in urban PM_{10} are not robust.
- Uncertainty in future sales projections of diesel vehicles (currently thought to be 12% higher than expected by 2010, which would increase urban PM_{10} by 15%). Predicted reductions reliant on effectiveness of new legislation (Euro IV and potential Euro V vehicle standards, applied to new vehicles sold after 2008 and 2010, respectively).
- A significant proportion of traffic-derived urban PM_{10} emissions (23% in 2001) do come from combustion alone. This includes brake and tyre wear emissions, which would not be affected by emission controls proposed currently and in the future.

In addition to these points, there are also uncertainties in climatic conditions that will prevail in the UK in the future and that may affect the weathering processes under discussion here. Climate change through radiative forcing will affect temperature ranges, wetting–drying and freeze–thaw cycles, etc., although the degree of such changes is unclear and will vary over the UK. In particular, the role of albedo may be even more significant owing to an observed increase in the strength of sunlight over Europe, which reversed a dimming trend up to 1985 (Wild et al. 2005).

Conclusion

It is clear that despite long-term declines in particulate pollution generally over the UK, stone buildings are still being soiled in our urban centres with traffic sources identified as the primary polluter. While acknowledging that soiling on buildings and associated damage mechanisms are the result of the inter-relationship of a number of complex factors, the experiments presented did provide an opportunity to isolate and examine the role of one of those factors.

Diesel-coated Portland Limestone was shown to exhibit higher surface temperatures than untreated stone when heated by solar radiation. This was caused by a reduction in albedo (hence, an increase in absorbed short-wave radiation) owing to the presence of diesel particulates on the stone surface. Intermediate temperature increases were also observed for coal particulate and for the coal–diesel combination. Associated with this were significant increases in early moisture loss on the diesel-treated samples.

It could be argued that trafficked areas of urban centres will still have elevated levels of particulate pollution in the near future and, that in association with climate and sunlight strength changes, building stones will face different and evolving environmental and pollution conditions from those previously experienced. There is clearly a need, therefore, for continuing research into the soiling of stone and other particulate-derived effects in our urban centres.

The authors would like to thank the School of Engineering, and the Built Environment and the School of Applied Sciences, at the University of Wolverhampton for their help and funding of this work.

References

AMOROSO, G. G. & FASSINA, V. 1983. *Stone Decay and Conservation, Atmospheric Pollution, Cleaning, Consolidation and Protection.* Elsevier, Oxford.

AQEG. 2005. *Particulate Matter in the United Kingdom.* Air Quality Expert Group. Defra, London.

ASHURST, J. & DIMES, G. 1999. *Conservation of Building and Decorative Stone.* Butterworth Heinemann, Oxford.

BONAZZA, A., BRIMBLECOMBE, P., GROSSI, C. M. & SABBIONI, C. 2007. Carbon in black crusts from the Tower of London. *Environmental Science and Technology*, **41**(12), 4199–4204.

BRIMBLECOMBE, P. & GROSSI, C. M. 2005. Aesthetic thresholds and blackening of stone buildings. *Science of the Total Environment*, **349**, 175–189.

DE BARQUIN, F. & DEREPPE, J. M. 1996. Drying of a white porous limestone monitored by NMR imaging. *Magnetic Resonance Imaging*, **14**(7/8), 941–943.

GROSSI, C. M. & BRIMBLECOMBE, P. 2004. Aesthetics of simulated soiling patterns on architecture. *Environmental Science and Technology*, **38**(14), 3971–3976.

HALL, K., LINGREN, B. S. & JACKSON, P. 2005. Rock albedo and monitoring of thermal conditions in respect of weathering: some expected and some unexpected results. *Earth Surface Processes and Landforms*, **30**(7), 801–811.

HAMILTON, R. S. & MANSFIELD, T. A. 1991. Airborne particulate elemental carbon: its sources, transport and contribution to dark smoke and soiling. *Atmospheric Environment*, **25A**(3/4), 715–723.

GOMEZ-HERAS, M., SMITH, B. J. & FORT, R. 2006. Surface temperature differences between minerals in crystalline rocks: implications for granular disaggregation of granites through thermal fatigue. *Geomorphology*, **78**, 236–249.

LEVINSON, R. & AKBARI, H. 2002. Effects of composition and composure on the solar reflectance of Portland cement concrete. *Cement and Concrete Research*, **32**, 1679–1698.

MCGREEVY, J. P. 1985. Thermal rock properties as controls on rock surface temperature maxima and possible implications for rock weathering. *Earth Surface Processes and Landforms*, **10**, 125–136

MCGREEVY, J. M., WARKE, P. A. & SMITH, B. J. 2000. Controls on stone temperatures and the benefits of interdisciplinary exchange. *Journal of the American Institute for Conservation*, **39**(2), 259–274.

PESAVA, P., AKSU, R., TOPRAK, S., HORVATH, H. & SEIDL, S. 1999. Dry deposition of particles to building surfaces and soiling. *Science of the Total Environment*, **235**, 25–35.

RODRIGUEZ-NAVARRO, C. & SEBASTIAN, E. 1996. Role of particulate matter from vehicle exhaust on porous building stones (limestone) sulfation. *Science of the Total Environment*, **187**(2), 79–91.

SEARLE, D. E. & MITCHELL, D. J. 2006. The effect of coal and diesel particulates on the weathering loss of Portland Limestone in an urban environment. *Science of the Total Environment*, **370**, 207–233.

SIMÃO, J., RUIZ-AGUDO, E. & RODRIGUEZ-NAVARRO, C. 2006. Effects of particulate matter from gasoline and diesel vehicle exhaust emissions on silicate stones sulphation. *Atmospheric Environment*, **40**, 6905–6917.

TOURNIER, B., JEANNETTE, D. & DESTRIGNEVILLE, C. 2000. Stone drying: an effective evaporating surface area. *In*: FASSINA, V. (ed.) *Proceedings of the 9th International Congress on Deterioration and Conservation of Stone*. Venice, June 19–24 2000. Elsevier, Amsterdam, 329–635.

WARKE, P. A. & SMITH, B. J. 1998. Effects of direct and indirect heating on the validity of rock weathering simulations studies and durability tests. *Geomorphology*, **22**, 247–357.

WARKE, P. A., SMITH, B. J. & MAGEE, R. W. 1996. Thermal response characteristics of stone: implications for weathering of soiled surfaces in urban environments. *Earth Surface Processes and Landforms*, **21**, 295–306.

WHITTOW, J. 1984. *Dictionary of Physical Geography*. Penguin, London.

WILD, M., GILGEN, H. ET AL. 2005. From dimming to brightening: decadal changes in solar radiation at Earth's surface. *Science*, **38**, 847–850.

The changing Maltese soil environment: evidence from the ancient cart tracks at San Pawl Tat-Tarġa, Naxxar

DEREK MOTTERSHEAD*, PAUL FARRES & ALASTAIR PEARSON

Department of Geography, University of Portsmouth, Buckingham Building, Lion Terrace, Portsmouth, Hampshire PO1 3HE, UK

*Corresponding author (e-mail: derek.mottershead@port.ac.uk)

Abstract: The historic cart ruts of Malta are incised into the underlying bedrock topography. Anomalous relationships between their routeways and the uneven terrain beneath suggest that they originated on a land surface different from that of today. An exposure close to a cart-rut location near Naxxar reveals evidence of limestone terrain development and its role in the evolution of the cart-rut patterns. Specifically, it reveals that cart trackways were most probably superimposed from a soil cover onto an underlying bedrock surface topographically different from the former soil surface. A model is developed demonstrating likely relationships between human activity, soil erosion and trackway evolution leading to the incision of the trackways into the bedrock.

The ancient rutted trackways incised into bare limestone surfaces form distinctive features of the Maltese landscape. They are commonly considered to be cart tracks, an interpretation recently reaffirmed by Mottershead *et al.* (2008). In the course of previous attempts to explain them authors have frequently invoked the theme of environmental change; several writers have, for instance, inferred that the removal of a soil cover by erosion was an integral factor in the formation of these features, implying that substantial environmental change has occurred in this way (Fenton 1918; Gracie 1954; Trump 1993; Hughes 1999). To date, however, firm site-specific local evidence to substantiate this hypothesis has been lacking.

The present paper, based largely on the San Pawl tat-Tarġa site near Naxxar supplemented by observations from San Ġwann, makes close and critical observation of the relationship between the tracks, the topography of the site and the immediately local rock surface. It investigates aspects of track form that are difficult to explain in relation to the contemporary features of the terrain which they traverse. Field evidence from an adjacent section reveals aspects of limestone terrain evolution and its relationship with a surface cover of unconsolidated materials and soils. These provide strong indications of environmental change in the form of soil cover loss, which are presented and evaluated along with their implications for local routeways.

Route selection on natural terrain

The existence of the tracks carved into rock exposed at the current ground surface does not necessarily mean that the original trackways were initiated on this rock surface. Simple field observation provides evidence that the original land surface on which they were formed was likely to have been substantially different in character from that of today. Furthermore, the land surface was apparently undergoing change as the trackways developed. Indeed, environmental change caused either by climatic variation at the broader scale, or human activities at the more local, is a common theme throughout the span of human history, especially in Malta (Hunt 1997).

The locations of the cart ruts in Malta (Fig. 1) are, at the broader scale, commonly related to regional routeways. Those at Naxxar, for example, are evidently a means of scaling the relatively steep slope of the Victoria Lines escarpment by means of large hairpin bends, and a climbing alignment commonly slightly oblique to the contours. In the same escarpment those at Binġemma funnel through a gap, evidently as a means of linking the lowlands to the north with the plateau beyond to the south. Within the broader constraint of the regional routeway, the traveller has a significant degree of choice as to the exact line she or he takes up the slope. This choice, on natural terrain and in the absence of a constructed road or track, is likely to be governed by immediately local factors such as inanimate obstacles in the form of crevices or rock steps, puddles, or living forms such as trees or thorny plants. In exactly the same way, a contemporary walker on rough or steep terrain would make immediately local decisions as to where to make a satisfactory footfall in taking the line of least difficulty.

From: SMITH, B. J., GOMEZ-HERAS, M., VILES, H. A. & CASSAR, J. (eds) *Limestone in the Built Environment: Present-Day Challenges for the Preservation of the Past*. Geological Society, London, Special Publications, **331**, 219–229. DOI: 10.1144/SP331.20 0305-8719/10/$15.00 © The Geological Society of London 2010.

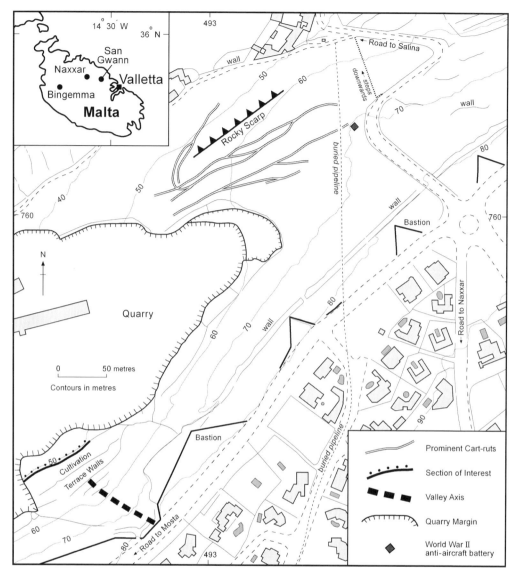

Fig. 1. Location map. Based on 1:2500 map of Malta, Sheet 4875, British Government Overseas Development Administration (Directorate of Overseas Surveys) 1971, series M8190 (DOS 052), with additional information from Google Earth, Virtual Earth and Plate VI in Zammit (1928).

There are features of the ruts at Naxxar that strongly suggest that the lines which they take across the ground are not the natural lines of least resistance in relation to the current terrain. At the location shown in Figure 2, for example, the ruts climb over a sharp step of some 50 cm up an escarpment formed by a resistant bed of rock. Climbing such a step, especially with a loaded vehicle, would have required much effort and unnecessarily so since a route of much lower gradient is present nearby, which offers a convenient alternative for avoiding the obstacle. If the present surface had been exposed at the time of track initiation, surely the most sensible action would have been to save energy, and potential damage to the vehicle, and take the readily available nearby route, which was both more comfortable and less energy-sapping.

A second feature at Naxxar, which throws light on the development of ruts through time, is the double bypass of a vertical shaft penetrating at

Fig. 2. Rut–bedrock discordance: ruts climb a 0.5 m step.

least 3 m down from the rock surface, which intersects a line of ruts (Fig. 3). In this case, a reasonable simplifying interpretation would be that the original straight line of the ruts drives straight across the shaft, which originated as a solution pipe in the underlying limestone. Evidently at that time the shaft would not have presented an obstacle to traffic, implying that it was not visible as such, or open to the surface. The development of a double bypass, however, suggests that the initial route became increasingly impassable as the trackway shifted incrementally away from the hazardous obstacle.

It is likely that other examples of bypass loops, which have previously been explained as engineered passing places, may now be reinterpreted in the light of the example above. Aerial photography of the loop illustrated in Figure 4, whose location is named by Zammit (1928) as Minsia, St Julian Heights and by Trump (2002, p. 282) as San Ġwann, provides a perspective that can be augmented by ground observations. The tracks here are

Fig. 3. Rut–bedrock discordance: tracks diverge left to bypass the open shaft on the right.

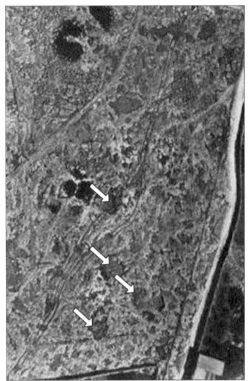

Fig. 4. Aerial view of a cart-rut bifurcation at San Ġwann. The darker-toned quasi-circular patches indicate depressions in the limestone surface (four of them indicated by arrows). The more direct left-hand branch passes through one of them; the longer looped SE branch to the right is interpreted as a subsequent bypass (north is at the top of the figure). Photograph reproduced from Zammit (1928) by kind permission of Antiquity Publications Ltd.

located on a limestone plateau surface evidently pocked with broad solution hollows. The shorter section of the loop is more clearly aligned with the onward extension of the track in both directions; it is shown in this image to transect a substantial depression in the rock surface more than 8 m long and up to 7 m across. Close examination of the southerly rut junction shows multiple rut tracks, with at least three clear rut pairs, and traces of more, that increasingly diverge from the more direct alignment (Fig. 5). Furthermore, the left-hand rut (3L in Fig. 5) of the most divergent trackway is incised into the right-hand rut of the most direct route (1R in Fig. 5). This cross-cutting relationship shows that the latter is the younger of the two ruts. This in turn suggests that the younger routes gradually diverged away from the original direct route, showing that the use of the latter post-dates the erosion of the original.

Fig. 5. Close-up of a junction with multiple cart ruts at San Ġwann, looking north. The bypass loop is to the right; the original track trends toward the large tree, which has colonized a hollow interpreted as solution pan. Each trackway is identified by left (L) and right (R) ruts. The left rut of track 3 is incised into the right rut of track 1. This cross-cutting relationship indicates the temporal relationship between the two trackways, of which track 3 is the younger. The trackway complex is interpreted as a temporal sequence notated 1–3, showing an incremental shift to the right through time as the trackway takes an increasingly wide line to avoid the obstacle posed by the solution hollow.

The features described above, then, imply discordances between the requirements for smooth passage of vehicular traffic and the terrain into which the rutted trackways are incised. They strongly suggest that the surface terrain over which the track-forming vehicles initially travelled differed significantly in nature from the current rocky terrain. Although direct evidence of a former soil cover on the site is lacking, the presence of such a cover would provide an explanation of how the trackways came to be established apparently independently of the bedrock terrain beneath. It then becomes pertinent to consider the natural soilscape (Hole 1978; Hole & Campbell 1985) of Malta prior to any modification by human activity, and the processes by which environmental change could have been effected.

The prehistoric soilscape of Malta

The slope bearing the cart ruts at San Pawl tat-Tarġa is a largely structural slope from which any previous soil cover has been removed, exposing the hard limestone surface of the Lower Coralline Limestone. A relatively resistant limestone exposed for an extended time period in a now Mediterranean climate is an environmental context that would almost inevitably have been covered by *terra rossa* materials in the past (Durn 2003). These materials generally vary in texture from clay to silty clay and are always reddish in colour, typically with hue redder than 5YR and chroma >3.5 in the Munsell soil notation (Bech *et al.* 1997), reflecting the preferential formation of the iron mineral hematite over goethite resulting from bisiallic type weathering (Schwertmann 1988; Fedoroff 1997; Yaalon 1997). It is normally accepted that the *terra rossa* material, from which the soil cover would have been derived, is the insoluble residue from the deep *in situ* weathering of the underlying carbonate rocks (Moresi & Mongelli 1988 and many others). However, a totally *in situ* origin has always been brought into question and thus many authors invoke airfall additions of external mineral components, such as wind-blown dust from the North African drylands, and volcanic fallout from Mediterranean sources such as Etna and Stromboli (Yaalon 1997; Durn 2003). Evidence for such additions can be found throughout the broader literature (Nihlen & Olson 1995 for example), and from modern satellite imagery (http://visibleearth.nasa.gov/858/S2000238.jpg, http://www.nasaimages.org/luna/servlet/detail/nasaNAS~10~10~86944~193535:Eruption-of-Sicily-s-Mt–Etna). An important observation made by Yaalon (1997) is that much of this added material will become integrated very quickly into the soil mass by bioturbation and other water-induced mixing mechanisms.

Malta, in general, fits into such an environmental context. Throughout Coralline Limestone exposures in NW Malta such small pockets of *terra rossa* materials can be found filling structural fissures in the rock, and also occasionally in small basins created by slightly deeper local weathering. It should be noted, however, that neither of these is common, and the majority of such slopes are now completely denuded of this type of material and its associated soil cover. Such rare *terra rossa* remnants as are found in Malta are best interpreted as the C or B_W horizons of a now completely eroded or truncated soil. Such a situation is far from unique in the Mediterranean, as Boero & Schwertmann (1989, p. 320) write about *terra rossa* soils:

> In general, it is rather difficult to find a complete terra rossa soil profile because erosion and deposition processes are commonly superimposed on karstic phenomena prevailing over long time spans.

One might expect, therefore, that the original soil cover would have been deep enough to allow the development of some form of *leptic rhodic cambisol* (FAO 2006). The idea that some clay migration may have also taken place is far from accepted, so it is unlikely that some sort of *luvisol* would have had

Fig. 6. Section showing the pits and fissures developed in the otherwise relatively planar surface of the underlying Coralline Limestone at San Pawl tat-Tarġa. The section is approximately 2 m high.

time to evolve. However, the development in most situations would be greater than the depth constraints placed on its characterization as some version of a *lepotosol*. The next thing to note is that *terra rossa* soils are often regarded as *palaeosols* of the relict form that owe more to weathering conditions of the Tertiary and Quaternary than they do to the Mediterranean environment of today (Bronger & Bruhn-Lobin 1997).

With specific reference to the San Pawl tat-Tarġa cart rut site, although the slope bearing the ruts is now completely bare of soil, a small exposure of the original *in situ terra rossa* material can be found nearby in the limestone quarry headcut at the base of the Victoria Lines escarpment (Fig. 6). This section remained visible throughout field visits spanning the period September 2004–March 2007.

The exposed section cuts across a small valley that is incised into a major escarpment and whose floor is covered with cultivation terraces. The concave nature of the valley encourages sediment transportation from opposing valley slopes and thus acts as a locus for sediment accumulation through surface wash and creep processes, and for subsequent soil formation. Here sediment accumulates more deeply than on slopes that are convex or planar in planform. It is also more likely to retain the sediment/soil cover should subsequent environmental conditions happen to become more conducive to erosion. The presence of agricultural terracing is indicative of intensive human usage of the environment, and serves as an additional sediment trap for material moved downslope by colluvial processes. It also serves to protect and preserve any older soil material beneath.

Limestone is a soluble rock and permeable by surface rainwater, which commonly seeps or flows through joints and fissures (Trudgill 1985; White 1988; Ford & Williams 1989). Solution will take place wherever acidic water comes into contact with the rock. Commonly a soil cover will bring acidic and relatively static soil moisture into contact with the limestone rockhead. Spatial variation in the rate of solution will cause local concentrations, and the consequent development of solution hollows and pipes. Dissolution of the soluble calcium carbonate will release any insoluble mineral components of the rock, which then simply accumulate as a residual blanket over the limestone, occupying any irregularities in the rockhead relief. Where a joint permits a flow of water to enter the limestone mass, the passage of water through the rock may remove any infilling residue, actively dissolving the limestone and enlarging an initial hollow or pipe into a substantial open shaft.

The natural state of the surface environment, prior to the effects of human activity on the landscape, would thus be a cover of soil and other surficial materials overlying the rockhead relief (the form of the rock surface), filling and burying the irregularities in the limestone surface, rendering them undetectable by eye at the ground surface. The rockhead relief, associated soils and other unconsolidated materials themselves provide evidence of the long-term evolution of the landscape, and the processes and conditions responsible. They also offer a possible analogue for the evolution of the adjacent surface environment.

The section shown in Figure 6 shows the boundary between the rock surface and the *terra rossa*, and gives a clear impression of the likely surficial cover–rock surface relationship on the nearby slope, the current site of cart ruts, before the loss of any soil cover. The section clearly exposes the rockhead relief, which presents a roughly planar rock surface topography, indented by shallow basins and hollows that retain remnants of a red stony soil. It is also pitted by solution pipes filled with unconsolidated earth surface materials, narrowing downwards and of more than 2 m in visible depth. The sequence of surficial and infill materials is well displayed in Figure 7, in which a shallow surface depression in the bedrock surface leads down to a fissure penetrating to at least 1.5 m; that is set out in Table 1.

Some basic observations were made on the unconsolidated material found in the solution pipe. Field observation shows that there is no evidence of sorting of the material or any evident bedding structures. This implies that it accumulated by slow accretion, rather than the action of flowing water.

In order to identify evidence of environmental change, analytical tests were carried out on the surficial materials present. A sample of the underlying limestone was slowly digested in acid in order to determine the concentration and nature of the insoluble weathering residue within the soluble limestone. Particle size distributions of the fine

Fig. 7. Close view of the soil overlying the rockhead and the solution pipe infill. The two soil layers and solution residue are identified, together with sample point locations. The survey tape is 1.4 m long.

Table 1. *The sequence of surficial materials at the sampled site in the quarry headcut at San Pawl tat-Tarġa*

Location	Thickness (m)	Material
Depression	0.30	Red silty clay soil with some stones
Depression	0.60	Red silty clay soil, very stony
Fissure	>0.90	Brown silty clay

fraction of five samples down the infill profile, plus the insoluble residue, were carried out with a Malvern laser particle size analyser, enabling 50 size classes below 64 μm to be identified (summarized into seven size classes in Table 2). Insoluble residue values of two samples of the local bedrock (the Xlendi Member of the Lower Coralline Limestone of Oligocene age) were derived by digestion of 20 g of powdered rock in 20% hydrochloric

Table 2. *Particle size fractions of the insoluble residue and pipe infill of the limestone*

Sample	Fine clay <0.001 mm	Clay <0.002 mm	Fine silt <0.016 mm	Silt <0.064 mm	Very fine sand <0.125 mm	Fine sand <1 mm	Sand >1 mm
1	21.00	21.17	35.52	14.59	6.84	0.31	0.61
2	12.08	17.93	46.38	18.26	3.64	0.28	1.42
3	11.92	15.29	40.00	25.04	6.60	1.00	0.17
4	12.98	17.79	43.81	17.10	4.68	2.17	1.46
5	11.32	19.24	47.38	17.25	2.63	1.54	0.65
Insoluble residue	11.57	19.67	48.44	17.64	2.69	0.00	0.00

acid, and yielded values of 1.28 and 1.42%, respectively, giving a mean value of 1.35%. Clay mineral analysis (Fig. 8) was undertaken with X-ray diffraction (XRD) to determine the mineralogical components of the clay fraction.

Particle size analysis reveals the proportions of the different grades of sediment in the samples. There is, overall, substantial commonality in the samples, with total sand fractions of less than 10%, a total silt fraction normally in the range 60–65% and a total clay fraction (with one exception) around 30%. Despite this commonality, however, there are subtle variations and three distinct subgroups can be identified. First, the insoluble residue has a very low sand content, and is unique in containing no material coarser than very fine sand. Secondly, samples 2–5 from within the pipe show great similarity in possessing total clay, silt and sand fractions, each within a very narrow range. Thirdly, sample 1, representing the current stony subsoil, possibly a cultivation layer, shows reduced silt and enhanced clay fractions in comparison to the others.

The similarity between the pipe infill (samples 2–5) and the insoluble residue strongly suggests that the infill comprises material mainly derived from long-term solution of the parent limestone. The presence of the very small amounts (1–2%) of fine and medium sand in the infill, and their total absence from the insoluble residue, may indicate a small amount of mixing during the process of pipe infilling. It may, however, simply be a function of the fact that the insoluble residue values are based on a single and not fully representative sample. Sample 1 shows depleted silt and enhanced clay fractions in comparison to the infill. This is interpreted as representing weathering of the silt fraction to produce clay materials. It is also possible that, as soil material on a cultivation terrace, it has been augmented and its composition therefore modified by soil imported from other areas, consistent with historic Maltese practices of soil conservation and enhancement.

XRD analysis (Fig. 8) reveals the mineralogical composition of the clay fraction of the insoluble residue of the limestone, the pipe infill and the surface soil. Two major conclusions can be drawn from the results. First, all of the XRD traces show a strong commonality, with illite and kaolinite as the dominant clay minerals. The strong similarity in clay mineralogy between the insoluble residue and the pipe infill implies that the latter is derived from the former.

The analytical data simply reflect much of the detail that is already known of *terra rossa* materials in the Mediterranean. However, whilst the infill clearly owes much of its characteristics to the weathering residue of the underlying limestone, its similarity to the weathering residue decreases towards the top of the profile. In addition, this particular site appears to exhibit a duplex soil in which the

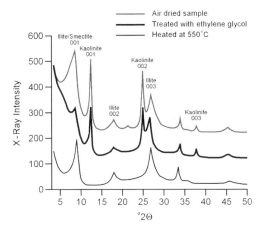

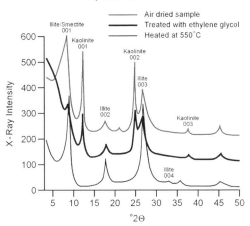

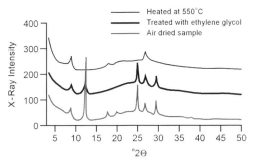

Fig. 8. XRD traces from San Pawl tat-Tarġa for samples 2, 4 and the insoluble residue. Note the commonality of the peaks in all three, implying that the two upper samples are derived from the latter material.

topmost material represents a later deposition phase in which the close proximity of the ancient and now degraded terrace wall created a soil trap for a combination of tillage soil, material delivered downslope by colluvial processes, and contributions of airfall deposits of aeolian and volcanic origin.

Environmental change: the evolving soilscape

There is currently little available direct evidence of the erosional history of Malta, and it is necessary to consider evidence from analogous regions elsewhere. For example, the classic 319 m core from Lake Ioannina in Greece has been intensively researched (Lawson et al. 2004; Wilson et al. 2008) and reveals considerable erosional activity in this upland basin from 5500 ^{14}C BP. A recent review by Butzer (2005) states that the most intense soil erosional phase was in the Bronze Age, and that this was then followed by a further series of phases of instability. The idea of erosion being initiated by clearance and followed by subsequent phases resulting from land use activities is also reported by Boardman & Bell (1982) and Brandt & Thornes (2002). Studies of erosional processes in the Mediterranean today clearly show that parameters characterizing land use and their interaction with vegetation performance dictate the magnitude and frequency of erosional activity (Kosmas et al. 2000). A question as yet not clearly evaluated is how much of these land use changes are themselves climatically induced and how much result from population pressure and/or technological developments in the use of the land.

The huge number of terrace walls on the Coralline Limestone slopes in Malta, many of them now abandoned, is a testament to the intensity of usage of these soils, and their potential for human-induced erosion throughout prehistory and historical time. Earlier phases of accelerated soil erosion are normally to be seen in the stratigraphic record preserved in deposition sites of Quaternary age. Malta, unlike much of the Mediterranean, has an extremely disjointed set of Quaternary deposits, insufficient to allow a detailed reconstruction of the erosional history of the slopes. The deposits reported by Pedley & Clarke (2002) around Mellieha Bay may, however, provide a small window into the past, although they have as yet received little detailed modern interpretation. They display a basal unit of *in situ terra rossa* material lying above a series of depositional phases punctuated by at least two stable periods, where soils again developed in the local surface material, and are also now buried by the most recent slope deposits.

The initiation of soil erosion in Malta is likely to be associated with the development of settled agriculture. Excavations at Skorba (Trump 2002) indicate human settlement in Malta associated with mixed farming, and dating from as early as 5000 BC. The development of temples and burial sites suggests a significantly expanding population whose existence would require increased agricultural production, implying woodland clearance and exposure of surface soils to erosion. On the evidence of a pollen record from Tal-Mejtin, Luqa, Trump (2002) states that the former tree cover had almost vanished by the Bronze Age (2500 BC). In contrast to the record of agricultural activity, there is no archaeological or environmental evidence of early wheeled vehicles on Malta, and any attempt to identify an appropriate chronological context must rely on a broader consideration of the evolution of wheeled vehicles. Recent evidence suggests that wheeled vehicles existed in Europe in the late sixth millennium BC, with the discovery of a wooden disc wheel 0.7 m in diameter and 5 cm thick, together with a 1.2 m-long supporting axle in Slovenia (Gasser 2003). This indicates the use of substantial load-bearing vehicles in the Mediterranean region as long as 7000 years ago. There is no specific evidence as to when they first reached Malta. There appears to be no technical or historical reason, however, why people of any of the early Maltese cultures could not have been the first local wheel users, and further contributed to the erosion of the original soils of Malta.

It is likely that the slope gradient and length plus the textural and structural properties of the potential A horizons in these soils allowed the material to be very quickly lost downslope. Once the surface vegetation cover at the study sites was broken by the passage of feet, hooves and wheels, the silty soils would be directly exposed to erosion agents. Their texture would render them easily erodible when wet, as they would readily slurry and flow away as sediment-laden overland flow during rainstorms, and be washed down joints, fissures and shafts in the limestone surface. When dry, the small soil particles would be readily transported away by winds. Erosion of such materials takes place far more rapidly than their replacement by the soil-forming processes of limestone solution. In fact, on the rutted slope at San Pawl tat-Tarġa it is easy to imagine very intense accelerated erosion by human and vehicular traffic with most of the soil material being lost over tens and hundreds rather than thousands of years, during which the thin soils of the hillslope were replaced by the clean washed bare limestone topography now present.

The building of terrace walls may well have diminished the erosion for a period of time but eventually these were abandoned and, as this took place and the walls were allowed to decay naturally, this in turn activated later, though less intense, truncation of the soil cover.

The origin and development of the cart ruts

The section described above, with the overlying materials still in place, represents an analogue for nearby rock surfaces from which any former unconsolidated cover is likely to have been removed, such as the more exposed planar slopes on which the rutted trackways are found at San Pawl tat-Tarġa. The topographic relations demonstrated between rockhead relief and the unconsolidated cover permit interpretation of surface terrain evolution through erosion of the overlying unconsolidated materials.

In an era of increasing human activity such soil would be readily erodible, and human activity itself would act as a significant eroding force, especially where it involved the passage of vehicular traffic. When the rockhead relief is covered by soils, it would be invisible to anyone traversing the surface above. Initially, the local choice of route over the landscape would have no need to take account of the underlying rock surface as the latter would be obscured by the overlying soil. Local routing decisions would be made on the basis, perhaps, of local topographic or vegetational constraints. If the soil were stripped away in thin layers, it is apparent that the nature of the rock surface would be only incrementally revealed as it emerged though the gradually thinning soils. Only as the soil eroded away would the bedrock obstacles become apparent, as individual high points began to appear through the soil cover. At that stage the true magnitude of bedrock obstacles, whether high or

Fig. 9. Modelled sequence of soil stripping to show the progressive emergence at the surface of the rockhead relief through times $t_0 - t_2$. The section is approximately 2 m high.

low, would not be identifiable, and there would be no need for substantial route change.

On first exposure of the bedrock surface following a general thinning of the overlying soil, a shaft would initially appear as an apparently soil-filled hollow in the bedrock surface, muddy in wet conditions and meriting avoidance with a bypass. Consequent upon further erosion through continuing use of the rutted trackway, the vertical shaft would become increasingly exposed, appearing recognizably deeper as the soil infill became further eroded, until the hollow became exposed as too deep to be traversable by vehicles. Its true nature would become increasingly apparent as the soil eroded further, revealing it to be a substantial rock-rimmed pit.

The multiple bypasses suggest that the first bypass itself became insufficient to avoid the exposed shaft, as its true width and depth as an obstacle became progressively more apparent. These facts favour the interpretation of the longer curving looped section as a subsequent bypass, as

Table 3. *Tabulation model of environmental change and the consequent development of ruts in bedrock*

Time	Nature of surface	Route selection	Surface development
t_0	Smooth ground surface over soil; occasional rises and dimples. Vegetation cover of varying density	Ease of trafficability strongly determined by vegetation cover. Route selection favours less densely vegetated areas on a uniform ground surface	Ruts develop in the silt/clay soil, with graded floors, and will attract subsequent traffic
t_1	Knolls and plateaux of bedrock begin to appear, separated by soil-filled hollows carrying rutted trackways	Traffic will tend to follow existing ruts on to the intermittent bedrock outcrops	The bedrock outcrops start to become rutted themselves. The hollows gradually deepen as the sticky clay is walked out by passing feet and wheels
t_2	Irregular surface of bedrock widely exposed, revealing sharp hollows and deep shafts	Rutted tracks become laterally uneven; deep shafts require bypassing	Bedrock surface hardens and fossilizes

it deviates from the general route alignment in taking a longer and less direct course in avoiding the solution hollow or shaft.

Figure 9 summarizes this sequence of stages in the stripping of the soil cover through times t_0-t_2, and the changing nature of the topographic surface as the soil was thinned and removed. Only at the present day, with all the former soil fill removed, is the nature of the bedrock topography fully revealed. This sequence implies that the initial land surface form of the rutted areas at San Pawl tat-Tarġa and San Ġwann was very different to that visible now, and, furthermore, that the local environment was continuing to change during, and probably because of, the continued usage of the rutted trackways.

This interpretation of the changing surface environment can be expressed by a simple model (Table 3). Here, it is shown how the changing nature of the surface is likely to influence local route selection as the formerly buried obstacles to traffic gradually become apparent, and then increase in severity. The vehicular footprint becomes superimposed onto the emergent bedrock, leading to the development of the rut patterns visible today, several features of which are clearly discordant to the bedrock surface. Once exposed at the surface, the bedrock becomes colonized by lichens, which in the case of less resistant rock such as the Tal-Pitkal Member of the Upper Coralline Limestone at Misraħ Għar il-Kbir forms a protective patina that fortifies the rutted surface form.

Conclusion

Remnants of a former soil cover are preserved on limestone exposed near the cart-rut site at San Pawl tat-Tarġa, Naxxar, and can be shown to offer an analogue for the development of the nearby bedrock surfaces that bear cart ruts. A progressive loss of soil cover permits the gradual superimposition of the rutted trackway patterns, which developed initially in soils, onto the underlying and formerly buried irregular karstic bedrock surfaces, of which the original carters would have been blissfully unaware. The newly exposed bedrock surface presents many obstacles to the passage of vehicles, and shows several discordances between trackway and local topography; in particular the emergence into visibility at the surface of shafts, and basins, and subsequent avoidance strategies in the form of bypasses.

Indeed, topographic irregularity of the rockhead form appears to be the norm at the observed sites. This is demonstrated by the existence of an almost horizontal and plane bare limestone surface on part of the San Ġwann site. This section, in the

Fig. 10. Exceptional uniform linear ruts at San Ġwann developed on an almost plane horizontal limestone surface devoid of obstacles.

absence of any bedrock obstacles or hazards, bears a pair of uncharacteristically straight and uniform ruts (Fig. 10). The exceptional nature of these ruts, in contrast to the curving, asymmetric, divergent and duplicate forms found elsewhere and developed in response to buried bedrock irregularities, emphasizes the abundance of the latter at the study sites.

Progressive soil erosion has created substantial change in the form and material character of the surface environment, which, it is argued here, is the key to explaining the origin of the cart ruts in their present form. The evidence presented here confirms the conjectures of previous authors in respect of erosion of a former soil cover (Fenton 1918; Gracie 1954; Trump 1993).

The erosion of a soil cover consequent upon increasing population pressure on the limited land resource of Malta is not confined to the local site described in this paper, but is typical of Mediterranean land use history (Hunt 1997) and likely to be an island-wide phenomenon. The fortuitous juxtaposition of the critical exposure and the cart-rut evidence enables the substantiation of this hypothesis, which is equally likely to apply to other similar sites. This is not to suggest that all Maltese cart ruts originated in this way, as those at Misrah Għar il-Kbir (Clapham Junction) differ in being specifically associated with large-scale local quarrying of unknown age. The general explanation offered here may, however, be widely applicable to the origin and development of those cart ruts that are associated with regional transport routes across Malta.

The University of Portsmouth assisted with field expenses. W. Johnson created Figures 1 and 8. D. Weights provided assistance and advice in respect of the XRD analyses and interpretation. G. Gabriel assisted with the laboratory analysis of soil materials. G. Gunston and J. Carden assisted with the fieldwork. R. Grima provided helpful advice and discussion. An anonymous reviewer offered helpful comments on an earlier version of this paper.

References

BECH, J., RUSTULLET, J., GARIGÓ, J., TÓBIAS, F. J. & MARTÍNEZ, R. 1997. The iron content of some red Mediterranean soils from northeast Spain and its pedological significance. *Catena*, **28**, 211–229.

BOARDMAN, J. & BELL, J. 1992. *Soil Erosion and Archaeology*. Wiley, Chichester.

BOERO, V. & SCHWERTMANN, U. 1989. Iron oxide mineralogy of Terra Rossa and its genetic implications. *Geoderma*, **44**, 319–327.

BRANDT, J. & THORNES, J. B. (eds) 2002. *Mediterranean Desertification and Land Use*. Wiley, Chichester.

BRONGER, A. & BRUHN-LOBIN, N. 1997. Paleopedology of Terrae rossae-Rhodoxeralfs from Quaternary calcarenites in NW Morocco. *Catena*, **28**, 279–295.

BUTZER, K. W. 2005. Environmental history in the Mediterranean world: cross-disciplinary investigation of cause-and-effect for degradation and soil erosion. *Journal of Archaeological Science*, **32**, 1773–1800.

DURN, D. 2003. Terra Rossa in the Mediterranean region: parent materials, composition and origin. *Geologia Croatica*, **56**(1), 83–100.

FEDOROFF, N. 1997. Clay illuviation in Red Mediterranean soils. *Catena*, **26**, 171–189.

FENTON, E. G. 1918. The Maltese cart-ruts. *Man*, **18**, 67–72.

FAO. 2006. *World Reference Base for Soil Resources 2006. A framework for International Classification, Correlation and Communication*. World Soil Resources Reports, **103**. Food and Agricultural Organization, Rome.

FORD, D. C. & WILLIAMS, P. 1989. *Karst Geomorphology and Hydrology*. Unwin Hyman, London.

GASSER, A. 2003. http://www.ukom.gov.si/eng/slovenia/background-information/oldest-wheel/.

GRACIE, H. S. 1954. The ancient cart-tracks of Malta. *Antiquity*, **28**, 91–98.

HOLE, F. D. 1978. An approach to landscape analysis with emphasis on soils. *Geoderma*, **21**, 1–13.

HOLE, F. D. & CAMPBELL, J. B. 1985. *Soil Landscape Analysis*. Rowman & Allanheld, Totowa, NJ.

HUGHES, K. J. 1999. Persistent features from a palaeolandscape: the ancient tracks of the Maltese Islands. *Geographical Journal*, **165**(1), 62–278.

HUNT, C. O. 1997. Quaternary deposits in the Maltese Islands: a microcosm of environmental change in Mediterranean lands. *Geo Journal*, **41**(2), 101–109.

KOSMAS, C., DAALATOS, N. G. & GERONTIDES, ST. 2000. The effect of land parameters on vegetation performance and degree of erosion under Mediterranean condition. *Catena*, **40**, 3–17.

LAWSON, I., FROGLEY, M., BRYANT, C., PREECE, R. & TZEDAKIS, P. 2004. The late glacial and Holocene environmental history of the Ioannina basin, Northwest Greece. *Quaternary Science Reviews*, **23**, 1599–1625.

MORESI, M. & MONGELLI, G. 1988. The relationship between the terra rossa and the carbonate-free residue of underlying limestones and dolostones in Apulia, Italy. *Clay Minerals*, **23**, 439–446.

MOTTERSHEAD, D., PEARSON, A. & SCHAEFER, M. 2008. The cart-ruts of Malta: an applied geomorphology approach. *Antiquity*, **82**(318), 1065–1079.

NIHLEN, T. & OLSON, S. 1995. Influence of eolian dust on soil formation in the Aegean area. *Zeitschrift für Geomorphologie*, **39**, 341–361.

PEDLEY, M. & CLARK, M. H. 2002. *Geological Itineraries in Malta and Gozo*. Enterprise Group (PEG), San Gwann, Matta.

SCHWERTMANN, U. 1988. Occurrence and formation of iron oxides in various pedoenvironments. *In*: STUCKI, J. W., GOODMAN, B. A. & SCHWERTMANN, U. (eds) *Iron Oxides and Clay Minerals*. NATO Advanced Study Institute. Reidel, Dordrecht, 267–308.

TRUDGILL, S. T. 1985. *Limestone Geomorphology*. Longman, London.

TRUMP, D. H. 1993. *Malta: An Archaeological Guide*. Progress Press, Valletta, Malta.

TRUMP, D. H. 2002. *Malta Prehistory and Temples*. Midsea Books, Malta.

WHITE, W. B. 1988. *Geomorphology and Hydrology of Karst Terrains*. Oxford University Press, Oxford.

WILSON, G. P., REED, J., LAWSON, I., FROGLEY, M., BRYANT, C., PREECE, R. & TZEDAKIS, P. 2008. Diatom response to the last glacial-interglacial transition in the Ioannina basin, northwest Greece: Implications for Mediterranean palaeoclimate reconstruction. *Quaternary Science Reviews* **27**, 428–440.

YAALON, D. H. 1997. Soils in the Mediterranean region: what makes them different? *Catena*, **28**, 157–169.

ZAMMIT, T. 1928. Prehistoric cart-tracks in Malta. *Antiquity*, **11**, 18–125.

Measurements of soiling and colour change using outdoor rephotography and image processing in Adobe Photoshop along the southern façade of the Ashmolean Museum, Oxford

MARY J. THORNBUSH

University of Oxford, Oxford University Centre for the Environment, Dyson Perrins Building, South Parks Road, Oxford OX1 3QY, UK

Present address: Lakehead University, Orillia Campus, Heritage Place, 1 Colborne Street West, Orillia, Ontario, L3V 7X5, Canada
(e-mail: mthornbu@lakeheadu.ca)

Abstract: This paper builds on work using Adobe Photoshop as image-processing software to obtain (histogram-based) quantification of camera-captured images. The analysis tracks cross-temporal surface colour change (in 2005 and 2007) at the façade-scale of a limestone building (cleaned between 2006 and 2007) by means of digital photographs obtained in repeat photographic surveys taken under different outdoor lighting conditions (of a clear sky v. overcast). The relevance of the study is to contribute to further research using the integrated digital photography and image processing (IDIP) method in an outdoor setting (O-IDIP) with differing levels of light at a façade- (building) scale necessary for assessments of soiling affecting decisions of maintenance and restoration. Calibration was performed using spectrophotometric data (acquired in the winter of 2006). Findings show that the calibrated method is able to measure change before v. after the cleaning of the southern façade, which was darker in 2005 (with a lower level of lightness), especially at the east elevation. Lightness (surface darkening or blackening) is more affected by outdoor lighting conditions than chromatic values along green–red (a) and blue–yellow (b) channels of colour. These findings confirm that there is more error associated with soiling measurements at the façade-scale (in an outdoor setting).

Patterns of soiling and decay have been examined at historical sites in Europe, leading to a vast literature. Fassina (1988) was one of the pioneers in the field, differentiating between rainwashed (white areas) v. unwashed (black areas) on the façade of historical monuments in Venice, Italy. He coined the term 'whitewashing' for vertical building surfaces that had been cleaned by rainwater of dirt adhering to stone in areas sheltered from rainfall, and noted that aerosols are transported by wind and condense on cold walls. Camuffo & Bernardi (1996) attributed blackening of the Ara Pacis, Rome to large-particle deposition in the near absence of wind and turbulence in a sheltered environment of a glass-panel building, where gravitational settling influenced horizontal, upward surfaces. Iñigo & Vicente-Tavera (2001) examined factors in micro-environments associated with the cloister of the Cathedral of Avila, Spain. They found that the reduced variability of temperature and relative humidity inside the cloister accounted for less degradation in interior zones; whereas, fracturing and granular disintegration led to decay on the outside of the cloister. Rainwater did not directly impinge inside the cloister, reducing frost damage and biodeterioration. These studies have shown that structures exposed to the natural (outdoor) environment are affected, both in terms of soiling and decay, by climatic variables, including temperature, humidity and rainfall.

Studies using laboratory specimens as well as building façades have deciphered the variables contributing towards patterns of soiling in particular. Pesava *et al.* (1999) found the greatest soiling towards the edges on the front and back sides of cubes. Deposition was lowest in the middle and increased towards the edges, particularly on the upwind edge. Windwashing was noted in the presence of strong gusts of wind even where there was rain protection. Soiling rates varied according to the main wind direction as well as surface roughness. Davidson *et al.* (2000) concluded that wind direction and rain intensity are responsible for soiling patterns at the Cathedral of Learning in Pittsburgh, Philadelphia, USA. Whereas soiling occurred at an annual timescale, the removal of soiling was observed to be decadal. Etyemezian *et al.* (2000) calculated rain fluxes to be consistent with soiling patterns also at the Cathedral of Learning. White areas corresponded with areas of high rain fluxes, whereas soiled areas received less rain. Wind speed, especially intermediate wind speeds,

From: SMITH, B. J., GOMEZ-HERAS, M., VILES, H. A. & CASSAR, J. (eds) *Limestone in the Built Environment: Present-Day Challenges for the Preservation of the Past.* Geological Society, London, Special Publications, **331**, 231–236. DOI: 10.1144/SP331.21 0305-8719/10/$15.00 © The Geological Society of London 2010.

affected these rain fluxes. Tang et al. (2004) more recently followed up on this research with field measurements, which included photographs of driving-rain gauges, showing differential levels of soiling. They attributed soiling patterns to the non-uniform distribution of wind-driven rain produced from interactions between wind, rainfall and building geometry. Walls receiving less wind-driven rain, on the eastern side of the cathedral, were soiled compared to those white and weathered walls on the western side that received higher volumes of wind-driven rain. These studies indicate that the external (outdoor) environment affects soiling patterns across building façades, which are governed by wind-driven rain (windwashing).

In recent years authors have turned to colour measurement of stone soiling (e.g. Prieto et al. 2004; Feliu et al. 2005; Franceschi et al. 2006), even incorporating video cameras (e.g. Erdogan 2000; Maurício & Figueiredo 2000). The Commission Internationale d'Eclairage (CIE) $L*a*b*$ system is particularly used extensively in the literature. Grossi et al. (2003), for instance, monitored changes in $L*$ (luminance or light reflectance) as a proxy measure for surface soiling of Spanish building stones, including samples of marble, granite, limestone and dolostone, in the urban environments of Burgos and Oviedo. They concluded that type of stone, exposure and climate (e.g. humidity and precipitation) all affect soiling. Durán-Suárez et al. (1995) used a colorimetric analysis in the assessment of restoring agents on whitening or darkening (loss of luminosity) of stone, which included consideration of chromatic change. Similarly, García-Talegon et al. (1998) monitored chromatic change associated with treatments and simulated ageing of stone. Samples darkened with treatment and lightened with ageing; whereas, untreated samples darkened with age. They also observed chromatic alterations of increased ($a*$) or reddening with age, and an increase in $b*$ or yellow/gold hues with treatment. Using CIE $L*a*b*$ co-ordinates, Urzi & Realini (1998) established a link between microorganisms and surface coloration. Realini et al. (1995) had previously used luminous reflectance in Milan, Italy to observe blackening of stone surfaces from the deposition of particulate matter rich in carbonaceous particles. This colorimetric research has revealed much about the soiling of exposed stone, including limestone, in a measurable way.

The aim of this paper is to contribute towards the use of the CIE $L*a*b*$ system in the measurement of outdoor surface colour change and soiling patterns across a building façade. This is not to provide a technical summary of the method, but rather to apply the already established method, which has been published as a new method to measure (calibrated) colour change using digital photographs, to measure façade-scale colour change. With the aid of field surveys based on repeat photography as a methodology (cf. Thornbush & Viles 2008), it is possible to acquire images that can then be processed using computer software. Images can be used to measure cross-temporal change on the surfaces of historical buildings as a non-destructive, inexpensive approach. Viles (1994), for instance, examined the Ashmolean Museum as a case study of stone decay in Oxford using photographs from 1882, 1900, 1963, 1976 and 1993 in a qualitative approach of stone decay and soiling patterns. Although more recently published research has focused on the laboratory application of integrated digital photography and image processing known as the IDIP method (Thornbush & Viles 2004a, b), this study extends it to the field at the façade- (or building) scale. Even though it has not been possible to quantify archival photographs due to various factors, such as a lack of photographic control (e.g. Thornbush & Viles 2005), close-up quality photographs have been successfully quantified using Adobe Photoshop (Thornbush & Viles 2007) as an accessible and easily used software. It is noteworthy that other programs are available to conduct a quantitative study (e.g. CorelDraw; ImageJ with a Plugin Color Inspector); however, this study contributes to the development of literature on the use of Adobe Photoshop as a software employed in an established method, namely the IDIP method in an outdoor environment (O-IDIP).

The Ashmolean Museum, built in the 1840s by Baker of Lambeth (Law 1998, p. 54), was restored in 1889 by H.W. Moore (p. 128). Its plinths have been recently replaced and so are not indicative of its soiling history. More recently, its southern façade was cleaned between 2006 and 2007, creating a cleaner (brighter and more colourful) appearance in 2007 than in 2005 – this can be seen; however, it is helpful to be able to quantify the visible change as a means to track changes (e.g. re-soiling) cross-temporally, if not cross-spatially. For example, as conveyed in Figure 1, there is greater traffic pollution due to congestion along St Giles Street (to the east of the building) than along Beaumont Street (along the southern façade) and, particularly, than along St John's Street (to the west of the building). One would, therefore, expect to see greater soiling towards the east along the southern façade of the Ashmolean Museum. However, a building-scale approach in an outdoor setting may complicate cross-spatial comparisons (affected by building geometry, changing weather, light conditions, etc.) and, for this reason, this study focuses on a cross-temporal comparison (rather than a cross-spatial comparison of the west v. east elevations) of parts of the southern façade. It is arguable that taking digital photographs of

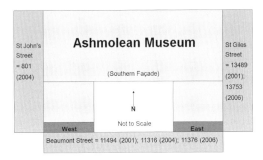

Fig. 1. Planform view of the general building structure, showing the west and east elevations along the southern building façade. Traffic counts (manual classified counts between 7 am and 7 pm from Table B of Oxfordshire County Council 2006 report of Oxfordshire traffic flows for all vehicles, pedal cycles and heavy goods vehicles) from 2001, 2004 and 2006 appear for corresponding streets.

the same parts of the building (with identical architectural features, including windows) at a similar time (of day and season) can render comparative measurements of soiling and colour change at the building-scale. In this study, 'soiling' refers to a darkening of the surface that leads to an overall lowered light reflectance of the building façade.

Materials and methods

Photographic sets captured for this study are of the west and east elevations along the southern façade of the Ashmolean Museum, facing onto Beaumont Street (see Fig. 1). All photographs were taken by the author with a freely-held digital camera between 0900 and 1200 in the spring. The first photographic survey took place in early March 2005 (under a clear sky), using a Nikon Coolpix 950 digital camera (which has a megapixel size of 1.2). Subsequent photographic surveys were carried out in April 2007, using a Nikon Coolpix S4 digital camera (with a megapixel size of 6), in the same manner across the street facing the southern building façade. This latter survey consisted of two components: under (a) a clear sky and (b) overcast conditions. Calibration measurements were acquired directly on the southern façade of the Ashmolean Museum in October 2006 with an X-Rite SP68 Sphere Spectrophotometer (after Thornbush 2008, p. 64, fig. 2).

An approach similar to that employed by Searle (2001) to examine the soiling of building façades in Bath, UK is relevant here, although he used Scion Image software for image processing. Using Adobe Photoshop (Version 7.0) it was possible to similarly measure image colouration, instead incorporating colour images in Lab Color rather than Grayscale images so that colour change could be assessed in addition to soiling. Windows in the building façade were not cleared to avoid user error (using the rectangular selection tool) because this would not have completely resolved the problem of glare and light scattering already evident in the outdoor environment and captured by the photographs. Results, then, are comparative cross-temporally, keeping in mind that windows are as they would naturally appear. Images were calibrated for brightness and contrast as well as colour levels using an average of three (points of) spectrophotometric measurements obtained directly from clean (rainwashed) surfaces of Portland stone near the SW balustrade (where mean $L^* = 77.39$, $a^* = 0.85$, $b^* = 7.25$, with standard deviation values of $L^* = 2.11$, $a^* = 0.15$, $b^* = 0.23$, respectively) accessible from street level. Portland stone was used in the calibration procedure because it is the only stone type in the composite (intricate) use of limestones along this façade that was naturally cleaned by rainwater (rainwashed) and represented in the area selected for image analysis. Many limestones used as building materials have a similar light cream/buff colour (with L^* c. 80, very low a^*, and slightly higher values of b^* of about 10). The digital photographs (all taken at a default resolution of 300 dpi) captured similar façade sections in all images. The colour measurements are of composite stonework, including Portland and Bath limestones, portraying a realistic assessment of intricate building façades. The data range between 0 and 255 across three channels of lightness ($L = 0\%$ for black and 100% for white), green–magenta ($a = 0\%$ towards green and 100% for magenta or red) and blue–yellow ($b = 0\%$ for blue and 100% for yellow). The method (of a hand-held camera as seen from the perspective of human perception, with selection of as similar an area as possible) is susceptible to human error. However, it is important for the approach to be human-based since decisions about soiling and restoration are affected by human perception of light v. dark. The use of CIE $L^*a^*b^*$ colour mode is mostly based on human perception of colour as it is a device-independent system, which is relevant for weathering studies.

Results

The colorimetric results for 2005 and 2007 appear in Figures 2 and 3, respectively. Mean L was lowest in 2005 at the SE elevation (36.32%), when standard deviation (SD) of L was greatest (28.12%). At the SW elevation, L values were lowest in 2005, including the mean (40.05%) and SD (27.39%). Lightness is greatest in 2007 at the SW elevation under the overcast condition (with a mean L of 63.46% and SD of 20.18%). In fact, all values of L are greater

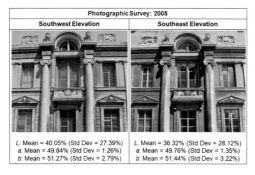

Fig. 2. Summary of results for the O-IDIP method for 2005 (taken under a clear sky).

at the SW elevation than the SE elevation, where the SD is usually greater.

Green–red values have a restricted range between 48.95% (at the SE elevation in 2007 under a clear sky condition) and 49.84% (at the SW elevation in 2005). The SD of a is low, ranging between 0.98% (at the SW elevation in 2007 under an overcast condition) and 1.35% (at the SE elevation in 2005). Blue–yellow values are also consistent, ranging between 51.27% (at the SW elevation in 2005) and 54.68% (at the SE elevation in 2007 under an overcast condition).

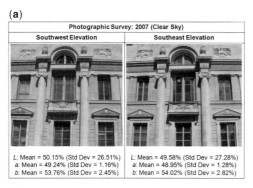

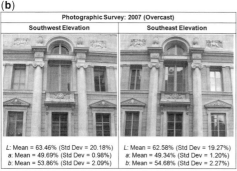

Fig. 3. Summary of the results for the O-IDIP method for 2007 under (**a**) a clear sky and (**b**) overcast conditions.

The SD of b is also low, ranging between 2.09% (at the SW elevation in 2007 under an overcast condition) and 3.22% (at the SE elevation in 2005).

Discussion

Spectrophotometric measurements for lightness of clean stone in this study ranged between 65.14 (Bath stone) and 77.39 (Portland stone). This compares well with previous studies; for instance, Fort et al. (2000) applied a spectrocolorimeter to the surface of cylindrical test specimens (each 50 mm in diameter) as well as to the façade of the Cathedral Nueva of Valladolid, Spain, attaining only somewhat higher L values of 70–80 (p. 443). Soiling accounted for a lightness reduction of 43.0, differing by 40.4 from rainwashed areas, from an original value of 79.3 (p. 444). In the current study, darkening was responsible for a similar reduction of 36.89% at the SW elevation and 41.96% at the SE elevation (differing from rainwashed stone, with an L value of 77.39%, respectively, by 48.25 and 53.07%). This could, in part, be a product of wind-driven rain (as in Tang et al. 2004), which comes from the SW in Oxford (Oxfordshire County Council 1995). Values for $a*$ ranged between 0.85 (Portland stone) and 5.56 (Bath stone), suggesting a greener rather than reddish coloration. The range for $b*$ was 7.25 (Portland stone) and 16.40 (Bath stone), indicating the more yellowish (buff) colouration of Bath stone. It has been possible to measure changes in soiling (in a quantitative fashion across a dark–light spectrum) as well as colour changes of Portland and Bath stones before and after cleaning. The results suggest a similar conclusion to Grossi et al. (2003) – that exposure and climate affect soiling. There is also some indication that cleaning restores a yellow colouration, which is the reverse of reddening (cf. García-Talegon et al. 1998). This has also been found to be the case with laser cleaning treatment in other studies (e.g. Pouli et al. 2006). Furthermore, limestone buildings, such as the Ashmolean, may suffer future yellowing with the accumulation of brownish organic deposits from pollution. Grossi et al. (2007) recently noted that sulphur dioxide (even at low concentrations in the atmosphere) can increase the chroma ($b*$) of limestones, yellowing the surface. They also discovered that the faster occurrence of change in $b*$ indicates earlier sulphation than darkening.

The SD of L values were consistently greater at the SE elevation than at the SW elevation and were notably reduced in the overcast condition (rather than with cleaning). A possible explanation for this is that SD values reflect photographic contrast, and are affected by shadows in the photographic surveys produced by the angle of the sun, its proximity and time of day. In the overcast condition, it was possible to reduce the SD of all values, including a and b values; whereas, cleaning only slightly improved (reduced) SD values, suggesting that shadows (rather than soiling patterns) are chiefly responsible for augmenting SD values, as they are reduced in the absence of direct sunlight (in the overcast condition). This finding limits any spatial measurements, although cross-temporal change can be assessed.

Caveats and further research

There are several problems with this study that should be considered. Architectural features associated with columns, for instance, can lead to shadows that distort the quantification of soiling (e.g. increasing the SD of L values, as at the SE elevation). It is best for direct solar radiation to be avoided in this approach to reduce, for instance, increased glare from windows. Calibration in this study was based on three points located on the balustrade at the SW elevation. Ideally, more samples should be taken. Another limitation is the limited range of observation of 2 years (2005 and 2007). Finally, different cameras were used (although both were Nikon Coolpix digital cameras).

Future research can provide a proper scale of severity for some of these problems. For instance, this approach should be performed on a plain façade devoid of architectural features or windows. However, the approach taken here helped to convey more common buildings with intricate architecture and windows. The use of different digital cameras of the same brand (Nikon Coolpix) should not produce too much of a difference in the results; however, this should be examined, especially as pixel dimensions and the number of pixels in the images differed by type of digital camera used – for instance, for lightness, pixel dimensions were greater (ranging between 3787 W × 3245 H and 4011 W × 3459 H) in 2007 than in 2005 (2186 W × 1856 H and 2187 W × 1885 H), and there were more pixels in 2007 (ranging between 12 288 815 and 13 874 049 pixels) than in 2005 (4 057 216 and 4 122 495 pixels). The calibration procedure varies from grey-scale calibration already tested by Thornbush (2008) and should be further examined. As most people regard a building as dirty with a reduction of the lightness parameter below 50% (Brimblecombe & Grossi 2005), it would be interesting to establish a quantitative link with thresholds in the perception of soiling using the O-IDIP method. Moreover, the SD values obtained in this approach could be used as an aid for future research attempting to measure this component of soiling perception (cf. Grossi & Brimblecombe 2004). In this way, there is also much potential here for the application of this method to measure change in biological colonization at various scales (e.g. of lichen growth), and this research is forthcoming.

Conclusions

There are several important contributions in this paper towards the development of a quantitative rephotography using Adobe Photoshop in an outdoor environment. First, the method (calibrated with spectrophotometric data) is capable of capturing the environmental effects of rainwashing and possibly pollution on spatially disparate locations (e.g. on the SW v. SE elevations), although it cannot disentangle the effects. Second, O-IDIP captured cross-temporal trends in soiling at the façade-scale (as with cleaning of the southern façade, which brightened and improved the colouration especially of yellow surfaces). For instance, both the SW and SE elevations had a more similar colouration after cleaning, which is an unlikely result of spatially produced differences. It is also noteworthy that both a and b (the chromatic values) were not as affected by outdoor lighting conditions (e.g. of a clear v. overcast sky) as L values, which are impacted by radiation as a measure of reflectance. Third, shadows can be effectively reduced by photographing on overcast days (as supported by previous work by Thornbush 2008), especially as it was found that many of the changes in SD values in an outdoor environment are attributable to cast shadows rather than soiling patterns resulting from contrasts established by the rainwashing of sections with protruding relief and the unwashing of recessed areas.

S. E. Thornbush provided field assistance. Thanks also to T. Yates at the Building Research Establishment for use of a spectrophotometer; H. Viles and T. Stevens also assisted me with this. D. Searle at the University of Wolverhampton offered very helpful suggestions and feedback.

References

BRIMBLECOMBE, P. & GROSSI, C. M. 2005. Aesthetic thresholds and blackening of stone buildings. *Science of the Total Environment*, **349**, 175–189.

CAMUFFO, D. & BERNARDI, A. 1996. Deposition of urban pollution on the Ara Pacis, Rome. *Science of the Total Environment*, **189/190**, 235–245.

DAVIDSON, C. I., TANG, W., FINGER, S., ETYEMEZIAN, V., STRIEGEL, M. F. & SHERWOOD, S. I. 2000. Soiling patterns on a tall limestone building: changes over 60 years. *Environmental Science and Technology*, **34**, 560–565.

DURÁN-SUÁREZ, J., GARCIÁ-BELTRÁN, A. & RODRÍGUEZ-GORDILLO, J. 1995. Colorimetric cataloguing of stone materials (biocalcarenite) and evaluation of the chromatic effects of different restoring agents. *Science of the Total Environment*, **167**, 171–180.

ERDOGAN, M. 2000. Measurement of polished rock surface brightness by image analysis method. *Engineering Geology*, **57**, 65–72.

ETYEMEZIAN, V., DAVIDSON, C. I., ZUFALL, M., DAI, W., FINGER, S. & STRIEGEL, M. 2000. Impingement of rain drops on a tall building. *Atmospheric Environment*, **34**, 2399–2412.

FASSINA, V. 1988. Stone decay of Venetian monuments in relation to air pollution. *In*: MARINOS, P. G. & KOUKIS, G. O. (eds) *Engineering Geology of Ancient Works, Monuments and Historical Sites*. A.A. Balkema, Rotterdam, 787–796.

FELIU, M. J., EDREIRA, M. C., MARTIN, J., CALLEJA, S. & ORTEGA, P. 2005. Study of various interventions in the façades of a historical building – methodology proposal, chromatic and material analysis. *Color Research and Application*, **30**, 382–390.

FORT, R., MINGARRO, F., LÓPEZ DE AZCONA, M. C. & RODRIGUEZ BLANCO, J. 2000. Chromatic parameters as performance indicators for stone cleaning techniques. *Color Research and Application*, **25**, 442–446.

FRANCESCHI, E., LETARDI, P. & LUCIANO, G. 2006. Colour measurements on patinas and coating system for outdoor bronze monuments. *Journal of Cultural Heritage*, **7**, 166–170.

GARCÍA-TALEGON, J., VICENTE, M. A., VICENTE-TAVERA, S. & MOLINA-BALLESTEROS, E. 1998. Assessment of chromatic changes due to artificial ageing and/or conservation treatments of sandstones. *Color Research and Application*, **23**, 46–51.

GROSSI, C. M. & BRIMBLECOMBE, P. 2004. Aesthetics of simulated soiling patterns on architecture. *Environmental Science and Technology*, **38**, 3971–3976.

GROSSI, C. M., BRIMBLECOMBE, P., ESBERT, R. M. & ALONSO, F. J. 2007. Color changes in architectural limestones from pollution and cleaning. *Color Research and Application*, **32**, 320–331.

GROSSI, C. M., ESBERT, R. M., DÍAZ-PACHE, F. & ALONSO, F. J. 2003. Soiling of building stones in urban environments. *Building and Environment*, **38**, 147–159.

IÑIGO, A. C. & VICENTE-TAVERA, S. 2001. Different degrees of stone decay on the inner and outer walls of a Cloister. *Building and Environment*, **36**, 911–917.

LAW, B. R. 1998. *Building Oxford's Heritage: Symm & Company from 1815*. Prelude Promotion, Oxford.

MAURÍCIO, A & FIGUEIREDO, C. 2000. Texture analysis of grey-tone images by mathematical morphology: a non-destructive tool for the quantitative assessment of stone decay. *Mathematical Geology*, **32**, 619–642.

OXFORDSHIRE COUNTY COUNCIL. 1995. *Oxfordshire's Environment. Environmental Monitoring Report, 1995*. Department of Planning and Property Services, Oxfordshire County Council, Oxford.

OXFORDSHIRE COUNTY COUNCIL. 2006. *Oxfordshire Traffic Counts 2006. Table B: Manual Classified Counts*. Oxfordshire County Council, Oxford. World wide web address: http://portal.oxfordshire.gov.uk/content/publicnet/council_services/roads_transport/traffic/traffic_monitoring/data-summaries/TableB-manual-classified-counts.pdf.

PESAVA, P., AKSU, R., TOPRAK, S., HORVATH, H. & SEIDL, S. 1999. Dry deposition of particles to building surfaces and soiling. *Science of the Total Environment*, **235**, 25–35.

POULI, P., TOTOU, G., FOTAKI, C., GASPARD, S., OUJJA, M., CASTILLEJO, M. & DOMINGO, C. 2006. A comprehensive study on the discoloration associated with laser cleaning of stonework. *In*: FORT, R., ÁLVAREZ DE BUERGO, M., GÓMEZ-HERAS, M. & VÁZQUEZ-CALVO, C. (eds) *Proceedings of the International Conference on Heritage, Weathering and Conservation*. A.A. Balkema Publishers, London, 687–692.

PRIETO, B., SILVA, B. & LANTES, O. 2004. Biofilm quantification on stone surfaces: comparison of various methods. *Science of the Total Environment*, **333**, 1–7.

REALINI, M., NEGROTTI, R., APPOLLONIA, L. & VAUDAN, D. 1995. Deposition of particulate matter on stone surfaces: an experimental verification of its effects on Carrara marble. *Science of the Total Environment*, **167**, 67–72.

SEARLE, D. E. 2001. *The Comparative Effects of Diesel and Coal Particulate Matter on the Deterioration of Hollington Sandstone and Portland Limestone*. PhD thesis, University of Wolverhampton.

TANG, W., DAVIDSON, C. I., FINGER, S. & VANCE, K. 2004. Erosion of limestone building surfaces caused by wind-driven rain: 1. field measurements. *Atmospheric Environment*, **38**, 5589–5599.

THORNBUSH, M. J. 2008. Grayscale calibration of outdoor photographic surveys of historical stone walls in Oxford, England. *Color Research and Application*, **33**, 61–67.

THORNBUSH, M. J. & VILES, H. A. 2004a. Integrated digital photography and image processing for the quantification of coloration on soiled limestone surfaces in Oxford, England. *Journal of Cultural Heritage*, **5**, 185–190.

THORNBUSH, M. J. & VILES, H. A. 2004b. Surface soiling pattern detected by integrated digital photography and image processing on exposed limestone in Oxford, England. *In*: SAIZ-JIMENEZ, C. (ed.) *Air Pollution and Cultural Heritage*. A.A. Balkema, London, 221–224.

THORNBUSH, M. J. & VILES, H. A. 2005. The changing façade of Magdalen College, Oxford: reconstructing long-term soiling patterns from archival photographs and traffic records. *Journal of Architectural Conservation*, **11**, 40–57.

THORNBUSH, M. J. & VILES, H. A. 2007. Photo-based decay mapping of replaced stone blocks on the boundary wall of Worcester College, Oxford. *In*: PŘIKRYL, R. & SMITH, B. J. (eds) *Building Stone Decay: From Diagnosis to Conservation*. Geological Society, London, Special Publications, **271**, 69–75.

THORNBUSH, M. J. & VILES, H. A. 2008. Photographic monitoring of soiling and decay of roadside walls, central Oxford, England. *Environmental Geology*, Special Issue, 1–11.

URZI, C. & REALINI, M. 1998. Colour changes of Noto's calcareous sandstone as related to its colonisation by microorganisms. *International Biodeterioration and Biodegradation*, **42**, 45–54.

VILES, H. A. 1994. *Time and Grime: Studies in the History of Building Stone Decay in Oxford*. School of Geography Research Paper, **50**.

Two-dimensional resistivity surveys of the moisture content of historic limestone walls in Oxford, UK: implications for understanding catastrophic stone deterioration

O. SASS[1] & H. A. VILES[2]*

[1]*Department of Geography and Regional Science, University of Graz, Heinrichstrasse 36, 8010 Graz, Austria*

[2]*School of Geography, OUCE, University of Oxford South Parks Road, Oxford, OX1 3QY, UK*

*Corresponding author (e-mail: heather.viles@ouce.ox.ac.uk)

Abstract: Catastrophic deterioration of limestone facades occurs where areas of stonework become rapidly hollowed out. It affects many historic buildings in Oxford, especially where soot-rich gypsum crusts have accumulated. In order to understand the processes of catastrophic deterioration we need to understand the microenvironmental conditions, especially the moisture distributions in the deteriorating walls. Geoelectric methods, in the form of two-dimensional (2D) resistivity surveys, have been used to study the distribution and amount of water stored in deteriorating limestone walls within the historic centre of Oxford. Fifteen vertical profiles, each 2–2.5 m in length, have been monitored at five sites using 50 medical electrodes and GeoTom equipment. Calculated moisture contents and distributions are presented for those profiles that extend up to 40 cm into the wall. The data indicate the diversity and complexity of moisture distributions within these often heterogeneous walls, which have also had long histories of decay and conservation. Replacement stone patches show consistently higher moisture conditions than the surrounding stone. Most profiles indicate the presence of wetter patches 5–10 cm behind the wall face under blackened crusts. Catastrophically decayed sections of profiles often exhibit wetter near-surface conditions than surrounding stonework, whilst areas with shallow but active decay are often much drier than surrounding crusted stone. In conclusion, the results give preliminary confirmation of a simple model of catastrophic decay and illustrate the complexity of moisture regimes in historic walls.

Knowledge of the amount and distribution of water held in porous masonry walls is of critical importance to understanding stone deterioration. Water is a key component of most deterioration processes, either directly (e.g. as a solvent) or indirectly (e.g. as it encourages growth of agents of biodeterioration). As part of a larger project on the catastrophic decay of building limestones within historic masonry walls in Oxford, UK, we are currently developing the use of two-dimensional (2D) resistivity surveys for assessing the distribution patterns of water within walls in order to test models of how catastrophic decay develops.

Geophysical methods, such as ground-penetrating radar or 2D resistivity surveys, provide non-invasive methods of sensing the water contents of porous media. They have been routinely used in a range of geological and archaeological applications over a long period of time, but only recently have geophysical survey techniques been applied to buildings and monuments. Leucci *et al.* (2006) were able to locate underground discontinuities and to estimate the volumetric water content under the pavement of an Italian crypt using ground-penetrating radar profiles. Sass & Viles (2006) used 2D resistivity surveys to verify the effect of soft wall capping on the moisture distribution within the stonework. Leucci *et al.* (2007) used a combined geophysical approach to investigate the moisture distribution within a single column. However, a number of challenges remain before any simple interpretation of resistivity profiles can be made because of the complex and heterogeneous nature of most masonry walls. As each wall is unique, there is an urgent need for more, and detailed, moisture data to find systematic patterns of water distribution and how they relate to weathering features, and vice versa.

Resistivity methods provide an indirect, approximate measure of moisture in porous media as they monitor the resistance of the material to the passage of an electric current. This resistance is directly influenced by water content, but the picture may also be complicated by the salt content and temperature of the water, and by the small-scale water, distribution within the pores

(Endrea & Knight 1992), all of which can affect the calibration between resistivity values and water content. A further factor needs to be considered, that is the porosity of the materials under observation. When the porosity of stone blocks is variable across and between profiles it is difficult to compare saturation levels, as the resistivity measurements tell us about absolute water content. Unless one carries out 'real time' wetting experiments in association with resistivity surveys it is impossible to know whether a given resistivity value represents, say, 50% saturation of a highly porous block or 100% saturation of a virtually non-porous block.

Limestone deterioration takes many forms, from dissolution of calcite and granular disintegration through to blistering and catastrophic decay (Robinson & Moses 2002; Török 2002). Most deterioration processes affecting limestone involve water in one way or another. Catastrophic decay of building limestones occurs when a single block, or sometimes a number of adjacent blocks, suffer from extreme active blistering or hollowing out. Preliminary observations of a wall in Oxford illustrated that, unlike in sandstones where catastrophic decay is isolated to individual blocks, a more contagious pattern is found in limestones (Smith & Viles 2006).

Understanding why some parts of walls exhibit only minor and shallow deterioration, whilst others are characterized by deep-seated, rapid and catastrophic decay, requires a detailed consideration of how such deterioration develops – for which understanding moisture distributions is a key task. The aims of this paper are, first, to use 2D resistivity surveys to provide a picture of internal moisture conditions (which are poorly understood to date) in a range of deteriorating walls within the historic centre of Oxford and, secondly, to use the observations to test a preliminary model of how catastrophic decay develops.

Conceptual model of the development of catastrophic decay

Catastrophic deterioration is widely observed within the historic centre of Oxford, predominantly but not exclusively, within the lower 1–1.5 m of the wall where capillary rise can operate (Viles 1993; Antill & Viles 1999). Often areas of five or more individual blocks are affected by the catastrophic decay, whilst there are also many walls that are characterized by individual blocks showing patchy rapid recession (what we call 'incipient catastrophic decay'). In some cases the deterioration appears to be actively progressing, whilst in others it appears to have been stabilized, often with further blackening and crust formation. Catastrophic decay is found on both old and newer stones within historic walls; on walls that are blackened and those which have been cleaned or were never soiled. However, as a general rule, we have observed catastrophic decay most frequently on walls that have suffered from extensive blackening and gypsum encrustation.

Smith & Viles (2006) present a simple conceptual model of how catastrophic decay develops on such limestone walls. The major components of the model are as follows.

(a) An initial phase of development of surface gypsum crusts, allied with weakening of the underlying stone. Gypsum crusts characteristically form on limestones in polluted atmospheres where sulphur dioxide reacts with limestone, often in the presence of soot and other particulates – forming a blackened, low-permeability crust. A near-surface zone within the underlying stone becomes softer and more permeable as a result of water becoming trapped inside the low-permeability crust and dissolving the calcite, probably enhanced by the presence of sulphur dioxide.
(b) Subsequent breaching of patches of the crust, leading to rapid weathering of the softer limestone underneath (in the form of blisters and incipient catastrophic decay).
(c) Either reformation of a surface gypsum crust through interaction of the stone surface with polluted air and particulates *or* rapid, catastrophic retreat of large areas of the surface. In turn such areas may become stabilized.

This series of events may be contrasted with what we hypothesize occurs in 'clean' areas without the development of these gypsum crusts. In this situation water can enter and leave the wall face easily and uniformly, and the stonework will suffer from gradual dissolution and granular disintegration. In both cases we assume that water enters the wall from two sources, that is, groundwater entering by capillary rise from the base of the wall, and driving rain impinging on the wall face. Where there is an effective damp course groundwater ingress will be minimized, but in many historic buildings water can enter easily from the ground.

If our model is correct we would expect to find at stage (a) a dry surface (where low permeability gypsum crusts are present) backed by a wetter zone (where water becomes trapped and where the stone becomes more porous and permeable). Where catastrophic decay is present (stage c) we might expect to find either wetter surface conditions within the hollows than further inside the wall (as porosity increases and the hollow microenvironment favours low evaporation rates in comparison with the surrounding wall face) or drier conditions in hollows stretching from the surface inwards (as water can now enter and exit freely without becoming trapped behind a low-permeability crust). At stage (b) of the model we might expect to find an intermediate state – with less clear gradients between a dry surface and wet near-surface interior.

Oxford stone and stone decay

Oxford's architectural and cultural heritage has, to a large extent, been written in Jurassic oolitic limestones. From the Medieval period onwards, the builders of Oxford have been creating buildings and monuments from locally-sourced stone and its more distantly quarried alternatives. Arkell (1947) provides a comprehensive summary of the history of Oxford stone, illustrating the early preference for Wheatley and Headington stones obtained from nearby villages, followed by the use of a range of stone types from the Cotswold Hills that were particularly needed for ashlar (e.g. Taynton and Milton). Over time, the search for suitable stone stretched to the Bath area (with a wide range of Bath stones used in Oxford from around 1820 onwards according to Arkell 1947), and up into Lincolnshire and Rutland (from where Clipsham Stone, which has proved to be highly durable in Oxford's environment, is sourced). The characteristically honey-toned colours of most stone used in Oxford means that the almost-white Portland Stone has not been used very commonly here, although there are notable examples where it has been used (e.g. the Ashmolean Museum and Queen's College). Within the last 100 years there has also been a move to use French limestones (including Savonnières), as cost and supply issues have made English limestone harder to obtain.

Stones used frequently in Oxford date from various periods within the Jurassic. Early Jurassic examples include those called 'Inferior oolite' – such as Clipsham and Guiting, which predate those called 'Great Oolite' (such as Taynton and Bath stones). The middle Jurassic is represented by the Corallian Limestone (such as Headington and Wheatley stones), whilst Portland Stone dates from the late Jurassic. These stones vary hugely in their petrology, although all are bioclastic, with a range of shell and fossil fragments, and many are dominated by ooliths. One of the most important stone types that was used within Oxford up until the nineteenth century is Headington Stone, quarried in a nearby village which is now a suburb of Oxford. Two distinctive stones were obtained from these quarries, that is, Headington Freestone used for ashlar walls, and Headington Hardstone used for plinths. Headington Freestone, especially the material quarried in later periods, has been found to be extremely vulnerable to polluted atmospheres, illustrating characteristic blackening and blistering. In contrast, Headington Hardstone, although often experiencing deep pitting, has proven to be extremely durable.

The Jurassic limestones used in Oxford's historic buildings have suffered from a long history of deterioration, repair and replacement. Extensive phases of repair and refacing occurred at the end of the nineteenth century and in the 1960s, when the Oxford Historic Buildings Fund paid for a large number of repairs (Oakeshott 1975). Histories of air pollution in Oxford indicate that from the mid-nineteenth century to the mid-twentieth century smoke and sulphur dioxide were serious problems, as a result of the burning of coal in domestic, college and other buildings within the city centre (Viles 1996). In recent years, the blanket of coal smoke has been replaced by more localized pollution from vehicles that predominantly affects roadside walls (Thornbush & Viles 2006). Thus, building stone deterioration today acts on stones with complex histories, which may well condition their response to present-day conditions.

Study sites

We have studied five sections of historic wall in Oxford city centre in detail in order to examine both the distribution of moisture within them and the nature and severity of decay. Working from west to east across the city, the first two sites are at Worcester College (one near the entrance and one on the roadside boundary wall). The next site is the so-called Wren Wall, which is adjacent to the Sheldonian Theatre, and just further east of this is the site at New College. In the extreme east of the city centre is the roadside boundary wall of Magdalen College in Longwall Street, referred to in this paper as 'Longwall'. The location of the sites is shown in Figure 1, alongside images of the New College walls studied and the geoelectrics survey in action at the Worcester College entrance site.

The *Worcester College entrance* site is part of the late eighteenth century façade of the building, constructed from Headington Stone (with Headington Freestone for the ashlar and Headington Hardstone in the plinth) with some patchy restoration in Clipsham Stone recorded from the 1960s. Four profiles have been studied in detail on the east-, north- and south-facing sections of the entrance. Each section forms the external wall of one of the main college buildings (e.g. chapel and dining hall). The *Worcester College boundary wall* site is an east-facing mixed rubble and block construction

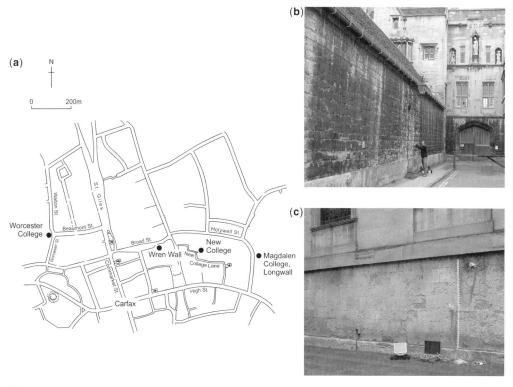

Fig. 1. (**a**) Location of the study sites. (**b**) Soot-encrusted and deteriorating stonework studied at New College. (**c**) Deteriorating stonework at the Worcester College entrance site with the resistivity survey in process.

wall that marks the edge of the college gardens, and is within 1 m or so of the busy road. The wall is composed of a patchwork of stones (including Headington, Weldon, Clipsham and Bath stones) and probably dates from a range of periods in the college's history. The particular section we monitored is in front of a building dating from 1939 and thus may have been restored around that same time – certainly the external face looks more regular in construction and homogeneous in stone type than the inner face and has clearly been refaced. Two profiles have been studied here, one on the roadside face and one exactly opposite on the inner face.

The *Wren Wall* is given this name as it is thought to be the only surviving masonry dating from the architectural work of Sir Christopher Wren in Oxford. His great building, the Sheldonian Theatre that was completed in 1669 of Headington Stone, was entirely refaced in between 1958 and 1961 in Clipsham Stone. However, the 'Wren Wall', thought to have been constructed at the same time to form a decorative boundary around the Sheldonian Theatre, survives largely untouched (with only some, probably, nineteenth century patchy repairs). According to the Royal Commission on Historical Monuments England (1939), this wall was built with niches in it to display the newly acquired Arundel marbles. It faces east and is well away from traffic. Two profiles have been recorded here.

Along *New College* lane, which now has almost no traffic, we have monitored a section of the external wall of the college cloisters that runs directly along the roadside. The cloisters date from the end of the fourteenth century (1380–1386 according to Arkell 1947) and were constructed of Headington Stone. Arkell noted in 1947 that these ancient stone walls had not been refaced at all for over five centuries and, indeed, although blackened still showed clear evidence of the original tool marks (Arkell 1947, p. 23, plate 5). Observing this same west-facing section of the wall today it is clear that it has hardly changed since 1947, whilst there appears to be much more active decay towards the base of the south-facing section. Three profiles have been measured on the south-facing section of the wall, and two on the west-facing section. The final site we monitored is part of the late fifteenth century boundary wall of

Magdalen College on Longwall Street, which includes the external walls of a projecting tower. This wall was most probably built of Headington Stone (Salter & Lobel 1954) and is now some 2 m away from a very busy road. Two profiles have been monitored here, one of the south-facing tower wall and one on the west-facing boundary wall.

Materials and methods

Two-dimensional resistivity surveys were carried out in October 2006 and July 2007 using a GeoTom device (Geolog2000, Augsburg, Germany) connected to two specially designed shielded cables, each of which had 25 clips – allowing a total of 50 electrodes to be used. Resistivity measurements were carried out by applying current into the sample through two current electrodes and measuring the resulting voltage difference at two potential electrodes. From the current and voltage values an apparent resistivity value was calculated. The relevant four electrodes for each single measurement were automatically selected using a switching unit. The result gave information on spatial averages of resistivities in a 2D section. The technique as we used it here was originally developed for use on rock walls within the natural environment (Sass 2003). As the electrical resistivity of massive or fissured rock is dependent upon its moisture content (Loke 1999), resistivity measurements allow a good estimation of moisture distribution. However, when adapting the technique for use on masonry walls we have to consider a range of potential complicating factors such as the variation in salt content and porosities between individual masonry units (i.e. individual stone blocks and mortar in joints, as well as the presence of voids) when we interpret and compare the resistivity profiles.

To avoid damage to the stonework, self-adhesive medical electrodes were stuck onto each wall in a vertical line, at either 4 or 5 cm intervals, allowing a profile of between 2 and 2.5 m to be measured. Leucci *et al.* (2007) performed a similar approach, however with a wider spacing and only 19 electrodes. At each site coupling tests were carried out to check for good electrical contact at all electrodes. The transitional resistivities between two adjacent electrodes were below 10 kΩ at rather moist sites and up to several 100 kΩ at drier sites. A Wenner configuration profile was then generated, involving 392 individual measurements. There are many different measurement configurations that can be run using the GeoTom, but we chose the Wenner configuration as it provides good resolution for surface-parallel structures and has a good signal to noise ratio (e.g. Reynolds 1997). In a Wenner array, for the first set of readings, the electrode spacing is one; then the second sequence of measurements is carried out with a spacing of two electrodes and so on. As the spacing increases, so the electrical field penetrates deeper into the stone. Penetration depths achieved by Wenner configurations are approximately one-sixth of the profile length, thus in this study 30 or 40 cm (depending on the electrode spacing). Only one spacing of the potential electrodes was measured in each separation of the current electrodes. We also tested dipole–dipole arrays, which are advantageous in terms of their better horizontal resolution; however, the penetration depth is lower and the measurements are susceptible to coupling problems of single electrodes.

Once the profiles were measured an inversion routine was run using the Res2Dinv software (Loke & Barker 1995) to calculate the 2D sections. We adapted the default settings to permit the maximum number of calculation nodes and enhanced the attenuation of single outliers as discussed in Sass (2003). A total of 15 profiles were created in all – six of which were on free-standing walls (Magdalen and Worcester College boundary walls and the Wren Wall), four on the external faces of buildings (Worcester College entrance sites) and five on walls with a semi-sheltered inside face (New College cloisters).

A range of adjunct measurements was also carried out to investigate the decay status of the stonework and to check the resistivity measurements. At all sites a visual assessment of the deterioration of each block of stone surveyed by the GeoTom profiles was made using simple decay survey techniques to help visually relate decay status to resistivity and moisture values. In order to check the GeoTom measurements we used a Protimeter to monitor the near-surface moisture values of a sample of blocks within each profile. The Protimeter is a small, easily available resistivity-based technique that records moisture in WME% units (i.e. % wood moisture equivalent) using two metal prongs held tightly against the stone surface. Although the Protimeter only measures surface water content it gives a rough and ready check on the Geotom values. In order to investigate the potential influence of variations in salt content on the resistivity values monitored by the Geotom (and thus to check the calibration used) we collected some data on the resistivity of pore waters from Oxford walls. We used the drilling of samples removed as part of restoration or other works from a range of historic Oxford buildings to obtain small powder samples. One gram of rock powder was then mixed thoroughly using a sonic bath and mechanical stirrer for 2 h with 100 ml of distilled water, filtered and measured (in micro-siemens

(μS) cm^{-1}) using a conductivity meter. These measurements gave an absolute upper bound on the likely salt content of pore waters, as under natural conditions pore waters would have much less opportunity to react with salts contained within the stone.

Once the 15 resistivity profiles were obtained a calibration curve derived from laboratory resistivity measurements on small limestone blocks with varying water content was used to convert the resistivities to moisture contents (% by weight) (Sass 1998). In order to simplify and standardize the presentation of results we then calculated the mean % moisture content for each profile (i.e. from all the 392 measurement points in each Wenner array) and calculated the mean moisture level for each site (Worcester College Entrance, Worcester College Boundary Wall, Magdalen College Longwall, Wren Wall and New College). Each profile was then recalculated and replotted with only the deviation from the mean indicated at each site. This procedure makes it possible to display the relatively wet and relatively dry areas of each profile using a consistent greyscale. The profiles of each site in terms of the relative internal moisture distribution can be directly compared. Caution should be exercised, however, in using the data to compare the degree of saturation of different blocks within one wall because of potential differences in porosity and salt content. Furthermore, direct comparisons of absolute moisture contents within profiles from different sites cannot be made because of the normalized presentation of the data (in terms of deviations from the mean at each site).

Results

General trends

Figure 2 shows the mean moisture values calculated from each of the 15 profiles. As can be seen, there is a large range – with Longwall 1 (a south-facing profile on the external wall of the projecting tower along the Magdalen College boundary wall) showing a mean of over 22% water content by weight, and the four sites around the Worcester College entrance having a mean of just over 1% by weight, and most of the other sites having mean values of between 2 and 10%. The maximum value of 22% is probably influenced by salts, as is discussed later. There is no clear correlation between the type of wall (i.e. free-standing, building, semi-sheltered) and the mean moisture content, although the wettest two profiles come

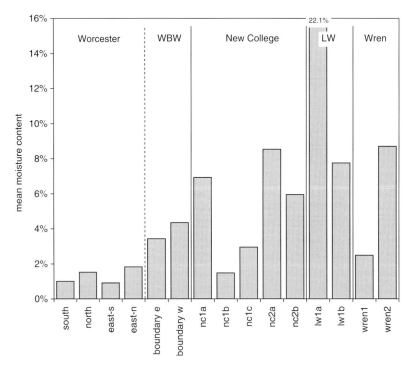

Fig. 2. Mean moisture values calculated for each of the 15 profiles.

from free-standing walls and the driest two profiles come from building walls. There is also no clear relationship between aspect and mean moisture content, although at the Worcester College entrance site the north-facing wall and the northern east-facing profile are wetter than the south and southern east-facing ones. Furthermore, the profiles New College 1b and New College 1c are markedly drier than the rest of the New College profiles, which we assign to wind funnelling rather than solar radiation (see New College results below for more details).

Open porosity values measured for a range of old and new samples of stone types used in Oxford are between about 13% (for new Stoke Ground base bed – a Bath stone) and 25% (for old stone, an unidentified Bath stone, from St John's College). It is quite likely that some of the highly weathered old Headington freestone in some of the measured profiles has even higher open porosity values. Thus, the range of absolute moisture content values calculated from the Geotom surveys is feasible (i.e. we do not get more water content than available porosity), validating our calibration.

Conductivities measured from powdered stone samples range from approximately 50 to 500 μS cm^{-1} – which would convert to resistivities of between 20 and 200 ohmmeter (Ωm). What this implies is that for two adjacent blocks of the same porosity and the same saturation values, if one had the lower salt content within the range and the other had a salt content within the upper part of the range there would be an order of magnitude difference in resistivity, and thus in the apparent moisture content. Thus, at least part of the variations in moisture we have recorded could be due to variations in salt content. However, as hypothesized earlier, we think the conductivities in natural pore waters would be much lower because of shorter residence and reaction times in comparison with the harsh regime of drilling powder samples and vigorously mixing them with distilled water. We thus assume that salinity variations play only a minor role and do not upset our resistivity–moisture calibration.

Looking at the moisture data in Figure 3, it is clear that there is a broad correlation between the surface wetness values in October 2007, as monitored on a sample of blocks with the Protimeter, and the average GeoTom moisture values. Anomalies are found at New College site 2A where the GeoTom records high moisture values, especially near the surface towards the base of the profile – not recorded in the Protimeter data. Interestingly, the Protimeter values for the Worcester College boundary wall sites are also relatively low in comparison with the mean GeoTom values, indicating that the immediate surface is drier than expected from the 2D survey results. As the Protimeter measurements have only been taken from spot samples within each GeoTom profile, and as the technique is known to have its limitations and drawbacks, the results should only be regarded as

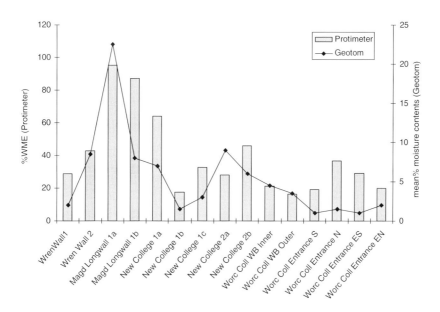

Fig. 3. Protimeter values (units of % wood moisture equivalent) from selected blocks and mean moisture values for each of the 15 profiles.

preliminary. Broadly speaking, however, the coincidence of the Protimeter and GeoTom values provide confirmation of the near-surface trends observed with the GeoTom.

Worcester College entrance

As discussed above, these four sites are all very dry compared with most of the other profiles. Figure 4 shows each profile plotted to illustrate the spatial distribution of deviations away from the mean moisture level calculated from all profiles measured at this site. All profiles but one (profile WE-east-N) started at the top of the plinth. The WE-east-N profile was achieved by adding 11 more electrodes on the plinth; two Wenner measurements (electrodes 1–50 and electrodes 12–61) were then merged to create the longer profile. All profiles reveal a drier top half and a wetter lower half. Profile WE-east-N illustrates the dense, non-porous nature of the Headington Hardstone plinth in the form of relatively low absolute moisture content here. For profiles WE-north and WE-south there is a clear zone of wetness around 5–10 cm behind the surface (notable in WE-south only after a period of simulated rainfall). At this depth, moisture is trapped as water infiltrates during rainfall but is then too deep to be affected by evaporation from the surface, while the deeper parts are influenced by drying from the interior side of the wall. Both the east-facing profiles (WE-east-S and WE-east-N) have a wetter zone just above the plinth, which correlates well with an observed zone of deterioration on both walls. Rainwater splash on the ledge of the plinth might contribute to enhanced water content (as well as a frequently used lawn sprinkler). At all sites the stonework is characterized by a brown–grey-coloured crust or patina, and extremely thin joints between the large stone blocks. All profiles contain a range of crisp blocks and also some with incipient, but stabilized, catastrophic decay, that is, breached and healed crusts and roughened surfaces. WE-north and WE-south both are both situated close to areas of catastrophic decay at the base of the ashlar that are now stable and recrusted. Only WE-east-S and WE-east-N contain blocks with active incipient catastrophic decay – which in both cases are found towards the base of the ashlar and characterized by wetter surface conditions as monitored by the GeoTom.

New College

Large variations in the mean water content of these profiles are illustrated in Figure 2. Figure 5

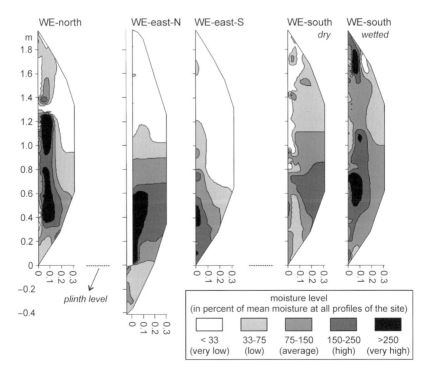

Fig. 4. Moisture profiles from the Worcester College entrance. Moisture levels are displayed as deviations from the mean of all profiles of the site (see legend).

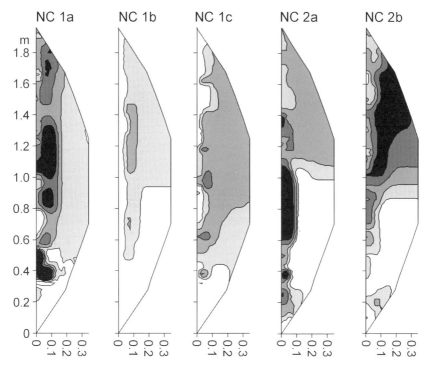

Fig. 5. Moisture profiles from the New College exterior cloister walls. Moisture levels are displayed as deviations from the mean of all profiles of the site (see the legend on Fig. 4).

illustrates the patterns of moisture content in terms of deviations from the mean moisture content calculated from each profile at this site. The wettest profiles in Figure 5 are those nearest the west corner of the wall, whilst those adjacent to the road junction are driest. We hypothesize that this spatial pattern relates to differing receipt of rain and drying winds. The roads in this area function as canyons, being surrounded by high walls, and the road junction appears to funnel wind onto the wall – which we hypothesize promotes frequent drying. Profiles NC1a, NC1b and NC1c all include the Headington Hardstone plinth, and all exhibit very dry conditions at the base. Wetter patches in a zone 5–10 cm behind the face are observed on profiles NC1a, NC1b and NC1c (as recorded at the Worcester College entrance sites). Profile NC1c is the least weathered, with most of the surface covered by a thick black crust and only small patches of blistering. A number of wet patches on the surface connected to wetter areas inside the stone seem to show some favoured infiltration paths of the rainwater. These do not correlate with mortar joints (which themselves here are highly encrusted with low-permeability black crusts) but may reflect an enhanced ingress of water within actively decaying patches where the crust has been breached. The most visibly weathered profile (NC1b), which is characterized by incipient catastrophic decay, is by far the driest. This may be a function of the generalized lack of black crust in this profile, with higher permeability weathered stone subject to more rapid drying and with large air-filled voids where flakes are detaching. Profile NC1a includes, at the base of the ashlar, a patch of active catastrophic decay that is reflected in higher surface and subsurface moisture content (to 15 cm inside the stone) than the adjacent blocks. However, a block subjected to patchy incipient catastrophic decay above, as with the large areas of profiles NC1b, is characterized by very dry surface conditions down to 10 cm inside the stone.

Profiles NC2a and NC2b are located on part of the wall without a clear plinth, but with an extensive mortar covering on the lower 20 cm or so. Both of these profiles also show a trend for drier conditions at the base – especially at depth within the wall. Interestingly, a replacement block in profile NC2a that is in apparently good condition but with a blackened surface has a very high moisture content. This may be due to higher porosity of the replacement stone type used in comparison with the original Headington Stone or to the lack of a thick gypsum crust on the 'young' surface in comparison with the highly encrusted medieval stonework. Profile NC2b contains a large area of active catastrophic decay around 1 m above pavement level, which is reflected by damper conditions stretching from the surface to about 10 cm inside the stone. There is no apparent explanation for the very wet patch behind the upper part of profile NC2b.

Wren Wall

The two Wren Wall profiles have a mean moisture content similar to the mean of all the New College profiles (see Fig. 2), but Wren 1 is much drier than Wren 2 (Fig. 6 – in which both Wren Wall profiles are plotted in terms of deviations away from the mean value calculated from both datasets). Both profiles illustrate extremely dry conditions in the interior of the wall, leading to particularly high moisture gradients from the surface inwards in Wren 2. Wren 1 exhibits wetter patches some 5–10 cm inside the stonework, as found in a number of other profiles, and also a dry plinth (as also noted in other profiles). A small replacement block on the plinth of Wren 1 is highlighted by much wetter near-surface conditions than the surrounding blocks – similar to the observations at New College profile 2A. Wren 2 is quite enigmatic as it shows such a generally wet surface and dry interior. One of the wet surface patches partially coincides with a large area of catastrophic decay in the middle of the plinth, but the others are located on stones with almost complete dark-grey–brown crust.

Longwall

The Magdalen College boundary wall along Longwall Street is by far the wettest site studied (Fig. 2). Both profiles are extremely wet at the base (Fig. 6 – in which both profiles are plotted in terms of deviations in moisture from the mean value calculated from both datasets) – perhaps because of the presence of raised gardens on the inner side of the wall although if this were the cause we might expect a more deep-seated wetness (cf. Longwall 2, which shows relatively dry conditions in the interior). The extremely high apparent moisture values at this site may be a result of two factors. First, this road experiences very high traffic levels, and salt has routinely been

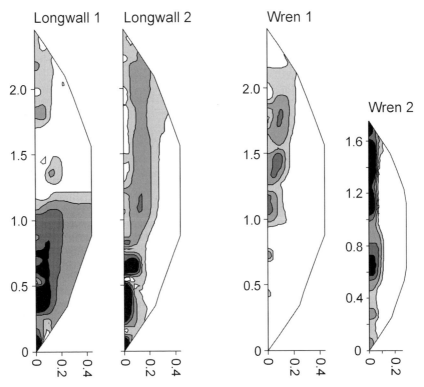

Fig. 6. Moisture profiles from the Magdalen College boundary wall on Longwall Street and from the Wren Wall. Moisture levels are displayed as deviations from the mean of all profiles of the respective sites (see the legend on Fig. 4).

applied to de-ice the road surfaces in winter (as well, perhaps, as the on road in front of the Worcester College boundary wall). No other sites will have been affected by road de-icing salt applications because of their location away from major traffic routes. Secondly, the wall was extensively cleaned with water washing to remove old impermeable black crusts approximately 18 months before the surveys took place. Longwall 2 exhibits wet patches 5–10 cm inside the face in the upper parts of the profile and is, thus, quite similar to a number of other profiles. The lower parts of Longwall 2, which show very high moisture levels from the surface to up to 20 cm inside the wall, are comprised of a series of replacement blocks, added during repairs to the wall in the early 1990s – illustrating another example of how replacement blocks seem to be much wetter than expected. Interestingly, the edges of the blocks appear to be, in contrast, extremely dry – and we suspect that this reflects the fact that in many parts of this profile the mortar joints have been washed out, leaving air-filled voids that will have very high resistivities.

Worcester College boundary wall

The mean moisture content of the Worcester College boundary wall is much lower than those found on the Magdalen College boundary wall along Longwall Street (Fig. 2), but higher than those found at the nearby Worcester College entrance sites. The inner, garden-facing and highly shaded section has a higher mean moisture value than the outer, roadside face. There is a clear wet zone towards the base of both sides of the wall (as illustrated in Fig. 7, which combines profiles from the inner and outer parts of the wall and plots values as deviations away from the mean value from the combined dataset). This wet zone extends to approximately 1 m above soil level on the garden side of the wall, and 60 cm above pavement level on the roadside. On both sides the wet zone clearly correlates with the location of the most deteriorated sections of the stonework. For example, the lower three courses of blocks on the road-facing side have lost their black crusts and are starting to deteriorate quite rapidly. It is worth noting that the pattern observed is not what would be expected where capillary rise dominates. Numerical simulations of such a situation would predict relatively moist conditions in the inside and evaporative drying from both sides. This may result from the presence of a rubble core within the wall (which also may explain the apparent dry conditions inside the Longwall and Wren Wall profiles) that may now have large, air-filled voids. Both profiles indicate that the wall is extremely dry in its upper sections, although there are a small number of

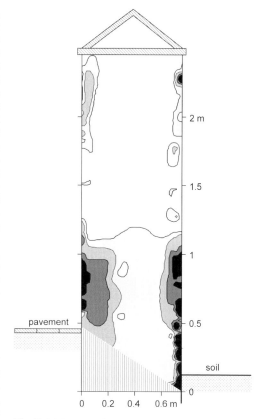

Fig. 7. Moisture profiles from either side of the Worcester College boundary wall. Moisture levels are displayed as deviations from the mean of all profiles of the site (see the legend on Fig. 4).

slightly wetter patches 5–10 cm inside the face in the higher parts of the profiles, as found at other sites

Discussion and conclusions

Our results indicate the value of resistivity surveys in non-invasive probing of the subsurface characteristics of historic masonry walls. Calibration using small samples provides a sensible picture of moisture values given the sorts of porosities known to occur in Oxford building stones. Cross-checking of our results against extreme values for salt content of pore waters (measured on drill powder samples) shows that variations in salinity may explain some of the variation in 'moisture' values. However, as salts are known to be important agents of deterioration on these walls, their influence on raising apparent moisture values, whilst affecting the calibration, is also reflected in their importance in creating 'hot spots' of decay.

Further investigations are now underway to improve the calibration of the GeoTom resistivity values for use on deteriorating walls.

The data presented indicate complex patterns of moisture distribution within historic limestone walls in the centre of Oxford. Some profiles show wetter sections towards the base of the wall (usually where a plinth is absent or where the plinth was not monitored), whilst others show wetter conditions at the top. A general finding is the role of plinths (where present) in reducing capillary rise from groundwater (as shown in the comparison between the New College 1 profiles and the Worcester College boundary wall profiles). The low porosity of these hard stones reduces the ingress of water. However, at several sites we noticed a wetter zone immediately above the top of the plinth (e.g. New College 1a) that may reflect the impact of the plinth in preventing rainwater from the upper parts of the wall 'draining out' below. This wetter zone often correlates with the occurrence of catastrophic decay – indicating that the plinth may encourage concentration of decay (as illustrated in Fig. 8 from the Worcester College entrance).

A further general trend exemplified by several profiles is for a zone of interconnected wet patches that occur between 5 and 10 cm inside the face. Such patches of heightened absolute moisture content could play a very important role in encouraging catastrophic decay, and seem to confirm our conceptual model that proposes such wet patches developing behind a low-permeability gypsum crust. It is particularly interesting that such patches appear even on the Longwall profiles, which have been recently water washed and the blackened crusts removed, and on New College profile 1b where large areas of the black crust have been removed by shallow deterioration.

Our dataset also indicates some complex relationships between catastrophically decayed patches (or incipiently catastrophic decay) and moisture levels. In many cases, actively

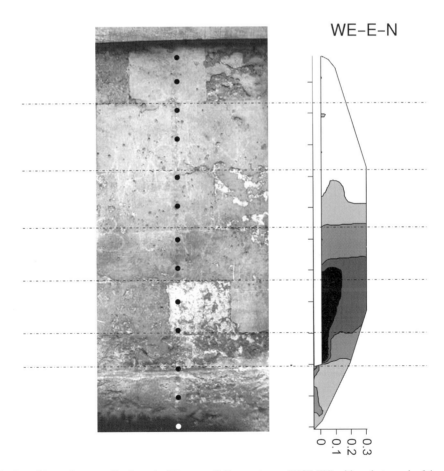

Fig. 8. One of the moisture profiles from the Worcester College entrance (WCE-EN) with a photograph of the façade at the same vertical scale illustrating the state of decay of individual blocks. Black circles show the location of every fifth electrode. Dotted horizontal lines indicate the position of mortar joints between the blocks.

deteriorating hollows are characterized by wetter surface and subsurface conditions than their surrounding blocks. However, where whole parts, or even small sections of, a profile are suffering from incipient decay (i.e. crust removed and shallow, sometimes patchy, active weathering occurs) drier than average conditions are characteristically found at the surface and below. Further investigations are now underway to explain these trends, but they seem to confirm our conceptual model. The sheer variety of mean moisture content and moisture distributions observed on these 15 profiles implies that, for historic walls with complex histories of decay and conservation, untangling the causes of, and managing the future of, rapid deterioration may be very complex.

References

ANTILL, S. J. & VILES, H. A. 1999. Deciphering the impacts of traffic on stone decay in Oxford: some preliminary observations from old limestone walls. *In*: JONES, M. S. & WAKEFIELD, R. D. (eds) *Aspects of Stone Weathering, Decay and Conservation*. Imperial College Press, London, 28–42.

ARKELL, W. J. 1947. *Oxford Stone*. Faber & Faber, London.

ENDREA, A. L. & KNIGHT, R. 1992. A theoretical treatment of the effect of microscopic fluid distribution on the dielectric properties of partially saturated rocks. *Geophysical Prospecting*, **40**, 307–324.

LEUCCI, G., CATALDO, R. & DE NUNZIO, G. 2006. Subsurface water-content identification in a crypt using GPR and comparison with microclimatic conditions. *Near Surface Geophysics*, **4**, 207–213.

LEUCCI, G., CATALDO, R. & DE NUNZIO, G. 2007. Assessment of fractures in porous materials with integrated geophysical methodologies. *Journal of Archaeological Science*, **34**, 222–232.

LOKE, M. H. 1999. *Electrical Imaging Surveys for Environmental and Engineering Studies – A Practical Guide to 2D and 3D Surveys*. Penang, Malaysia.

LOKE, M. H. & BARKER, R. D. 1995. Least-squares deconvolution of apparent resistivity pseudosections. *Geophysics*, **60**, 1682–1690.

OAKESHOTT, W. F. (ed.) 1975. *Oxford Stone Restored*. Oxford University Press, Oxford.

REYNOLDS, J. M. 1997. *An Introduction to Applied and Environmental Geophysics*. Wiley, Chichester.

ROBINSON, D. A. & MOSES, C. A. 2002. Rapid, asymmetric weathering of a limestone obelisk in a coastal environment: Telscombe Cliffs, Brighton, UK. *In*: PRIKRYL, R. & VILES, H. A. (eds) *Understanding and Managing Stone Decay*. Karolinum Press, Prague, 147–160.

ROYAL COMMISSION ON HISTORICAL MONUMENTS ENGLAND. 1939. *An Inventory of the Historical Monuments in the City of Oxford*. HMSO, London.

SALTER, H. E. & LOBEL, M. D. (eds) 1954. *A History of the County of Oxford. Volume 3: The University of Oxford*. Victoria County Histories. Oxford University Press, London, 144–162.

SASS, O. 1998. *Die Steuerung von Steinschlagmenge und -verteilung durch Mikroklima, Gesteinsfeuchte und Gesteinseigenschaften im westlichen Karwendelgebirge (Bayerische Alpen)*. Münchener Geographische Abhandlungen B, **29** (PhD thesis).

SASS, O. 2003. Moisture distributions in rockwalls derived from 2D-resistivity measurements. *Zeitschrift für Geomorphologie Supplement Band*, **132**, 51–69.

SASS, O. & VILES, H. A. 2006. How wet are these walls? Testing a novel technique for measuring moisture in ruined walls. *Journal of Cultural Heritage*, **7**, 257–263.

SMITH, B. J. & VILES, H. A. 2006. Rapid, catastrophic decay of building limestones: thoughts on causes, effects and consequences. *In*: FORT, R., ALVAREZ DE BUERGO, M., GOMEZ-HEREZ, M. & VÁZQUEZ-CALVO, C. (eds) *Heritage, Weathering and Conservation*. Taylor & Francis, London, 191–198.

THORNBUSH, M. J. & VILES, H. A. 2006. Changing patterns of soiling and microbial growth on building stone in Oxford, England after implementation of a major traffic scheme. *Science of the Total Environment*, **367**, 203–211.

TÖRÖK, A. 2002. Oolitic limestone in a polluted atmospheric environment in Budapest: weathering phenomena and alterations in physical properties. *In*: SIEGESMUND, S., WEISS, T. & VOLLBRECHT, A. (eds) *Natural Stone, Weathering Phenomena, Conservation Strategies and Case Studies*. Geological Society, London, Special Publications, **205**, 363–379.

VILES, H. A. 1993. The environmental sensitivity of blistering of limestone walls in Oxford, England: a preliminary study. *In*: THOMAS, D. S. G. & ALLISON, R. J. (eds) *Landscape Sensitivity*. Wiley, Chichester, 309–326.

VILES, H. A. 1996. 'Unswept stone, besmeer'd by sluttish time' – air pollution and building stone decay in Oxford, 1790–1960. *Environment and History*, **2**, 359–372.

Index

abrasion resistance 132–133, 134, 184, 189
acid rain 2, 54–55
additives
    application techniques 74, 147, 149, 153
    cleaning of Santa Engrácia, Lisbon 186
    crystallization and salt damage reduction 72–75, 93–102
    historic restoration of Tudela Cathedral 197
    see also consolidated porous limestones; restoration
Adobe Photoshop 231–236
ageing, simulated 232
aggregates
    in historic lime mortars 119, 121, 122
    tuffeau powder for eggshell lime mortar 140–145
    use for concrete, Cyprus 127–135
    use of Coralline limestone, Malta 23
albedo, effect of particulates on 211, 215–216, 217
Alexandria (Egypt) 62
algae/alginates 108, 110, 200–201
Algeria 27, 34
alveolar weathering (honeycombing) 3–4, 5, 7, 15
'Ammonitico Rosso' 30
anhydrite 130
anisotropy in carbonate rocks 40, 42
anthropogenic effects on building stone 54–55
Arab period, Malta 19
aragonite 130, 203, 205, 206
artificial stone as replacement 113–117
Assisi (Italy) 183
Athens (Greece) 13, 29
atmospheric pollution 4, 54–55, 57, 58, 61
    and Orléans tuffeau 103, 104, 108–110
    Oxford (UK) 238, 239
atomic absorption spectrometry (AAS/AES) 184
Austria 147, 148
autogenous healing, in mortars 124
Avila Cathedral (Spain) 231

backscatter electron imaging (BSE), lime mortars 121, 140
bafflestone, Germany 41
Bath (UK) 4, 233
Bath Limestone 231–236, 239, 240, 243
bedding, effects of 4, 161–162, 165, 167, 195
Belgium 13, 203–208
binder evolution, effects of climate on 119–126
bindstone, Germany 41
biocalcarenite, effects of borax on 93–102
bioclastic limestones, 3, 4, 33, 34, 203–208, 239,
    see also fossil content; reef limestones; shell limestones
biodeterioration of monumental stone 79–92, 114
biofilms 2, 5, 6, 79–92
biogenic limestone, replacement with new local stone 113, 114, 115–117
biological activity/growth 6, 79–92, 108, 110, 115, 190, 228
biomicrites 129–135, 171–182, 195–202
biomicrosparite-biomicrosparrudite 187
biomicrudite 187
biomineralization, and borax 97

biopelintrasparite/grainstone 157–169
biopelsparite 187
biosparite aggregates for concrete, Cyprus 127–135
biosparrudite, research prior to restoration, Spain 171–182
bird droppings 191
black crusts/patina, 114, 115, 197, 232, 244, 245, 247, 248
    see also gypsum; particulates; soiling
bleaching 56
blistering, historic walls, Oxford (UK) 5–6, 238, 245
borax, and salt damage reduction 72–75, 93–102
boundstone, Germany 41, 48
Bratislava (Slovak Republic) 148
Brazil 80
breccias 30, 32, 33, 187
Bronze Age, Malta 226
Budapest (Hungary) 5, 6, 8, 147, 148
Bulgaria 183
Burgos (Spain) 232
bursting (salt crystallization) test 15, 61, 69

Cadiz (Spain) 80, 172
Caen limestones, Calvados (France) 165
calc-sinter, Germany 39
calcarenites 3, 203–208
Calcário Ança 79–92
Calcário Lioz 79–92, 186, 187, 188–190
calcic lime characterization 139
calciclastic limestones, pore structure and durability 157–169
calcite
    aggregates for concrete 130
    effects of borax on salt crystallization 95, 97–101
    euhedral 123, 124, 125
    in hen eggshells 138
    in Maastricht limestone 203, 204
    mineralogy and weathering 42, 45, 53, 55
    recrystallization 5, 130
    sparry/sparitic 130, 203, 204
calcium sulphate see gypsum
Calvados (France), Caen limestones 165
Campanile limestone decay, Tudela Cathedral 195–202
capillary absorption, 205, 172, 176, 105–6,
    see also water absorption/uptake
capillary rise 69–70, 73, 113
    Campanile limestone, Tudela Cathedral 196–197, 201
    historic walls, Oxford (UK) 238, 247, 248
carbonic acid 53, 54, 55
carboxylic acid 74
Carrara (Italy) 29, 30–31
cart tracks, Malta 219–229
Carthage (Tunisia) 13
catastrophic decay 4, 5, 237–249
    conceptual model 238, 248–249
    'incipient' 238, 244, 245, 248–249
cave dwellings, Malta 15, 17–18
cavernous hollows 3–4, 5, 8, 238, 239, 249
cement, 23, 181, see also Portland cement

chemical analysis
    Campanile limestone, Tudela Cathedral 196–201
    particulates 210
    Santa Engrácia stones, Lisbon 184, 187
    and stone replacement 113
    weathered tuffeau 108–110
cherts 130
Chlorella 82, 87
chlorides, 70, 122, 124, *see also* salt crystallization; sodium chloride
chlorite 130, 197
Chlorophyla 82
chlorophyll *a* monitoring and determination 81–82, 88–90
chromatic parameters *see* colorimetry
cladding 1, 55, 57, 157
classifications
    fractures 42
    German carbonate rocks (Folk/Dunham) 40–42
    Köppen–Geiger climate classification 120
    stylolites 42
clay mineralogy
    aggregates for concrete 130
    of Campanile limestone, Tudela Cathedral 197, 200
    of Coralline Limestone, Malta 225
    and hydric expansion in sandstones 200
    and hygric expansion 55
    and stylolites 57
    tuffeau 110, 140
cleaning of historic stone, 186, 191, 232, 247, 248, *see also* additives
climate, 119–126, 231–232, *see also* weathering
climate change 8, 217
Clipsham Stone 239, 240
coal pollution 209–218, 239
Coimbra (Portugal) 80
colorimetry 172, 184, 231–236
colour changes, in primary bioreceptivity experiment, 85–86, *see also* black crusts/patina; orange edging; yellowing
Commission Internationale d'Eclairage (CIE), L*a*b system 232, 233
compactness, Real Alcázar limestones, Seville 172, 176
concrete 23, 68, 127–135, 191
consolidated porous limestones 147–155
cooling experiments, and salt damage 70–71
Corallian limestones, Oxford (UK) 239
Coralline limestones, Malta 13–25, 219–229
cracking, Santa Engrácia, Lisbon, 190, *see also* dessication
Croatia 27, 29, 33
crustal layer 5, 55
    Coralline Limestone, Malta 228
    historic walls, Oxford (UK) 244, 245, 246, 247, 248
    Maastricht limestone 203, 204, 205–207
    see also black crusts/patina; gypsum efflorescence/crusts
cryptoflorescence 64
crystal morphology, in binders 122, 123, 124–125
crystallization pressure 67–68, 72, 73–74
crystallization of salts *see* salt crystallization
Cyanobacteria 82, 86
Cyprus, limestone aggregates for concrete 127–135
Czech Republic 147, 148

damp, rising *see* capillary rise
de-icing salts 5, 61, 246–247
decay
    features in German carbonate rocks 53–57
    Maastricht limestone, Tongeren Cathedral 203–208
    rates and localities, Real Alcázar limestones, Seville 175–178
    'switching off' of rapid decay 5–6, 7
    see also catastrophic decay; weathering
'deep salts' 5, 7
deformation, Santa Engrácia stones, Lisbon 191
denaturing gradient gel electrophoresis (DGGE) 80–81
density
    consolidation trials on porous limestones 148, 149, 151
    eggshell lime and tuffeau powder mortar 140
    German carbonate rocks 49
    Portuguese limestones 159, 167
    Real Alcázar limestones, Seville 172, 176
    Santa Engrácia stones, Lisbon 184, 189
    weathered tuffeau 107–108
dessication cracks, historic lime mortars, 121, 123, 124, *see also* drying
detergents, Santa Engrácia, Lisbon 186
diesel particulates 209–218
differential degradation, Santa Engrácia, Lisbon 191
differential scanning calorimetry (DSC) 68, 70
digital image processing 231–236
dissolution processes, 2–8, 238, *see also* weathering
dolomite (mineral) 42, 53, 55, 187
dolomites 42, 48
dolomitic marble 30
dolostones 27, 30, 31, 37, 232
drilling resistance 5, 205, 206
dry deposition 7
drying
    effects on historic walls, Oxford (UK) 245
    and salt damage 63–65, 68
    tests on eggshell lime and tuffeau powder mortar 141–142, 144
    see also dessication; hydric expansion; moisture
durability, 203, 157–69, *see also* deformation; hardness; strength
dusts 2, 6, 8, 55, 57, 58
dynamic mobility analyser (DMA) 68, 71

efflorescence
    on biogenic limestone 114
    Campanile limestone, Tudela Cathedral 196, 201
    effects of borax on 93, 98, 99
    in historic lime mortars 125
    research into 63–65, 66, 69, 70, 73
    Santa Engrácia, Lisbon 184
    see also gypsum; thenardite
eggshell lime and tuffeau powder mortar 137–145
Egypt 13, 27, 29, 33–34, 62
elasticity, consolidation trials on porous limestones 153
environmental change 6, 8, 219–229
environmental effects on binder evolution 119–126
environmental SEM (ESEM) 65, 68, 72, 73, 94–95, 96, 97
environmental X-ray diffraction 69
epitaxial growth, borax and 97
epoxy resins 70, 197

epsomite 63, 93–102
erosional processes, Malta 219–229
exfoliation, on biogenic limestone 114
extracellular polymeric substances (EPS) 79, 87

fabric 39–42, 149
false onyx 34
fillers in historic lime mortars 125
fire, effect on historic lime mortars 125
fire damage, German carbonate rocks 53, 56
fissures 5, 45, 176, 190, 191
flakiness of limestones for aggregates 129, 132–133, 134
flaking 4, 5, 190, 196, 245
floatstone, Germany 41, 48
fluid dynamics and transport *see* water absorption/uptake
fluorescence technique, determination of chlorophyll *a* 81–82, 88–90
fossil content, 1, 4, 45–48, 187, 203, 173–5, *see also* bioclastic limestones
fossil fuel combustion 2
fractures 39, 42, 68, 167, 191
    Campanile limestone, Tudela Cathedral 196, 197, 200–201
framestone, Germany 41, 48
franca stone 80
France
    carbonate rocks 27, 32–33, 140, 147, 165, 239
    eggshell lime and tuffeau powder mortar 137–145
    weathering effects on tuffeau 103–111
freeze–thaw action 2, 4, 5, 184, 188
freezing, in concrete 68
frost action, on travertine 115
frost heave, in ice 72
frost resistance 50, 55, 57, 203
fungal growth 6, 108, 110

geophysical surveys 237–249
German carbonate rocks 37–59
glauconite 47, 110, 175
Globigerina Limestone, Malta 3, 13–25, 62
goethite 222
Gozo 15, 16, 21–22
grainstones 29, 41, 47–48, 57
    aggregates for concrete, Cyprus 129–135
    mechanical properties of Maastricht limestone 203–208
    pore structure and durability 157–169
    silicic acid ester treatment 147–155
Granada (Spain) 61, 80, 93–102
granite, light reflectance monitoring 232
granularity 4, 5, 7, 192, 238
gravimetric analysis, Santa Engrácia stones, Lisbon 184
Greece 13, 21, 27, 29, 32, 113–117, 147
ground penetrating radar, Italy 237
groundwater contamination 5
gypsum, in capillary-rise experiment 70
gypsum efflorescence/crusts 2, 4–5, 6, 7–8, 103–111, 209
    German limestones 45, 47, 55, 57–58
    historic walls, Oxford (UK) 238, 245
    permeability 238
    Santa Engrácia, Lisbon 190–191, 192
gypsum, rendering, Real Alcázar limestones, Seville 181

halite 70
hardness 1, 172, 177
Headington Stone 239, 240, 243, 244, 245
heating-cooling cycles, 8, 70–71, 209–218, *see also* thermal expansion
hematite 222
historical photographs, Tudela Cathedral 197, 200
Hollington Sandstone, effect of particulates on 209–218
hollows, cavernous *see* cavernous hollows
honeycombing (alveolar weathering) 3–4, 5, 7, 15
humic acid 53
humidity, and decay of Real Alcázar limestones, Seville 179
Hungary 5, 6, 8, 147–155
hydric expansion, Campanile limestone, Tudela Cathedral 196, 197, 199–201
hydrostatic pressure, and salt damage 67–68
hydrous dilation tests, eggshell lime and tuffeau powder mortar 141–142, 144
hygric expansion, German carbonate rocks, 51–52, 55, *see also* crystallization pressure

ice crystal growth 72
Iceland spar 95, 97–101
ICP-OES 108, 139
illite 197, 225
imbibition *see* water absorption/uptake
infrared techniques, Santa Engrácia, Lisbon 184, 185
intrabiomicrite 131
intrapelmicrosparite/grainstone 157–169
Ireland (Northern), mortar samples 119–126
iron, 6, 48, 110, 198, see also *terra rossa*
Israel 13, 27, 29, 33
Istrian Limestone 1, 3
Italy
    carbonate rocks 13, 21, 27–31, 79–92, 147
    ground penetrating radar 237
    primary bioreceptivity experiment 79–92
    see also Assisi; Carrara; Milan; Naples; Pisa; Rome; Salento; Venice

Jurassic limestones, 30, 31, 33, 34, 39, 239, *see also* oolitic limestones

kaolinite 225
karst 2, 3, 7, 33

L*a*b system (CIE) 232, 233
landscape evolution, Malta 219–229
Latvia, mortar samples 119–126
lead 108, 110
Lecce (Italy) 80
lichens 228
light reflectance monitoring 232
lime mortars, historic 119–126
limestone terrain evolution, Malta 219–229
linear variable differential transformer (LVDT) 71
Lioz stone *see* Calcário Lioz
Lisbon (Portugal) 80, 183–193
Logos-Edessa (Greece) 114–117
London (UK) 1, 2

Maastricht limestone, mechanical properties 203–208
Macael (Spain) 29
magnesite 123
magnesium 187, 198, 199, 200, 201
magnesium sulphate/epsomite 63, 69–71, 93–102
Mallorca, cavernous hollows 3
Malta 3, 13–25, 62, 147, 219–229
marbles 13, 19, 30
    characteristics 3, 55, 200
    light reflectance monitoring 232
    'Ruivina', Portugal 187, 190
    use of term 13, 27, 31, 32
marine limestones, France 33
mechanical properties *see* physical/mechanical
    properties
mechanical resistance
    eggshell lime and tuffeau powder mortar 142, 143, 144
    weathered tuffeau 104–105
Mediterranean region 27–35, 79–92
mercury intrusion porosimetry (MIP)
    Campanile limestone, Tudela Cathedral 197, 198–199
    consolidation trials on porous limestones 148
    eggshell lime and tuffeau powder mortar 140
    Maastricht limestone 204
    Portuguese limestones 158, 162–164
    Real Alcázar limestones, Seville 172, 176
    weathered tuffeau 106–108
mercury pycnometry, Maastricht limestone 204
micas 187
micrite 175
micritic calcite, Maastricht limestone 203, 204
micritic limestones 31, 45, 46, 53
    aggregates for concrete 127–135
    silicic acid ester treatment 147–155
    see also biomicrites
microbial colonization of monumental stone, 79–92, *see also* biofilms; biological activity
microkarst 56
microscopy, 69, 72, 121, 122, 123, 171, *see also* scanning electron microscopy
microspar 127–135
Milan (Italy) 232
mineralogy
    Campanile limestone, Tudela Cathedral 197–198
    German carbonate rocks 42, 45, 53, 55, 57
    historic lime mortars 122, 123, 124, 125
    limestone for aggregates 130–131
    Santa Engrácia stones, Lisbon 187
    weathered tuffeau 108–110
mirabilite 63, 66, 69, 71, 73, 93–102
'Moca Creme', Portuguese limestone 157–169
modelling 65–67, 73, 238–239, 249
modifiers of supersaturation and crystal growth 72–74, 93–102
moisture
    content, historic walls, Oxford (UK) 237–249
    and historic lime mortars 123–125
    particulate-induced loss 209–218
    see also water absorption/uptake
molecular analysis of natural biofilm 80–81
molecular modelling 73
Moravia (Czech Republic) 148
Morocco 27, 34

Morón (Spain) 172
mortars 1, 113, 137
    black crusts/patina on joints of 245
    hydraulic, Malta 19, 20, 22
    Santa Engrácia, Lisbon 191
    see also lime mortars; Portland cement; restoration mortars
mudstones 29, 39, 41, 42–45, 57

Naples (Italy) 13
Netherlands 203
niter 191–192
nitrogen oxides, and acid rain 2, 54
Northern Ireland, mortar samples 119–126
nucleation and crystal growth 62–63, 68, 70
    modifiers 72–73, 74, 93–102

O-IDIP (outdoor integrated digital image processing) 231–236
oolitic limestones 3, 4, 6, 7, 8
    Germany 38, 39
    Mediterranean region 30, 31, 33
    Oxford (UK) 239
    silicic acid ester treatment 147–155
opal-cristobalite tridymite (opal-CT) 108, 140
optical emission spectrometry with inductively coupled plasma (ICP-OES) 108, 139
orange edging matter, weathered tuffeau 104, 105, 106, 108, 110
organic (humic) acid 53
organic matter 175, 179, 200–201
Orléans (France) 103, 104, 108, 110
Oviedo (Spain) 232
Oxford (UK)
    façade-scale colorimetry, Ashmolean Museum 231–236
    limestones used in historic buildings 1, 4, 5–6, 7, 239
    moisture content of historic limestone walls 237–249

packstones 41, 45–47, 127–135, 147–155
Pakhna Formation, Cyprus 128, 130–131
palomera stone *see* Piedra Escúzar
particle shape, limestones for aggregates 129
particle size
    Coralline Limestone, Malta 223–225
    eggshell lime and calcic lime 139
particulates 2, 5, 8, 108, 209–218, 232
pelmicrite 187
permeability 3, 238
petrography
    Campanile limestone, Tudela Cathedral 197–198
    limestones for aggregates 128, 130–131
    Maastricht limestone, Tongeren Cathedral 203–204
    Portuguese limestones 157, 158, 159
    Real Alcázar limestones, Seville 171–175
    Santa Engrácia stones, Lisbon 184, 187
phosphonates, and salt damage reduction 73, 74
phosphorus, in weathered tuffeau 110
photographic records 171, 197, 200, 231–236
photosynthetic monitoring 81, 85–86

physical/mechanical properties
    Campanile limestone, Tudela Cathedral 196, 197, 199–201
    consolidated porous limestones 148–153
    German carbonate rocks 39–42, 49–53
    limestones for aggregates 128–133
    Maastricht limestone 203–208
    Real Alcázar limestones, Seville 171–176
    Santa Engrácia stones, Lisbon 184, 188
    stone replacement and artificial stone 113, 114, 115–116
Piedra Escúzar (Piedra Franca; Piedra Palomera) 79–92, 171–182
Piedra San Cristobal 79–92
Piedra Tosca 171–182
Pietra di Lecce 79–92
Pisa (Italy) 183
pitting 7, 196
Pittsburgh Cathedral of Learning (USA) 231–232
plasters, and pore size 124
point count analysis, lime mortars 121
polishing machines 186
pollution *see* acid rain; atmospheric pollution; coal pollution; dust; particulates; sulphur dioxide pollution; traffic pollution; urban environment
polymers, 70, 74, 79, 87–88, 197, *see also* silicic acid ester trials
pore clogging 62, 63–65, 68, 70, 72, 73
pore network/spacing/structure
    and durability of Portuguese limestones 157–169
    German carbonate rocks 39, 42
    in historic lime mortars 122–123, 124, 125
    and salt damage 68
    weathered tuffeau 106–108
pore size
    Campanile limestone, Tudela Cathedral 197, 198–199, 200
    consolidated porous limestones 148, 149, 151–153
    Maastricht limestone, Tongeren Cathedral 204, 205, 206
    Real Alcázar limestones, Seville 176–177
    and stone replacement 114, 115
porosity 1, 5, 7
    Campanile limestone, Tudela Cathedral 198, 199, 200
    consolidated porous limestones 147–155
    deteriorating historic walls, Oxford (UK) 237–238, 241–242, 243, 247
    eggshell lime and tuffeau powder mortar 140–141
    German carbonate rocks 42–48, 49, 51, 55
    historic lime mortars 122–123, 124, 125
    limestones for aggregates 130, 131, 134
    Maastricht limestone, Tongeren Cathedral 204, 205, 206
    Portuguese limestones 159–162, 164, 167
    primary bioreceptivity experiment 84, 85
    Real Alcázar limestones, Seville 172, 175, 176
    Santa Engrácia stones, Lisbon 184, 188
    and stone replacement 114
    weathered tuffeau 107–108
    see also mercury intrusion porosimetry (MIP)
Portland cement 22, 23, 137, 195, 198, 200, 201
Portland Limestone 2, 167, 239
    effect of particulates on 209–218

façade-scale colorimetry 231–236
portlandite 53, 139, 141
Portugal 79–92, 157–169, 183–193
potassium chloride 70
potassium nitrate 191–192
pozzolana 116, 117
prehistoric sites, Malta 15, 16–18, 219–229
pressure, 51–52, 55, *see also* crystallization pressure
principal component analysis (PCA) 82
Protimeter 241, 243
pseudosparite 175
El Puerto de Santa María (Spain) 172
pulverization, Campanile limestone, Tudela Cathedral 197
pycnometry, Maastricht limestone 204
pyrite, in weathered tuffeau 110

quarries
    Algeria 34
    Croatia 33
    Egypt 34
    France 32–33, 140
    Greece 29, 32
    Israel 33
    Italy 29, 30–31, 80
    Malta 15, 21, 228
    Netherlands 203
    Portugal 80, 158, 183, 190
    Spain 29, 31, 80, 172, 197
    Tunisia 34
    Turkey 32
    United Kingdom 239

radiant heat, and particulates 211, 215–216, 217
radiocarbon dating 119
rainwater, 231, 232, 233, 234, 235, 244, 245, 248
    see also acid rain
reef limestones 37, 39, 127–135
rendering 181, 186
replacement stone 4, 113–117
    artificial stone 113–117
    in historic walls, Oxford (UK) 239, 240, 245, 247
    and pore size distribution 114, 115
resistance, mechanical
    eggshell lime and tuffeau powder mortar 142, 143, 144
    weathered tuffeau 104–105
resistance to fragmentation, limestones for aggregates 129
resistivity surveys, moisture content of historic walls, Oxford (UK) 237–249
restoration 4, 232
    mortars 137–145, 171, 181, 197
    prior research, Real Alcázar, Seville 171–182
    see also additives; replacement stone
Rhodes (Greece) 114, 116–117
rising damp *see* capillary rise
roads, de-icing salts, 5, 61, 246–247, *see also* traffic pollution
Romans 18–19, 137
Rome (Italy), Ara Pacis 231
rudstone 29, 41, 48

Salento (Italy) 80
salt crystallization 2, 3, 4, 5, 8, 61–77, 114
    Campanile limestone, Tudela Cathedral 195, 197, 199, 200–201
    German carbonate rocks 51, 55, 57
    Globigerina limestones 15
    and historic lime mortars 124, 125
    Portuguese limestones 157–158, 164–167
    prevention methods 72–74, 93–102
    salt content of pore water, historic walls, Oxford (UK) 241–242, 243, 247
    Santa Engrácia stones, Lisbon 190–192
    test (bursting test) 8, 15, 51, 69
    see also weathering
sandstones
    consolidation treatment 153
    effect of particulates on 209–218
    hydric expansion 200
    rapid retreat 5, 6, 7
    replacement stone, Tudela Cathedral 195, 197, 198, 200, 201
saturation coefficient, 3, see also supersaturation
Savonnières limestone 239
scaling 5, 6, 7, 8, 56, 190
scanning electron microscopy (SEM) 82, 87–88, 107, 108, 172, 210
    environmental SEM 65, 68, 72, 73, 94–95, 96, 97
    of mortars 121, 122, 123, 124, 140–141, 144
    and porosity 149, 153, 159
sealing using wax, Santa Engrácia, Lisbon 186
seepage evaporation, Santa Engrácia, Lisbon 191
self-healing in mortars 124
'Semi-rijo', Portuguese limestone 157–169
Seville (Spain) 80, 171–182
shales, German 39
shell limestones, 29, 37, 38–39, 45–48,
    see also bioclastic limestones
shrinkage see dessication
Sibbe (Holland), source for Maastricht limestone 203
silicic acid ester trials, porous limestones 147–155
Slovakia 147, 148
smectite 110, 198, 200
smoke pollution 239
sodium chloride 61, 69, 70, 130
sodium sulphate, 69–73, 93–102, see also mirabilite; thenardite
soil environment changes, Malta 219–229
soiling 7, 172–173, 216–217, 231–236
solar radiation 211, 215–216, 217, 235
solution features, Malta, 221, 223–225, 226,
    see also dissolution
soot, 55, 57, 238, see also particulates
sorptivity testing 70
Spain, 13, 21, 27, 28, 29, 30, 31, 79–92, see also Avila; Burgos; Cadiz; Granada; Macael; Morón; Oviedo; Puerto de Santa María; Seville; Tarragona; Tudela; Utrera; Valladolid
spalling, Santa Engrácia, Lisbon 190
sparite calcite cement, Tudela Cathedral limestone 198
sparitic calcite, Maastricht limestone 203, 204
sparitic limestones, Germany 52, 53, 55
spectrometry 88–90, 172, 184, 233–235
spectrophotometry 88–90, 172, 233–235
speleothems 29, 30

strength
    Campanile limestone, Tudela Cathedral 195, 200
    eggshell lime and tuffeau powder mortar 142, 143, 144
    German carbonate rocks 50–51, 56
    porous limestones 148, 149–151, 153
    Santa Engrácia stones, Lisbon 184, 188, 189
    and stone replacement 113, 114, 115
    weathered tuffeau 104–105
stress, and salt damage 68
stress history/'memory' (of stone) 2
stromatolithes 45
stylolites 29, 33, 39, 42, 45, 48, 56, 57
subflorescence 63–65, 69, 73
sulphur 2, 4–5, 8, 54, 58, 61
    in gypsum crust on tuffeau 108–110
    in historic lime mortars 122
    see also gypsum; magnesium sulphate; sodium sulphate
sulphur dioxide pollution 108, 234, 238, 239
supersaturation 62–63, 65–67, 68
    modifiers 72–73, 93–102
surface hardness, Real Alcázar limestones, Seville 172, 177
surfactants, and salt damage reduction 73, 74
synchrotron measurements 69, 70, 71

Tarragona (Spain) 62–63
temperature cycles see cooling; freeze-thaw; frost; heating; thermal expansion
terra rossa materials, Malta 222–223, 225, 226
terrain evolution, Malta 219–229
texture, German limestones 42
thenardite 69, 70, 71, 73, 190, 191, 192
thermal energy, effect of particulates on 211, 215–216, 217
thermal expansion, 52–53, 189, see also heating
thermogravimetric analysis 106, 138
thermoporoelasticity 71
Tongeren Cathedral (Belgium) 203–208
tosca limestone, Réal Alcázar, Seville 171–182
traffic pollution 209, 210, 216, 217, 232, 239, 246
travertine 29–30, 31, 34, 39, 49, 51
    replacement with artificial stone 113, 114–117
Tudela Cathedral (Spain), Campanile limestone decay 195–202
tufa 29–30, 39, 48
tuffeau (stone) 103–111, 137–138, 142–144
tuffeau powder and eggshell lime mortar 137–145
tuffs, consolidation treatment 152, 153
Tunisia 13, 27, 34
Turkey 13, 27, 32

ultrasonic velocity 5, 148, 149–151, 153, 172, 177
underground locations, Malta 13, 17–18, 19, 20
United Kingdom, 1, 2, 4, 13, 61, 147, 233,
    see also Oxford
United States of America 119–126, 231–232
urban environment, effect on tuffeau, 103–111,
    see also traffic pollution
Usseau, Vienne (France) 140
Utrecht (Netherlands) 203
Utrera (Spain) 172

vacuum consolidation 147, 149, 153
Valladolid (Spain), Cathedral Nueva 234
Valletta (Malta) 13, 15, 20–21
vaterite 53
vehicular pollution *see* traffic pollution
Venice (Italy) 1, 231
videocamera recording of colour changes and soiling 232
Vienna (Austria) 147, 148

wackestones 29, 41, 45, 57, 127–135, 195, 197
warping experiments 68, 70, 71–72
water absorption/uptake 3
   limestones for aggregates 129, 132, 133, 134
   Portuguese limestones 158, 159–162, 164
   in primary bioreceptivity experiment 84–85
   properties of German carbonate rocks 49–50
   Santa Engrácia stones, Lisbon 184, 188
   and stone replacement 113, 114
   weathered tuffeau 105–106
   see also capillary absorption; moisture; porosity
water transfer, eggshell lime and tuffeau powder mortar 141–142, 143, 144
wax sealing, Santa Engrácia stones, Lisbon 186
weathering 2–8
   and approaches to stone replacement 113–117
   of Campanile limestone, Spain 195–202
   of Coralline limestones, Malta 219–229
   effects on binder evolution 119–126
   effects on tuffeau 103–111
   of German limestones 45–48, 53–57
   of Globigerina limestones 15
   laboratory simulations of salt weathering 4
   of Maastricht limestone, Belgium 203–208
   properties of limestones for aggregates 129, 132–133
   Real Alcázar limestones, Spain 171–182
   role of particulates 209–218
   Santa Engrácia, Lisbon 190–192
   see also decay; salt crystallization
Weldon Stone 240
Wenner configuration profiling 241–242
wetting–drying 2, 4, 8, 209–218
wheel, early use of 226
wind drying, effects on historic walls, Oxford (UK) 245
windwashing 231, 232

X-ray diffraction (XRD) analysis
   Campanile limestone 197
   Coralline Limestone, Malta 225
   effects of borax on salt crystallization 94–95, 97
   of eggshells for lime mortar 138
   and stone replacement 113
   using environmental XRD 69
   of weathered tuffeau 106, 108
X-ray dispersive spectrometry (EDS) 172
X-ray fluorescence (XRF) 70, 113, 184, 197

yellowing, from sulphur dioxide 234